Means Residential Repair & Remodeling Costs

Contractor's Pricing Guide 2004

Senior Editor
Robert W. Mewis, CCC

Contributing Editors
Barbara Balboni
Robert A. Bastoni
John H. Chiang, PE
Robert J. Kuchta
Robert C. McNichols
Melville J. Mossman, PE
John J. Moylan
Jeannene D. Murphy
Stephen C. Plotner
Michael J. Regan
Eugene J. Spencer
Marshall J. Stetson
Phillip R. Waier, PE

Senior Engineering Operations Manager
John H. Ferguson, PE

Vice President, Product Management
Roger J. Grant

President
Curtis B. Allen

Sales Director
John M. Shea

Production Manager
Michael Kokernak

Production Coordinator
Marion E. Schofield

Technical Support
Thomas J. Dion
Jonathan Forgit
Mary Lou Geary
Gary L. Hoitt
Paula Reale-Camelio
Robin Richardson
Kathryn S. Rodriguez
Sheryl A. Rose

Book & Cover Design
Norman R. Forgit

RSMeans

Means Residential Repair & Remodeling Costs

Contractor's Pricing Guide 2004

- Quick Costs to Help You Estimate

- New, Easy-to-Use Format . . . Organized the Way You Build

- Hundreds of Work Items with Material, Labor & Equipment

$39.95 per copy. (In United States).
Price subject to change without prior notice.

Reed Construction Data

Copyright © 2003

Construction Publishers & Consultants
63 Smiths Lane
Kingston, MA 02364-0800
(781) 422-5000

Printed in the United States of America

10 9 8 7 6 5 4 3 2 1

ISSN 1542-9156

ISBN 0-87629-717-3

Foreword

RSMeans is a product line of Reed Construction Data, a leading provider of construction information, products, and services in North America and globally. Reed Construction Data's project information products include more than 100 regional editions, national construction data, sales leads, and local plan rooms in major business centers. Reed Construction Data's PlansDirect provides surveys, plans and specifications. The First Source suite of products consists of *First Source for Products,* SPEC-DATA™, MANU-SPEC™, CADBlocks, Manufacturer Catalogs and First Source Exchange (www.firstsourceexchange.com) for the selection of nationally available building products. Reed Construction Data also publishes ProFile, a database of more than 20,000 U.S. architectural firms. RSMeans provides construction cost data, training, and consulting services in print, CD-ROM and online. Reed Construction Data, headquartered in Atlanta, is owned by Reed Business Information (www.cahners.com), a leading provider of critical information and marketing solutions to business professionals in the media, manufacturing, electronics, construction and retail industries. Its market-leading properties include more than 135 business-to-business publications, over 125 Webzines and Web portals, as well as online services, custom publishing, directories, research and direct-marketing lists. Reed Business Information is a member of the Reed Elsevier plc group (NYSE: RUK and ENL)—a world-leading publisher and information provider operating in the science and medical, legal, education and business-to-business industry sectors.

Our Mission

Since 1942, RSMeans has been actively engaged in construction cost publishing and consulting throughout North America.

Today, over 50 years after the company began, our primary objective remains the same: to provide you, the construction and facilities professional, with the most current and comprehensive construction cost data possible.

Whether you are a contractor, an owner, an architect, an engineer, a facilities manager, or anyone else who needs a fast and reliable construction cost estimate, you'll find this publication to be a highly useful and necessary tool.

Today, with the constant flow of new construction methods and materials, it's difficult to find the time to look at and evaluate all the different construction cost possibilities. In addition, because labor and material costs keep changing, last year's cost information is not a reliable basis for today's estimate or budget.

That's why so many construction professionals turn to RSMeans. We keep track of the costs for you, along with a wide range of other key information, from city cost indexes . . . to productivity rates . . . to crew composition . . . to contractor's overhead and profit rates.

RSMeans performs these functions by collecting data from all facets of the industry, and organizing it in a format that is instantly accessible to you. From the preliminary budget to the detailed unit price estimate, you'll find the data in this book useful for all phases of construction cost determination.

The Staff, the Organization, and Our Services

When you purchase one of RSMeans' publications, you are in effect hiring the services of a full-time staff of construction and engineering professionals.

Our thoroughly experienced and highly qualified staff works daily at collecting, analyzing, and disseminating comprehensive cost information for your needs. These staff members have years of practical construction experience and engineering training prior to joining the firm. As a result, you can count on them not only for the cost figures, but also for additional background reference information that will help you create a realistic estimate.

The RSMeans organization is always prepared to help you solve construction problems through its five major divisions: Construction and Cost Data Publishing, Electronic Products and Services, Consulting Services, Insurance Services, and Educational Services. Besides a full array of construction cost estimating books, Means also publishes a number of other reference works for the construction industry. Subjects include construction estimating and project and business management; special topics such as HVAC, roofing, plumbing, and hazardous waste remediation; and a library of facility management references.

In addition, you can access all of our construction cost data through your computer with RSMeans CostWorks 2004 CD-ROM, an electronic tool that offers over 50,000 lines of Means detailed construction cost data, along with assembly and whole building cost data. You can also access Means cost information from our Web site at www.rsmeans.com

What's more, you can increase your knowledge and improve your construction estimating and management performance with a Means Construction Seminar or In-House Training Program. These two-day seminar programs offer unparalleled opportunities for everyone in your organization to get updated on a wide variety of construction-related issues.

RSMeans also is a worldwide provider of construction cost management and analysis services for commercial and government owners and of claims and valuation services for insurers.

In short, RSMeans can provide you with the tools and expertise for constructing accurate and dependable construction estimates and budgets in a variety of ways.

Robert Snow Means Established a Tradition of Quality That Continues Today

Robert Snow Means spent years building his company, making certain he always delivered a quality product.

Today, at RSMeans, we do more than talk about the quality of our data and the usefulness of our books. We stand behind all of our data, from historical cost indexes... to construction materials and techniques... to current costs.

If you have any questions about our products or services, please call us toll-free at 1-800-334-3509. Our customer service representatives will be happy to assist you or visit our Web site at www.rsmeans.com

Table of Contents

How the Book is Built: An Overview

A Powerful Construction Tool

You have in your hands one of the most powerful construction tools available today. A successful project is built on the foundation of an accurate and dependable estimate. This book will enable you to construct just such an estimate.

For the casual user the book is designed to be:

- quickly and easily understood so you can get right to your estimate
- filled with valuable information so you can understand the necessary factors that go into the cost estimate

For the regular user, the book is designed to be:

- a handy desk reference that can be quickly referred to for key costs
- a comprehensive, fully reliable source of current construction costs, so you'll be prepared to estimate any project
- a source book for project cost, product selections, and alternate materials and methods

To meet all of these requirements we have organized the book into the following clearly defined sections.

Unit Price Section

The cost data has been divided into 23 sections representing the order of construction. Within each section is a listing of components and variations to those components applicable to that section. Costs are shown for the various actions that must be done to that component.

Reference Section

This section includes information on Location Factors, and a listing of Abbreviations.

Location Factors: Costs vary depending upon regional economy. You can adjust the "national average" costs in this book to over 930 major cities throughout the U.S. and Canada by using the data in this section.

Abbreviations: A listing of the abbreviations used throughout this book, along with the terms they represent, is included.

Index

A comprehensive listing of all terms and subjects in this book to help you find what you need quickly when you are not sure where it falls in the order of construction.

The Scope of This Book

This book is designed to be as comprehensive and as easy to use as possible. To that end we have made certain assumptions and limited its scope in three key ways:

1. We have established material prices based on a "national average."
2. We have computed labor costs based on a 7 major region average of open shop wage rates.
3. We have targeted the data for projects of a certain size range.

Project Size

This book is intended for use by those involved primarily in Residential Repair and Remodeling construction costing between $10,000-$100,000.

With reasonable exercise of judgment the figures can be used for any building work. For other types of projects, such as new home construction or commercial buildings, consult the appropriate Means publication for more information.

How to Use the Book: The Details

What's Behind the Numbers? The Development of Cost Data

The staff at RSMeans continuously monitors developments in the construction industry in order to ensure reliable, thorough and up-to-date cost information.

While *overall* construction costs may vary relative to general economic conditions, price fluctuations within the industry are dependent upon many factors. Individual price variations may, in fact, be opposite to overall economic trends. Therefore, costs are continually monitored and complete updates are published yearly. Also, new items are frequently added in response to changes in materials and methods.

Costs—$ (U.S.)

All costs represent U.S. national averages and are given in U.S. dollars. The Means Location Factors can be used to adjust costs to a particular location. The Location Factors for Canada can be used to adjust U.S. national averages to local costs in Canadian dollars.

Material Costs

The RSMeans staff contacts manufacturers, dealers, distributors, and contractors all across the U.S. and Canada to determine national average material costs. If you have access to current material costs for your specific location, you may wish to make adjustments to reflect differences from the national average.

Included within material costs are fasteners for a normal installation. RSMeans engineers use manufacturers' recommendations, written specifications and/or standard construction practice for size and spacing of fasteners. Adjustments to material costs may be required for your specific application or location. Material costs do not include sales tax.

Labor Costs

Labor costs are based on the average of open shop wages from across the U.S. for the current year. Rates along with overhead and profit markups are listed on the inside back cover of this book.

- If wage rates in your area vary from those used in this book, or if rate increases are expected within a given year, labor costs should be adjusted accordingly.

Labor costs reflect productivity based on actual working conditions. These figures include time spent during a normal workday on tasks other than actual installation, such as material receiving and handling, mobilization at site, site movement, breaks, and cleanup.

Productivity data is developed over an extended period so as not to be influenced by abnormal variations and reflects a typical average.

Equipment Costs

Equipment costs include not only rental costs, but also operating costs such as fuel, oil, and routine maintenance. Equipment and rental rates are obtained from industry sources throughout North America—contractors, suppliers, dealers, manufacturers, and distributors.

Factors Affecting Costs

Costs can vary depending upon a number of variables. Here's how we have handled the main factors affecting costs.

Quality—The prices for materials and the workmanship upon which productivity is based represent sound construction work. They are also in line with U.S. government specifications.

Overtime—We have made no allowance for overtime. If you anticipate premium time or work beyond normal working hours, be sure to make an appropriate adjustment to your labor costs.

Productivity—The labor costs for each line item are based on working an eight-hour day in daylight hours in moderate temperatures. For work that extends beyond normal work hours or is performed under adverse conditions, productivity may decrease.

Size of Project—The size, scope of work, and type of construction project will have a significant impact on cost. Economies of scale can reduce costs for large projects. Unit costs can often run higher for small projects. Costs in this book are intended for the size and type of project as previously described in "How the Book Is Built: An Overview." Costs for projects of a significantly different size or type should be adjusted accordingly.

Location—Material prices in this book are for metropolitan areas. However, in dense urban areas, traffic and site storage limitations may increase costs. Beyond a 20-mile radius of large cities, extra trucking or transportation charges may also increase the material costs slightly. On the other hand, lower wage rates may be in effect. Be sure to consider both these factors when preparing an estimate, particularly if the job site is located in a central city or remote rural location.

In addition, highly specialized subcontract items may require travel and per diem expenses for mechanics.

Other factors—
- season of year
- contractor management
- weather conditions
- building code requirements
- availability of:
    - adequate energy
    - skilled labor
    - building materials
- owner's special requirements/ restrictions
- safety requirements
- environmental considerations

General Conditions—The extreme right-hand column of each chart gives the "Totals." These figures contain the installing contractor's O&P (in other words, the contractor doing the work). Therefore, it is necessary for a general contractor to add a percentage of all subcontracted items. For a detailed breakdown of O&P see the inside back cover of this book.

Unpredictable Factors—General business conditions influence "in-place" costs of all items. Substitute materials and construction methods may have to be employed. These may affect the installed cost and/or life cycle costs. Such factors may be difficult to evaluate and cannot necessarily be predicted on the basis of the job's location in a particular section of the country. Thus, where these factors apply, you may find significant, but unavoidable cost variations for which you will have to apply a measure of judgment to your estimate.

Final Checklist

Estimating can be a straightforward process provided you remember the basics. Here's a checklist of some of the items you should remember to do before completing your estimate.

Did you remember to . . .

- factor in the Location Factor for your locale
- take into consideration which items have been marked up and by how much
- mark up the entire estimate sufficiently for your purposes
- include all components of your project in the final estimate
- double check your figures to be sure of your accuracy
- call RSMeans if you have any questions about your estimate or the data you've found in our publications

Remember, RSMeans stands behind its publications. If you have any questions about your estimate . . . about the costs you've used from our books . . . or even about the technical aspects of the job that may affect your estimate, feel free to call the RSMeans editors at 1-800-334-3509.

Unit Price Section

Table of Contents

Unit Price Section

Table of Contents

How to Use the Unit Price Pages

The prices in this publication include all overhead and profit mark-ups for the installing contractor. If a general contractor is involved, an additional 10% should be added to the cost shown in the "Total" column.

The costs for material, labor and any equipment are shown here. Material costs include delivery, and a 10% mark-up. No sales taxes are included. Labor costs include all mark-ups as shown on the inside back cover of this book. Equipment costs also include a 10% mark-up.

The cost in the total column should be multiplied by the quantity of each item to determine the total cost for that item. The sum of all item's total costs becomes the project's total cost.

The book is arranged into 23 categories that follow the order of construction for residential projects. The category is shown as white letters on a black background at the top of each page.

A brief specification of the cost item is shown here.

Graphics representing some of the cost items are found at the beginning of each category.

The cost data is arranged in a tree structure format. Major cost sections are left justified and are shown as bold headings within a grey background. Cost items and sub-cost items are found under the bold headings. The complete description should read as "bold heading, cost item, sub-cost item."
For example, the identified item's description is "Beam/Girder, Solid Wood Beam, 2″ x 8″."

Each action performed to the cost item is conveniently listed to the right of the description. See the explanation below for the scope of work included as part of the action.

The unit of measure upon which all the costs are based is shown here. An abbreviations list is included within the reference section found at the back of this book.

Rough Frame / Structure

Rough Framing

Beam / Girder

	Unit	Material	Labor	Equip.	Total	Specification
Solid Wood Beam						
2″ x 8″						
Demolish	L.F.		.62		.62	Cost includes material and labor to install 2″ x 8″ beam.
Install	L.F.	.88	.97		1.85	
Demolish and Install	L.F.	.88	1.59		2.47	
Clean	L.F.	.10	.20		.30	
Paint	L.F.	.12	1.49		1.61	
Minimum Charge	Job		157		157	
2″ x 10″						
Demolish	L.F.		.77		.77	Cost includes material and labor to install 2″ x 10″ beam.
Install	L.F.	1.19	1.05		2.24	
Demolish and Install	L.F.	1.19	1.82		3.01	
Clean	L.F.	.11	.25		.36	
Paint	L.F.	.15	.92		1.07	
Minimum Charge	Job		157		157	
2″ x 12″						
Demolish	L.F.		.93		.93	Cost includes material and labor to install 2″ x 12″ beam.
Install	L.F.	1.62	1.14		2.76	
Demolish and Install	L.F.	1.62	2.07		3.69	
Clean	L.F.	.13	.29		.42	
Paint	L.F.	.18	2.21		2.39	
Minimum Charge	Job		157		157	
4″ x 8″						
Demolish	L.F.		3.16		3.16	Cost includes material and labor to install 4″ x 8″ beam.
Install	L.F.	3.67	1.40	.67	5.74	
Demolish and Install	L.F.	3.67	4.56	.67	8.90	
Clean	L.F.	.11	.25		.36	
Paint	L.F.	.15	.92		1.07	
Minimum Charge	Job		157		157	
4″ x 10″						
Demolish	L.F.		3.96		3.96	Cost includes material and labor to install 4″ x 10″ beam.
Install	L.F.	4.60	1.47	.71	6.78	
Demolish and Install	L.F.	4.60	5.43	.71	10.74	
Clean	L.F.	.13	.29		.42	
Paint	L.F.	.18	1.07		1.25	
Minimum Charge	Job		157		157	
4″ x 12″						
Demolish	L.F.		4.75		4.75	Cost includes material and labor to install 4″ x 12″ beam.
Install	L.F.	5.50	1.56	.75	7.81	
Demolish and Install	L.F.	5.50	6.31	.75	12.56	
Clean	L.F.	.15	.33		.48	
Paint	L.F.	.18	2.21		2.39	
Minimum Charge	Job		157		157	

Action	Scope
Demolish	Includes removal of the item and hauling the debris to a truck, dumpster, or storage area.
Install	Includes the installation of the item with normal placement and spacing of fasteners.
Demolish and Install	Includes the demolish action and the install action.
Reinstall	Includes the reinstallation of a removed item.
Clean	Includes cleaning of the item.
Paint	Includes up to two coats of finish on the item (unless otherwise indicated).
Minimum Charge	Includes the minimum labor or the minimum labor and equipment cost. Use this charge for small quantities of material.

Job Costs

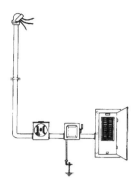

Scaffolding	Temporary Electrical Service	Air Compressor

Construction Fees		Unit	Material	Labor	Equip.	Total	Specification
Asbestos Test	Minimum Charge	Job		181		181	Minimum charge for testing.
Architectural	Minimum Charge	Day	480			480	Minimum labor charge for architect.
Engineering	Minimum Charge	Day	250			250	Minimum labor charge for field engineer.

Construction Permits		Unit	Material	Labor	Equip.	Total	Specification
Job (less than 5K)	Minimum Charge	Job	110			110	Permit fees, job less than 5K.
Job (5K to 10K)	Minimum Charge	Job	220			220	Permit fees, job more than 5K, but less than 10K.
Job (10K to 25K)	Minimum Charge	Job	550			550	Permit fees, job more than 10K, but less than 25K.
Job (25K to 50K)	Minimum Charge	Job	1100			1100	Permit fees, job more than 25K, but less than 50K.
Job (50K to 100K)	Minimum Charge	Job	2200			2200	Permit fees, job more than 50K, but less than 100K.
Job (100K to 250K)	Minimum Charge	Job	5500			5500	Permit fees, job more than 100K, but less than 250K.
Job (250K to 500K)	Minimum Charge	Job	11000			11000	Permit fees, job more than 250K, but less than 500K.
Job (500K to 1M)	Minimum Charge	Job	22000			22000	Permit fees, job more than 500K, but less than 1 million.

Job Costs

Trade Labor		Unit	Material	Labor	Equip.	Total	Specification
Carpenter							
Daily (8 hours)							
	Install	Day		315		315	Daily labor rate for one carpenter.
	Minimum Charge	Job		157		157	
Weekly (40 hours)							
	Install	Week		1575		1575	Weekly labor rate for one carpenter.
	Minimum Charge	Job		157		157	
Drywaller							
Daily (8 hours)							
	Install	Day		315		315	Daily labor rate for one carpenter.
	Minimum Charge	Job		157		157	
Weekly (40 hours)							
	Install	Week		1575		1575	Weekly labor rate for one carpenter.
	Minimum Charge	Job		157		157	
Roofer							
Daily (8 hours)							
	Install	Day		288		288	Daily labor rate for one skilled roofer.
	Minimum Charge	Job		144		144	
Weekly (40 hours)							
	Install	Week		1450		1450	Weekly labor rate for one skilled
	Minimum Charge	Job		144		144	roofer.
Painter							
Daily (8 hours)							
	Install	Day		276		276	Daily labor rate for one painter.
	Minimum Charge	Job		138		138	
Weekly (40 hours)							
	Install	Week		1375		1375	Weekly labor rate for one painter.
	Minimum Charge	Job		138		138	
Mason							
Daily (8 hours)							
	Install	Day		320		320	Daily labor rate for one skilled mason.
	Minimum Charge	Job		160		160	
Weekly (40 hours)							
	Install	Week		1600		1600	Weekly labor rate for one skilled
	Minimum Charge	Job		160		160	mason.
Electrician							
Daily (8 hours)							
	Install	Day		345		345	Daily labor rate for an electrician.
	Minimum Charge	Job		172		172	
Weekly (40 hours)							
	Install	Week		1725		1725	Weekly labor rate for an electrician.
	Minimum Charge	Job		172		172	
Plumber							
Daily (8 hours)							
	Install	Day		345		345	Daily labor rate for a plumber.
	Minimum Charge	Job		172		172	
Weekly (40 hours)							
	Install	Week		1725		1725	Weekly labor rate for a plumber.
	Minimum Charge	Job		172		172	
Common Laborer							
Daily (8 hours)							
	Install	Day		230		230	Daily labor rate for a common
	Minimum Charge	Job		115		115	laborer.

Job Costs

Trade Labor		Unit	Material	Labor	Equip.	Total	Specification
Weekly (40 hours)							
	Install	Week		1150		1150	Weekly labor rate for a common
	Minimum Charge	Job		115		115	laborer.

Temporary Utilities		Unit	Material	Labor	Equip.	Total	Specification
Power Pole							
Job Site Electricity							
	Install	Month	33			33	Includes average cost of electricity per
	Minimum Charge	Job		172		172	month.
Water Service							
Job Site Water							
	Install	Month	62.50			62.50	Includes average costs for water for
	Minimum Charge	Month	62.50			62.50	one month.
Telephone							
	Install	Month	224			224	Includes average cost for local and
	Minimum Charge	Month	224			224	long distance telephone service for
							one month.
Chemical Toilet							
	Install	Month			248	248	Monthly rental of portable toilet.
	Minimum Charge	Day			12.40	12.40	
Office Trailer							
Place and Remove							
	Install	Job	234			234	Includes cost to a contractor to deliver,
	Minimum Charge	Job		125		125	place and level an office trailer and
							remove at end of job.
Storage Container							
	Install	Job	155			155	Includes cost to a contractor to deliver
	Minimum Charge	Job		125		125	and place a storage container and
							remove at end of job.
Fencing							
	Install	L.F.	1.80			1.80	Includes material costs for temporary
	Minimum Charge	L.F.	275			275	fencing.
Railing							
	Install	L.F.	7.15	8.90	.47	16.52	Includes material and labor to install
	Minimum Charge	L.F.	275			275	primed steel pipe.

Equipment Rental		Unit	Material	Labor	Equip.	Total	Specification
Flatbed Truck							
Daily Rental							
	Install	Day			134	134	Daily rental of flatbed truck.
	Minimum Charge	Job			134	134	
Weekly Rental							
	Install	Week			670	670	Weekly rental of flatbed truck.
	Minimum Charge	Job			134	134	
Monthly Rental							
	Install	Month			2675	2675	Monthly rental of flatbed truck.
	Minimum Charge	Job			134	134	

Equipment Rental		Unit	Material	Labor	Equip.	Total	Specification
Dump Truck							
Daily Rental							
	Install	Day			360	360	Daily rental of dump truck.
	Minimum Charge	Day			360	360	
Weekly Rental							
	Install	Week			1800	1800	Weekly rental of dump truck.
	Minimum Charge	Day			360	360	
Monthly Rental							
	Install	Month			7150	7150	Monthly rental of dump truck.
	Minimum Charge	Day			360	360	
Forklift							
Daily Rental							
	Install	Day			240	240	Daily rental of forklift.
	Minimum Charge	Day			240	240	
Weekly Rental							
	Install	Week			1200	1200	Weekly rental of forklift.
	Minimum Charge	Day			240	240	
Monthly Rental							
	Install	Month			4800	4800	Monthly rental of forklift.
	Minimum Charge	Day			240	240	
Bobcat							
Daily Rental							
	Install	Day			158	158	Daily rental of wheeled, skid steer
	Minimum Charge	Day			158	158	loader.
Weekly Rental							
	Install	Week			790	790	Weekly rental of wheeled, skid steer
	Minimum Charge	Day			158	158	loader.
Monthly Rental							
	Install	Month			3150	3150	Monthly rental of wheeled, skid steer
	Minimum Charge	Day			158	158	loader.
Air Compressor							
Daily Rental							
	Install	Day			57.50	57.50	Daily rental of air compressor.
	Minimum Charge	Day			57.50	57.50	
Weekly Rental							
	Install	Week			288	288	Weekly rental of air compressor.
	Minimum Charge	Day			57.50	57.50	
Monthly Rental							
	Install	Month			1150	1150	Monthly rental of air compressor.
	Minimum Charge	Day			57.50	57.50	
Jackhammer							
Daily Rental							
	Install	Day			12.30	12.30	Daily rental of air tool, jackhammer.
	Minimum Charge	Day			12.30	12.30	
Weekly Rental							
	Install	Week			61.50	61.50	Weekly rental of air tool, jackhammer.
	Minimum Charge	Day			12.30	12.30	
Monthly Rental							
	Install	Month			246	246	Monthly rental of air tool,
	Minimum Charge	Day			12.30	12.30	jackhammer.
Concrete Mixer							
Daily Rental							
	Install	Day			99	99	Daily rental of concrete mixer.
	Minimum Charge	Day			99	99	

Job Costs

Equipment Rental		Unit	Material	Labor	Equip.	Total	Specification
Weekly Rental							
	Install	Week			495	495	Weekly rental of concrete mixer.
	Minimum Charge	Day			99	99	
Monthly Rental							
	Install	Month			1975	1975	Monthly rental of concrete mixer.
	Minimum Charge	Day			99	99	
Concrete Bucket							
Daily Rental							
	Install	Day			14.55	14.55	Daily rental of concrete bucket.
	Minimum Charge	Day			14.55	14.55	
Weekly Rental							
	Install	Week			73	73	Weekly rental of concrete bucket.
	Minimum Charge	Day			14.55	14.55	
Monthly Rental							
	Install	Month			291	291	Monthly rental of concrete bucket.
	Minimum Charge	Day			14.55	14.55	
Concrete Pump							
Daily Rental							
	Install	Day			790	790	Daily rental of concrete pump.
	Minimum Charge	Day			790	790	
Weekly Rental							
	Install	Week			3950	3950	Weekly rental of concrete pump.
	Minimum Charge	Day			790	790	
Monthly Rental							
	Install	Month			15800	15800	Monthly rental of concrete pump.
	Minimum Charge	Day			790	790	
Generator							
Daily Rental							
	Install	Day			28.50	28.50	Daily rental of generator.
	Minimum Charge	Day			28.50	28.50	
Weekly Rental							
	Install	Week			143	143	Weekly rental of generator.
	Minimum Charge	Day			28.50	28.50	
Monthly Rental							
	Install	Month			575	575	Monthly rental of generator.
	Minimum Charge	Day			28.50	28.50	
Welder							
Daily Rental							
	Install	Day			70.50	70.50	Daily rental of welder.
	Minimum Charge	Day			70.50	70.50	
Weekly Rental							
	Install	Week			350	350	Weekly rental of welder.
	Minimum Charge	Day			70.50	70.50	
Monthly Rental							
	Install	Month			1400	1400	Monthly rental of welder.
	Minimum Charge	Day			70.50	70.50	
Sandblaster							
Daily Rental							
	Install	Day			17.20	17.20	Daily rental of sandblaster.
	Minimum Charge	Day			17.20	17.20	
Weekly Rental							
	Install	Week			86	86	Weekly rental of sandblaster.
	Minimum Charge	Day			17.20	17.20	

Job Costs

Equipment Rental

		Unit	Material	Labor	Equip.	Total	Specification
Monthly Rental							
	Install	Month			345	345	Monthly rental of sandblaster.
	Minimum Charge	Day			17.20	17.20	
Space Heater							
Daily Rental							
	Install	Day			16.30	16.30	Daily rental of space heater.
	Minimum Charge	Day			16.10	16.10	
Weekly Rental							
	Install	Day			81.50	81.50	Weekly rental of space heater.
	Minimum Charge	Day			16.10	16.10	
Monthly Rental							
	Install	Day			325	325	Monthly rental of space heater.
	Minimum Charge	Day			16.10	16.10	

Move / Reset Contents

		Unit	Material	Labor	Equip.	Total	Specification
Small Room							
	Install	Room		28		28	Includes cost to remove and reset the
	Minimum Charge	Job		115		115	contents of a small room.
Average Room							
	Install	Room		33.50		33.50	Includes cost to remove and reset the
	Minimum Charge	Job		115		115	contents of an average room.
Large Room							
	Install	Room		46.50		46.50	Includes cost to remove and reset the
	Minimum Charge	Job		115		115	contents of a large room.
Extra Large Room							
	Install	Room		73.50		73.50	Includes cost to remove and reset the
	Minimum Charge	Job		115		115	contents of a extra large room.
Store Contents							
Weekly							
	Install	Week	152			152	Storage fee, per week.
Monthly							
	Install	Month	455			455	Storage fee, per month.

Cover and Protect

		Unit	Material	Labor	Equip.	Total	Specification
Cover / Protect Walls							
	Install	SF Wall		.09		.09	Includes labor costs to cover and
	Minimum Charge	Job		115		115	protect walls.
Cover / Protect Floors							
	Install	SF Flr.		.09		.09	Includes labor costs to cover and
	Minimum Charge	Job		115		115	protect floors.

Temporary Bracing

		Unit	Material	Labor	Equip.	Total	Specification
Shim Support Piers							
	Install	Ea.	3.31	90.50		93.81	Includes labor and material to install
	Minimum Charge	Job		271		271	shims support for piers.

Job Costs

Jack and Relevel

Jack and Relevel		Unit	Material	Labor	Equip.	Total	Specification
Single Story							
	Install	Ea.	1.22	1.08		2.30	Includes labor and equipment to jack
	Minimum Charge	Job		271		271	and relevel single-story building.
Two-story							
	Install	Ea.	1.22	1.81		3.03	Includes labor and equipment to jack
	Minimum Charge	Job		271		271	and relevel two-story building.

Dumpster Rental		Unit	Material	Labor	Equip.	Total	Specification
Small Load							
	Install	Week			465	465	Dumpster, weekly rental, 2 dumps, 5
	Minimum Charge	Week			465	465	yard container.
Medium Load							
	Install	Week			560	560	Dumpster, weekly rental, 2 dumps, 10
	Minimum Charge	Week			560	560	yard container.
Large Load							
	Install	Week			985	985	Dumpster, weekly rental, 2 dumps, 30
	Minimum Charge	Week			985	985	yard container.

Debris Hauling		Unit	Material	Labor	Equip.	Total	Specification
Per Ton							
	Install	Ton		71.50	39	110.50	Includes hauling of debris by trailer or
	Minimum Charge	Ton		535	291	826	dump truck.
Per Cubic Yard							
	Install	C.Y.		24	12.95	36.95	Includes hauling of debris by trailer or
	Minimum Charge	Job		535	291	826	dump truck.
Per Pick-up Truck Load							
	Install	Ea.		128		128	Includes hauling of debris by pick-up
	Minimum Charge	Job		535	291	826	truck.

Dump Fees		Unit	Material	Labor	Equip.	Total	Specification
Per Cubic Yard							
	Install	C.Y.	6.95			6.95	Dump charges, building materials.
	Minimum Charge	C.Y.	6.95			6.95	
Per Ton							
	Install	Ton	55.50			55.50	Dump charges, building materials.
	Minimum Charge	Ton	55.50			55.50	

Scaffolding		Unit	Material	Labor	Equip.	Total	Specification
4' to 6' High							
	Install	Month	51.50			51.50	Includes monthly rental of scaffolding,
	Minimum Charge	Month	51.50			51.50	steel tubular, 30" wide, 7' long, 5' high.
7' to 11' High							
	Install	Month	61.50			61.50	Includes monthly rental of scaffolding,
	Minimum Charge	Month	61.50			61.50	steel tubular, 30" wide, 7' long, 10' high.

Job Costs

Scaffolding		Unit	Material	Labor	Equip.	Total	Specification
12' to 16' High							
	Install	Month	71			71	Includes monthly rental of scaffolding,
	Minimum Charge	Month	71			71	steel tubular, 30" wide, 7' long, 15' high.
17' to 21' High							
	Install	Month	95			95	Includes monthly rental of scaffolding,
	Minimum Charge	Month	95			95	steel tubular, 30" wide, 7' long, 20' high.
22' to 26' High							
	Install	Month	105			105	Includes monthly rental of scaffolding,
	Minimum Charge	Month	105			105	steel tubular, 30" wide, 7' long, 25' high.
27' to 30' High							
	Install	Month	114			114	Includes monthly rental of scaffolding,
	Minimum Charge	Month	114			114	steel tubular, 30" wide, 7' long, 30' high.

Construction Clean-up		Unit	Material	Labor	Equip.	Total	Specification
Initial							
	Clean	S.F.		.29		.29	Includes labor for general clean-up.
	Minimum Charge	Job		115		115	
Progressive							
	Clean	S.F.		.14		.14	Includes labor for progressive site/job
	Minimum Charge	Job		115		115	clean-up on a weekly basis.

Job Preparation

Building Demolition		Unit	Material	Labor	Equip.	Total	Specification
Wood-framed Building							
1 Story Demolition							
	Demolish	S.F.		1.07	1.32	2.39	Includes demolition of 1 floor only light wood-framed building, including hauling of debris. Dump fees not included.
	Minimum Charge	Job		1700	2100	3800	
2nd Story Demolition							
	Demolish	S.F.		1.09	1.35	2.44	Includes demolition of 2nd floor only light wood-framed building, including hauling of debris. Dump fees not included.
	Minimum Charge	Job		1700	2100	3800	
3rd Story Demolition							
	Demolish	S.F.		1.42	1.75	3.17	Includes demolition of 3rd floor only light wood-framed building, including hauling of debris. Dump fees not included.
	Minimum Charge	Job		1700	2100	3800	
Masonry Building							
	Demolish	S.F.		1.52	1.89	3.41	Includes demolition of one story masonry building, including hauling of debris. Dump fees not included.
	Minimum Charge	Job		1700	2100	3800	
Concrete Building							
Unreinforced							
	Demolish	S.F.		3.12	.97	4.09	Includes demolition of a one story non-reinforced concrete building, including hauling of debris. Dump fees not included.
	Minimum Charge	Job		1700	2100	3800	
Reinforced							
	Demolish	S.F.		3.57	1.11	4.68	Includes demolition of a one story reinforced concrete building, including hauling of debris. Dump fees not included.
	Minimum Charge	Job		1700	2100	3800	
Steel Building							
	Demolish	S.F.		2.03	2.51	4.54	Includes demolition of a metal building, including hauling of debris. Dump fees not included. No salvage value assumed for scrap metal.
	Minimum Charge	Job		1700	2100	3800	

Job Preparation

General Clean-up		Unit	Material	Labor	Equip.	Total	Specification
Remove Debris							
	Clean	S.F.		.70	.13	.83	Includes removal of debris caused by the loss including contents items and loose debris.
	Minimum Charge	Job		350	67	417	
Muck-out							
	Clean	S.F.		.45		.45	Includes labor to remove 2" of muck and mud after a flood loss.
	Minimum Charge	Job		115		115	
Deodorize / Disinfect							
	Clean	SF Flr.		.15		.15	Includes deodorizing building with application of mildicide and disinfectant to floors and walls - will vary depending on the extent of deodorization required.
	Minimum Charge	Job		115		115	
Clean Walls							
Heavy							
	Clean	S.F.		.11		.11	Includes labor and material for heavy cleaning (multiple applications) with detergent and solvent.
	Minimum Charge	Job		115	.	115	
Clean Ceiling							
Light							
	Clean	S.F.		.18		.18	Includes labor and material to clean light to moderate smoke and soot.
	Minimum Charge	Job		115		115	
Heavy							
	Clean	S.F.		.28		.28	Includes labor and material to clean heavy smoke and soot.
	Minimum Charge	Job		115		115	

Flood Clean-up		Unit	Material	Labor	Equip.	Total	Specification
Emergency Service Call							
After Hours / Weekend							
	Minimum Charge	Job		153		153	Includes minimum charges for deflooding clean-up after hours or on weekends.
Grey Water / Sewage							
	Minimum Charge	Job		115		115	Includes minimum charges for deflooding clean-up.
Water Extraction							
Clean (non-grey)							
	Clean	SF Flr.	.02	.46		.48	Includes labor to remove clean uncontaminated water from the loss.
	Minimum Charge	Job		115		115	
Grey (non-solids)							
	Clean	SF Flr.	.02	.92		.94	Includes labor to remove non-solid grey water from the loss.
	Minimum Charge	Job		115		115	
Sewage / Solids							
	Clean	SF Flr.	.04	1.15		1.19	Includes labor to remove water from the loss that may contain sewage or solids.
	Minimum Charge	Job		115		115	

Job Preparation

Flood Clean-up		Unit	Material	Labor	Equip.	Total	Specification
Mildicide Walls							
Topographical							
	Clean	SF Wall		.21		.21	Clean-up of deflooding including
	Minimum Charge	Job		115		115	topographical mildicide application.
Injection Treatments							
	Clean	SF Flr.	25	3.53		28.53	Clean-up of deflooding including
	Minimum Charge	Job		115		115	mildicide injection.
Carpet Cleaning							
Uncontaminated							
	Clean	SF Flr.		.14		.14	Includes labor to clean carpet.
	Minimum Charge	Job		115		115	
Contaminated							
	Clean	SF Flr.		.23		.23	Includes labor to clean contaminated
	Minimum Charge	Job		115		115	carpet.
Stairway							
	Clean	Ea.	.10	2.70		2.80	Includes labor to clean contaminated
	Minimum Charge	Job		115		115	carpet on a stairway.
Carpet Treatment							
Disinfect / Deodorize							
	Clean	SF Flr.		.14		.14	Includes disinfecting and deodorizing
	Minimum Charge	Job		115		115	of carpet.
Disinfect For Sewage							
	Clean	SF Flr.		.37		.37	Includes labor to disinfect and
	Minimum Charge	Job		115		115	deodorize carpet exposed to sewage.
Mildicide							
	Clean	SF Flr.	.01	.14		.15	Includes labor and materials to apply
	Minimum Charge	Job		115		115	mildicide to carpet.
Thermal Fog Area							
	Clean	SF Flr.	.03	.35		.38	Includes labor and materials to
	Minimum Charge	Job		115		115	perform thermal fogging during
							deflooding clean-up.
Wet Fog / ULV Area							
	Clean	SF Flr.	.02	.20		.22	Includes labor and materials to
	Minimum Charge	Job		115		115	perform wet fogging / ULV during
							deflooding clean-up.
Steam Clean Fixtures							
	Clean	Ea.	.01	23		23.01	Includes labor and materials to steam
	Minimum Charge	Job		115		115	clean fixtures during deflooding
							clean-up.
Pressure Wash Sewage							
	Clean	SF Flr.		2.09		2.09	Includes labor to pressure wash
	Minimum Charge	Job		115		115	sewage contamination.
Airmover							
	Minimum Charge	Day			16.10	16.10	Daily rental of air mover.

Job Preparation

Flood Clean-up

Flood Clean-up		Unit	Material	Labor	Equip.	Total	Specification
Dehumidifier							
Medium Size							
	Minimum Charge	Day			33.50	33.50	Daily rental of dehumidifier, medium.
Large Size							
	Minimum Charge	Day			66.50	66.50	Daily rental of dehumidifier, large.
Air Purifier							
HEPA Filter-Bacteria							
	Minimum Charge	Ea.	195			195	HEPA filter for air purifier.

Asbestos Removal

Asbestos Removal		Unit	Material	Labor	Equip.	Total	Specification
Asbestos Analysis							
	Minimum Charge	Job		181		181	Minimum charge for testing.
Asbestos Pipe Insulation							
1/2" to 3/4" Diameter							
	Demolish	L.F.	.35	4.69		5.04	Minimum charge to remove asbestos.
	Minimum Charge	Job		2900		2900	Includes full Tyvek suits for workers changed four times per eight-hour shift with respirators, air monitoring and supervision by qualified professionals.
1" to 3" Diameter							
	Demolish	L.F.	1.08	14.55		15.63	Minimum charge to remove asbestos.
	Minimum Charge	Job		2900		2900	Includes full Tyvek suits for workers changed four times per eight-hour shift with respirators, air monitoring and supervision by qualified professionals.
Asbestos-based Siding							
	Demolish	S.F.	2.40	32.50		34.90	Minimum charge to remove asbestos.
	Minimum Charge	Job		2900		2900	Includes full Tyvek suits for workers changed four times per eight-hour shift with respirators, air monitoring and supervision by qualified professionals.
Asbestos-based Plaster							
	Demolish	S.F.	1.85	25		26.85	Minimum charge to remove asbestos.
	Minimum Charge	Job		2900		2900	Includes full Tyvek suits for workers changed four times per eight-hour shift with respirators, air monitoring and supervision by qualified professionals.
Encapsulate Asbestos Ceiling							
	Install	S.F.	.29	.14		.43	Includes encapsulation with penetrating sealant sprayed onto ceiling and walls with airless sprayer.
	Minimum Charge	Job		2900		2900	
Scrape Acoustical Ceiling							
	Demolish	S.F.	.07	.82		.89	Minimum charge to remove asbestos.
	Minimum Charge	Job		2900		2900	Includes full Tyvek suits for workers changed four times per eight-hour shift with respirators, air monitoring and supervision by qualified professionals.

Job Preparation

Asbestos Removal

		Unit	Material	Labor	Equip.	Total	Specification
Seal Ceiling Asbestos	Install	S.F.	1.10	2.74		3.84	Includes attaching plastic cover to ceiling area and sealing where asbestos exists, taping all seams, caulk or foam insulation in cracks and negative air system.
	Minimum Charge	Job		2900		2900	
Seal Wall Asbestos	Install	S.F.	1.10	2.43		3.53	Includes attaching plastic cover to wall area and sealing where asbestos exists, taping all seams, caulk or foam insulation in cracks and negative air system.
	Minimum Charge	Job		2900		2900	
Seal Floor Asbestos	Install	S.F.	1.76	4.86		6.62	Includes attaching plastic cover to floor area and sealing where asbestos exists, taping all seams, caulk or foam insulation in cracks and negative air system.
	Minimum Charge	Job		2900		2900	
Negative Air Vent System	Install	Day			43.50	43.50	Daily rental of negative air vent system.
	Minimum Charge	Week			218	218	
HEPA Vacuum Cleaner	Install	Day			18.25	18.25	Daily rental of HEPA vacuum, 16 gallon.
	Minimum Charge	Week			91	91	
Airless Sprayer	Install	Day			40	40	Daily rental of airless sprayer.
	Minimum Charge	Week			201	201	
Decontamination Unit	Install	Day			133	133	Daily rental of decontamination unit.
	Minimum Charge	Week			665	665	
Light Stand	Install	Day			10.30	10.30	Daily rental of floodlight w/tripod.
	Minimum Charge	Week			51.50	51.50	

Room Demolition

		Unit	Material	Labor	Equip.	Total	Specification
Strip Typical Room	Demolish	S.F.		2.30		2.30	Includes 4 labor hours for general demolition.
	Minimum Charge	Job		128		128	
Strip Typical Bathroom	Demolish	S.F.		3.06		3.06	Includes 4 labor hours for general demolition.
	Minimum Charge	Job		128		128	
Strip Typical Kitchen	Demolish	S.F.		2.94		2.94	Includes 4 labor hours for general demolition.
	Minimum Charge	Job		128		128	
Strip Typical Laundry	Demolish	S.F.		2.62		2.62	Includes 4 labor hours for general demolition.
	Minimum Charge	Job		128		128	

Job Preparation

Selective Demolition		Unit	Material	Labor	Equip.	Total	Specification
Concrete Slab							
4″ Unreinforced							
	Demolish	S.F.		1.54	.23	1.77	Includes labor and equipment to
	Minimum Charge	Job		655	98	753	remove unreinforced concrete slab and haul debris to truck or dumpster.
6″ Unreinforced							
	Demolish	S.F.		2.18	.33	2.51	Includes labor and equipment to
	Minimum Charge	Job		655	98	753	remove unreinforced concrete slab and haul debris to truck or dumpster.
8″ Unreinforced							
	Demolish	S.F.		3.58	.54	4.12	Includes labor and equipment to
	Minimum Charge	Job		655	98	753	remove unreinforced concrete slab and haul debris to truck or dumpster.
4″ Reinforced							
	Demolish	S.F.		1.93	.29	2.22	Includes labor and equipment to
	Minimum Charge	Job		655	98	753	remove reinforced concrete slab and haul debris to truck or dumpster.
6″ Reinforced							
	Demolish	S.F.		2.90	.43	3.33	Includes labor and equipment to
	Minimum Charge	Job		655	98	753	remove reinforced concrete slab and haul debris to truck or dumpster.
8″ Reinforced							
	Demolish	S.F.		4.75	.71	5.46	Includes labor and equipment to
	Minimum Charge	Job		655	98	753	remove reinforced concrete slab and haul debris to truck or dumpster.
Brick Wall							
4″ Thick							
	Demolish	S.F.		2.14	.32	2.46	Includes labor and equipment to
	Minimum Charge	Job		655	98	753	remove brick wall and haul debris to a central location for disposal.
8″ Thick							
	Demolish	S.F.		2.94	.44	3.38	Includes labor and equipment to
	Minimum Charge	Job		655	98	753	remove brick wall and haul debris to a central location for disposal.
12″ Thick							
	Demolish	S.F.		3.92	.59	4.51	Includes labor and equipment to
	Minimum Charge	Job		655	98	753	remove brick wall and haul debris to a central location for disposal.
Concrete Masonry Wall							
4″ Thick							
	Demolish	S.F.		2.37	.36	2.73	Includes minimum labor and
	Minimum Charge	Job		655	98	753	equipment to remove 4″ block wall and haul debris to a central location for disposal.
6″ Thick							
	Demolish	S.F.		2.94	.44	3.38	Includes minimum labor and
	Minimum Charge	Job		655	98	753	equipment to remove 6″ block wall and haul debris to a central location for disposal.

Job Preparation

Selective Demolition		Unit	Material	Labor	Equip.	Total	Specification
8" Thick							
	Demolish	S.F.		3.62	.54	4.16	Includes minimum labor and
	Minimum Charge	Job		655	98	753	equipment to remove 8" block wall
							and haul debris to a central location
							for disposal.
12" Thick							
	Demolish	S.F.		5.20	.78	5.98	Includes minimum labor and
	Minimum Charge	Job		655	98	753	equipment to remove 12" block wall
							and haul debris to a central location
							for disposal.
Tree Removal							
Small, 4" - 8"							
	Demolish	Ea.		82.50	60.50	143	Minimum charge, tree removal.
	Minimum Charge	Job		115		115	
Medium, 9" - 12"							
	Demolish	Ea.		123	91	214	Minimum charge, tree removal.
	Minimum Charge	Job		115		115	
Large, 13" - 18"							
	Demolish	Ea.		148	109	257	Minimum charge, tree removal.
	Minimum Charge	Job		375	122	497	
Very Large, 19" - 24"							
	Demolish	Ea.		185	137	322	Minimum charge, tree removal.
	Minimum Charge	Job		375	122	497	
Stump Removal							
	Demolish	Ea.		50.50	113	163.50	Includes minimum charges for a site
	Minimum Charge	Job		209	63.50	272.50	demolition crew and backhoe /
							loader.

Excavation

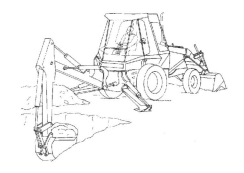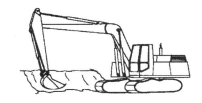

Backhoe/Loader - Wheel Type	Tractor Loader - Wheel Type	Backhoe - Crawler Type

Excavation		Unit	Material	Labor	Equip.	Total	Specification
Footings							
Hand Trenching							
	Install	L.F.		1.44		1.44	Includes hand excavation for footing
	Minimum Charge	Job		57.50		57.50	or trench up to 10″ deep and 12″ wide.
Machine Trenching							
	Install	L.F.		.54	.23	.77	Includes machine excavation for
	Minimum Charge	Job		268	114	382	footing or trench up to 10″ deep and 12″ wide.
Trenching w / Backfill							
Hand							
	Install	L.F.		2.30		2.30	Includes hand excavation, backfill and
	Minimum Charge	Job		57.50		57.50	compaction.
Machine							
	Install	L.F.		2.68	7.20	9.88	Includes machine excavation, backfill
	Minimum Charge	Job		115	32.50	147.50	and compaction.

Backfill		Unit	Material	Labor	Equip.	Total	Specification
Hand (no compaction)							
	Install	L.C.Y.		13.50		13.50	Includes backfill by hand in average
	Minimum Charge	Job		115		115	soil without compaction.
Hand (w / compaction)							
	Install	E.C.Y.		27		27	Includes backfill by hand in average
	Minimum Charge	Job		115		115	soil with compaction.
Machine (no compaction)							
	Install	L.C.Y.		2.68	1.14	3.82	Includes backfill, by 55 HP wheel
	Minimum Charge	Job		268	114	382	loader, of trenches from loose material piled adjacent without compaction.
Machine (w / compaction)							
	Install	L.C.Y.		7.15	3.05	10.20	Includes backfill, by 55 HP wheel
	Minimum Charge	Job		268	114	382	loader, of trenches from loose material piled adjacent with compaction by vibrating tampers.

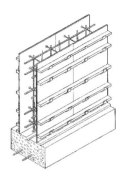

Concrete Footing and Formwork

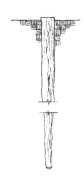

Wood Pile

Piles	Unit	Material	Labor	Equip.	Total	Specification
Treated Wood, 12″ Butt						
To 30′						
Demolish	L.F.		3.63	1.10	4.73	Includes material, labor and
Install	V.L.F.	9.75	3.49	2.76	16	equipment to install up to 30′ long
Demolish and Install	V.L.F.	9.75	7.12	3.86	20.73	treated wood piling.
Paint	L.F.	.37	1.58		1.95	
Minimum Charge	Job		2175	1725	3900	
30′ to 40′						
Demolish	L.F.		3.63	1.10	4.73	Includes material, labor and
Install	V.L.F.	9.50	3.12	2.47	15.09	equipment to install 30′ to 40′ long
Demolish and Install	V.L.F.	9.50	6.75	3.57	19.82	treated wood piling.
Paint	L.F.	.37	1.58		1.95	
Minimum Charge	Job		2175	1725	3900	
40′ to 50′						
Demolish	L.F.		3.63	1.10	4.73	Includes material, labor and
Install	V.L.F.	10.20	3.03	2.40	15.63	equipment to install 40′ to 50′ long
Demolish and Install	V.L.F.	10.20	6.66	3.50	20.36	treated wood piling.
Paint	L.F.	.37	1.58		1.95	
Minimum Charge	Job		2175	1725	3900	
50′ to 60′						
Demolish	L.F.		3.63	1.10	4.73	Includes material, labor and
Install	V.L.F.	11.30	2.73	2.16	16.19	equipment to install 50′ to 60′ long
Demolish and Install	V.L.F.	11.30	6.36	3.26	20.92	treated wood piling.
Paint	L.F.	.37	1.58		1.95	
Minimum Charge	Job		2175	1725	3900	
60′ to 80′						
Demolish	L.F.		3.63	1.10	4.73	Includes material, labor and
Install	V.L.F.	19.35	3.02	2.63	25	equipment to install 60′ to 80′ long
Demolish and Install	V.L.F.	19.35	6.65	3.73	29.73	treated wood piling.
Paint	L.F.	.37	1.58		1.95	
Minimum Charge	Job		2175	1725	3900	
Untreated Wood, 12″ Butt						
To 30′						
Demolish	L.F.		3.63	1.10	4.73	Includes material, labor and
Install	V.L.F.	6.60	3.49	2.76	12.85	equipment to install up to 30′ long
Demolish and Install	V.L.F.	6.60	7.12	3.86	17.58	untreated wood piling.
Paint	L.F.	.37	1.58		1.95	
Minimum Charge	Job		2175	1725	3900	

Foundation

Piles		Unit	Material	Labor	Equip.	Total	Specification
30' to 40'							
	Demolish	L.F.		3.63	1.10	4.73	Includes material, labor and
	Install	V.L.F.	6.60	3.12	2.47	12.19	equipment to install 30' to 40' long
	Demolish and Install	V.L.F.	6.60	6.75	3.57	16.92	untreated wood piling.
	Paint	L.F.	.37	1.58		1.95	
	Minimum Charge	Job		2175	1725	3900	
40' to 50'							
	Demolish	L.F.		3.63	1.10	4.73	Includes material, labor and
	Install	V.L.F.	6.60	3.03	2.40	12.03	equipment to install 40' to 50' long
	Demolish and Install	V.L.F.	6.60	6.66	3.50	16.76	untreated wood piling.
	Paint	L.F.	.37	1.58		1.95	
	Minimum Charge	Job		2175	1725	3900	
50' to 60'							
	Demolish	L.F.		3.63	1.10	4.73	Includes material, labor and
	Install	V.L.F.	6.70	2.73	2.16	11.59	equipment to install 50' to 60' long
	Demolish and Install	V.L.F.	6.70	6.36	3.26	16.32	untreated wood piling.
	Paint	L.F.	.37	1.58		1.95	
	Minimum Charge	Job		2175	1725	3900	
60' to 80'							
	Demolish	L.F.		3.63	1.10	4.73	Includes material, labor and
	Install	V.L.F.	8.35	2.60	2.05	13	equipment to install 60' to 80' long
	Demolish and Install	V.L.F.	8.35	6.23	3.15	17.73	untreated wood piling.
	Paint	L.F.	.37	1.58		1.95	
	Minimum Charge	Job		2175	1725	3900	
Precast Concrete							
10" Square							
	Demolish	L.F.		3.71	1.13	4.84	Includes material, labor and
	Install	V.L.F.	8.30	3.12	2.47	13.89	equipment to install a precast,
	Demolish and Install	V.L.F.	8.30	6.83	3.60	18.73	prestressed, 40' long, 10" square
	Paint	L.F.	.37	1.58		1.95	concrete pile.
	Minimum Charge	Job		2175	1725	3900	
12" Square							
	Demolish	L.F.		3.98	1.21	5.19	Includes material, labor and
	Install	V.L.F.	10.45	3.21	2.54	16.20	equipment to install a precast,
	Demolish and Install	V.L.F.	10.45	7.19	3.75	21.39	prestressed, 40' long, 12" square
	Paint	L.F.	.37	1.58		1.95	concrete pile.
	Minimum Charge	Job		2175	1725	3900	
14" Square							
	Demolish	L.F.		4.07	1.24	5.31	Includes material, labor and
	Install	V.L.F.	12.30	3.64	2.88	18.82	equipment to install a precast,
	Demolish and Install	V.L.F.	12.30	7.71	4.12	24.13	prestressed, 40' long, 14" square
	Paint	L.F.	.37	1.58		1.95	concrete pile.
	Minimum Charge	Job		2175	1725	3900	
16" Square							
	Demolish	L.F.		4.28	1.30	5.58	Includes material, labor and
	Install	V.L.F.	19.45	3.90	3.08	26.43	equipment to install a precast,
	Demolish and Install	V.L.F.	19.45	8.18	4.38	32.01	prestressed, 40' long, 16" square
	Paint	L.F.	.37	1.58		1.95	concrete pile.
	Minimum Charge	Job		2175	1725	3900	
18" Square							
	Demolish	L.F.		4.64	1.41	6.05	Includes material, labor and
	Install	V.L.F.	23.50	4.88	4.25	32.63	equipment to install a precast,
	Demolish and Install	V.L.F.	23.50	9.52	5.66	38.68	prestressed, 40' long, 18" square
	Paint	L.F.	.37	1.58		1.95	concrete pile.
	Minimum Charge	Job		2175	1725	3900	

Foundation

Piles

	Unit	Material	Labor	Equip.	Total	Specification
Cross Bracing						
Treated 2″ x 6″						
Demolish	L.F.		.75		.75	Cost includes material and labor to
Install	L.F.	.84	4.37		5.21	install 2″ x 6″ lumber for cross-bracing
Demolish and Install	L.F.	.84	5.12		5.96	of foundation piling or posts.
Reinstall	L.F.		3.50		3.50	
Clean	S.F.		.29		.29	
Paint	L.F.	.17	1.06		1.23	
Minimum Charge	Job		540		540	
Treated 2″ x 8″						
Demolish	L.F.		.75		.75	Cost includes material and labor to
Install	L.F.	1.05	4.52		5.57	install 2″ x 8″ lumber for cross-bracing
Demolish and Install	L.F.	1.05	5.27		6.32	of foundation piling or posts.
Reinstall	L.F.		3.62		3.62	
Clean	S.F.		.29		.29	
Paint	L.F.	.22	1.10		1.32	
Minimum Charge	Job		540		540	
Treated 2″ x 10″						
Demolish	L.F.		.77		.77	Cost includes material and labor to
Install	L.F.	1.30	4.68		5.98	install 2″ x 10″ lumber for
Demolish and Install	L.F.	1.30	5.45		6.75	cross-bracing of foundation piling or
Reinstall	L.F.		3.74		3.74	posts.
Clean	S.F.		.29		.29	
Paint	L.F.	.26	1.13		1.39	
Minimum Charge	Job		540		540	
Treated 2″ x 12″						
Demolish	L.F.		.93		.93	Cost includes material and labor to
Install	L.F.	1.99	4.84		6.83	install 2″ x 12″ lumber for
Demolish and Install	L.F.	1.99	5.77		7.76	cross-bracing of foundation piling or
Reinstall	L.F.		3.87		3.87	posts.
Clean	S.F.		.29		.29	
Paint	L.F.	.31	1.15		1.46	
Minimum Charge	Job		540		540	
Metal Pipe						
6″ Concrete Filled						
Demolish	L.F.		5.15	1.57	6.72	Includes material, labor and
Install	V.L.F.	12	4.55	3.60	20.15	equipment to install concrete filled
Demolish and Install	V.L.F.	12	9.70	5.17	26.87	steel pipe pile.
Paint	L.F.	.29	1.58		1.87	
Minimum Charge	Job		2175	1725	3900	
6″ Unfilled						
Demolish	L.F.		5.15	1.57	6.72	Includes material, labor and
Install	V.L.F.	10.25	4.16	3.29	17.70	equipment to install unfilled steel pipe
Demolish and Install	V.L.F.	10.25	9.31	4.86	24.42	pile.
Paint	L.F.	.29	1.58		1.87	
Minimum Charge	Job		2175	1725	3900	
8″ Concrete Filled						
Demolish	L.F.		5.15	1.57	6.72	Includes material, labor and
Install	V.L.F.	11.65	4.75	3.75	20.15	equipment to install concrete filled
Demolish and Install	V.L.F.	11.65	9.90	5.32	26.87	steel pipe pile.
Paint	L.F.	.29	1.58		1.87	
Minimum Charge	Job		2175	1725	3900	
8″ Unfilled						
Demolish	L.F.		5.15	1.57	6.72	Includes material, labor and
Install	V.L.F.	10.85	4.37	3.45	18.67	equipment to install unfilled steel pipe
Demolish and Install	V.L.F.	10.85	9.52	5.02	25.39	pile.
Paint	L.F.	.29	1.58		1.87	
Minimum Charge	Job		2175	1725	3900	

Foundation

Piles		Unit	Material	Labor	Equip.	Total	Specification
10" Concrete Filled							
	Demolish	L.F.		5.30	1.62	6.92	Includes material, labor and
	Install	V.L.F.	15.15	4.85	3.84	23.84	equipment to install concrete filled
	Demolish and Install	V.L.F.	15.15	10.15	5.46	30.76	steel pipe pile.
	Paint	L.F.	.29	1.58		1.87	
	Minimum Charge	Job		2175	1725	3900	
10" Unfilled							
	Demolish	L.F.		5.30	1.62	6.92	Includes material, labor and
	Install	V.L.F.	13.55	4.37	3.45	21.37	equipment to install unfilled steel pipe
	Demolish and Install	V.L.F.	13.55	9.67	5.07	28.29	pile.
	Paint	L.F.	.29	1.58		1.87	
	Minimum Charge	Job		2175	1725	3900	
12" Concrete Filled							
	Demolish	L.F.		5.30	1.62	6.92	Includes material, labor and
	Install	V.L.F.	17.65	5.25	4.16	27.06	equipment to install concrete filled
	Demolish and Install	V.L.F.	17.65	10.55	5.78	33.98	steel pipe pile.
	Paint	L.F.	.29	1.58		1.87	
	Minimum Charge	Job		2175	1725	3900	
12" Unfilled							
	Demolish	L.F.		5.30	1.62	6.92	Includes material, labor and
	Install	V.L.F.	16.70	4.60	3.63	24.93	equipment to install unfilled steel pipe
	Demolish and Install	V.L.F.	16.70	9.90	5.25	31.85	pile.
	Paint	L.F.	.29	1.58		1.87	
	Minimum Charge	Job		2175	1725	3900	
Steel H Section							
H Section 8 x 8 x 36#							
	Demolish	L.F.		3.60	1.09	4.69	Includes material, labor and
	Install	V.L.F.	9.50	3.41	2.70	15.61	equipment to install steel H section
	Demolish and Install	V.L.F.	9.50	7.01	3.79	20.30	pile.
	Paint	L.F.	.29	1.58		1.87	
	Minimum Charge	Job		2175	1725	3900	
H Section 10 x 10 x 57#							
	Demolish	L.F.		3.94	1.20	5.14	Includes material, labor and
	Install	V.L.F.	15.05	3.58	2.83	21.46	equipment to install steel H section
	Demolish and Install	V.L.F.	15.05	7.52	4.03	26.60	pile.
	Paint	L.F.	.29	1.58		1.87	
	Minimum Charge	Job		2175	1725	3900	
H Section 12 x 12 x 74#							
	Demolish	L.F.		4.07	1.24	5.31	Includes material, labor and
	Install	V.L.F.	19.85	4.30	3.75	27.90	equipment to install steel H section
	Demolish and Install	V.L.F.	19.85	8.37	4.99	33.21	pile.
	Paint	L.F.	.29	1.58		1.87	
	Minimum Charge	Job		2175	1725	3900	
H Section 14 x 14 x 89#							
	Demolish	L.F.		4.28	1.30	5.58	Includes material, labor and
	Install	V.L.F.	24	4.70	4.10	32.80	equipment to install steel H section
	Demolish and Install	V.L.F.	24	8.98	5.40	38.38	pile.
	Paint	L.F.	.29	1.58		1.87	
	Minimum Charge	Job		2175	1725	3900	

Foundation Post		Unit	Material	Labor	Equip.	Total	Specification
Wood Foundation Post							
4″ x 4″							
	Demolish	L.F.		.58	.18	.76	Cost includes material and labor to
	Install	L.F.	.92	3.83		4.75	install 4″ x 4″ treated wood post by
	Demolish and Install	L.F.	.92	4.41	.18	5.51	hand in normal soil conditions.
	Reinstall	L.F.		3.06		3.06	
	Minimum Charge	Job		540		540	
4″ x 6″							
	Demolish	L.F.		.58	.18	.76	Cost includes material and labor to
	Install	L.F.	1.39	3.83		5.22	install 4″ x 6″ treated wood post by
	Demolish and Install	L.F.	1.39	4.41	.18	5.98	hand in normal soil conditions.
	Reinstall	L.F.		3.06		3.06	
	Minimum Charge	Job		540		540	
6″ x 6″							
	Demolish	L.F.		.58	.18	.76	Cost includes material and labor to
	Install	L.F.	2.07	3.83		5.90	install 6″ x 6″ treated wood post by
	Demolish and Install	L.F.	2.07	4.41	.18	6.66	hand in normal soil conditions.
	Reinstall	L.F.		3.06		3.06	
	Minimum Charge	Job		540		540	
Water Jet							
	Demolish	Ea.		68		68	Includes material, labor and
	Install	Ea.	54.50	525		579.50	equipment to water jet a 6″ x 6″ wood
	Demolish and Install	Ea.	54.50	593		647.50	post under an existing building. Cost
	Reinstall	Ea.		524.80		524.80	may vary depending upon location
	Minimum Charge	Job		460		460	and access.
Shim							
	Install	Ea.	3.03	54		57.03	Includes labor and material to trim or
	Minimum Charge	Job		540		540	readjust existing foundation posts
							under an existing building. Access of
							2 - 4 foot assumed.
Cross-bracing							
Treated 2″ x 6″							
	Demolish	L.F.		.75		.75	Cost includes material and labor to
	Install	L.F.	.84	4.37		5.21	install 2″ x 6″ lumber for cross-bracing
	Demolish and Install	L.F.	.84	5.12		5.96	of foundation piling or posts.
	Reinstall	L.F.		3.50		3.50	
	Clean	S.F.		.29		.29	
	Paint	L.F.	.17	1.06		1.23	
	Minimum Charge	Job		540		540	
Treated 2″ x 8″							
	Demolish	L.F.		.75		.75	Cost includes material and labor to
	Install	L.F.	1.05	4.52		5.57	install 2″ x 8″ lumber for cross-bracing
	Demolish and Install	L.F.	1.05	5.27		6.32	of foundation piling or posts.
	Reinstall	L.F.		3.62		3.62	
	Clean	S.F.		.29		.29	
	Paint	L.F.	.22	1.10		1.32	
	Minimum Charge	Job		540		540	
Treated 2″ x 10″							
	Demolish	L.F.		.77		.77	Cost includes material and labor to
	Install	L.F.	1.30	4.68		5.98	install 2″ x 10″ lumber for
	Demolish and Install	L.F.	1.30	5.45		6.75	cross-bracing of foundation piling or
	Reinstall	L.F.		3.74		3.74	posts.
	Clean	S.F.		.29		.29	
	Paint	L.F.	.26	1.13		1.39	
	Minimum Charge	Job		540		540	

Foundation

Foundation Post

	Unit	Material	Labor	Equip.	Total	Specification
Treated 2" x 12"						Cost includes material and labor to
Demolish	L.F.		.93		.93	install 2" x 12" lumber for
Install	L.F.	1.99	4.84		6.83	cross-bracing of foundation piling or
Demolish and Install	L.F.	1.99	5.77		7.76	posts.
Reinstall	L.F.		3.87		3.87	
Clean	S.F.		.29		.29	
Paint	L.F.	.31	1.15		1.46	
Minimum Charge	Job		540		540	

Concrete Footing

	Unit	Material	Labor	Equip.	Total	Specification
Continuous						
12" w x 6" d						
Demolish	L.F.		1.11	.90	2.01	Includes material and labor to install
Install	L.F.	3.11	2.59	.02	5.72	continuous reinforced concrete footing
Demolish and Install	L.F.	3.11	3.70	.92	7.73	poured by chute cast against earth
Minimum Charge	Job		625		625	including reinforcing, forming and
						finishing.
12" w x 12" d						
Demolish	L.F.		2.14	1.73	3.87	Includes material and labor to install
Install	L.F.	7.35	5.15	.04	12.54	continuous reinforced concrete footing
Demolish and Install	L.F.	7.35	7.29	1.77	16.41	poured by chute cast against earth
Minimum Charge	Job		625		625	including reinforcing, forming and
						finishing.
18" w x 10" d						
Demolish	L.F.		2.78	2.25	5.03	Includes material and labor to install
Install	L.F.	6.80	6.35	.04	13.19	continuous reinforced concrete footing
Demolish and Install	L.F.	6.80	9.13	2.29	18.22	poured by chute cast against earth
Minimum Charge	Job		625		625	including reinforcing, forming and
						finishing.
24" w x 24" d						
Demolish	L.F.		4.63	3.75	8.38	Includes material and labor to install
Install	L.F.	20	20.50	.15	40.65	continuous reinforced concrete footing
Demolish and Install	L.F.	20	25.13	3.90	49.03	poured by chute cast against earth
Minimum Charge	Job		625		625	including reinforcing, forming and
						finishing.
Stem Wall						
Single Story						
Demolish	L.F.		10.30	8.35	18.65	Includes material and labor to install
Install	L.F.	9.90	54.50	44	108.40	stem wall 24" above and 18" below
Demolish and Install	L.F.	9.90	64.80	52.35	127.05	grade for a single story structure.
Minimum Charge	Job		625		625	Includes 12" wide and 8" deep
						footing.
Two Story						
Demolish	L.F.		10.30	8.35	18.65	Includes material and labor to install
Install	L.F.	12.30	78	63.50	153.80	stem wall 24" above and 18" below
Demolish and Install	L.F.	12.30	88.30	71.85	172.45	grade typical for a two story structure,
Minimum Charge	Job		625		625	includes 18" wide and 10" deep
						footing.

Foundation

Concrete Slab		Unit	Material	Labor	Equip.	Total	Specification
Reinforced							
4″							
	Demolish	S.F.		1.93	.29	2.22	Cost includes material and labor to
	Install	S.F.	1.62	.41		2.03	install 4″ reinforced concrete
	Demolish and Install	S.F.	1.62	2.34	.29	4.25	slab-on-grade poured by chute
	Clean	S.F.	.03	.26		.29	including forms, vapor barrier, wire
	Paint	S.F.	.12	.33		.45	mesh, 3000 PSI concrete, float finish,
	Minimum Charge	Job		143		143	and curing.
6″							
	Demolish	S.F.		2.90	.43	3.33	Cost includes material and labor to
	Install	S.F.	2.24	.61	.01	2.86	install 6″ reinforced concrete
	Demolish and Install	S.F.	2.24	3.51	.44	6.19	slab-on-grade poured by chute
	Clean	S.F.	.03	.26		.29	including forms, vapor barrier, wire
	Paint	S.F.	.12	.33		.45	mesh, 3000 PSI concrete, float finish,
	Minimum Charge	Job		143		143	and curing.
8″							
	Demolish	S.F.		4.75	.71	5.46	Cost includes material and labor to
	Install	S.F.	2.87	.80	.01	3.68	install 8″ reinforced concrete
	Demolish and Install	S.F.	2.87	5.55	.72	9.14	slab-on-grade poured by chute
	Clean	S.F.	.03	.26		.29	including forms, vapor barrier, wire
	Paint	S.F.	.12	.33		.45	mesh, 3000 PSI concrete, float finish,
	Minimum Charge	Job		143		143	and curing.
Unreinforced							
4″							
	Demolish	S.F.		1.54	.23	1.77	Cost includes material and labor to
	Install	S.F.	1.06	.71	.01	1.78	install 4″ concrete slab-on-grade
	Demolish and Install	S.F.	1.06	2.25	.24	3.55	poured by chute, including forms,
	Clean	S.F.	.03	.26		.29	vapor barrier, 3000 PSI concrete, float
	Paint	S.F.	.12	.33		.45	finish, and curing.
	Minimum Charge	Job		143		143	
6″							
	Demolish	S.F.		2.18	.33	2.51	Cost includes material and labor to
	Install	S.F.	1.55	.72	.01	2.28	install 6″ concrete slab-on-grade
	Demolish and Install	S.F.	1.55	2.90	.34	4.79	poured by chute, including forms,
	Clean	S.F.	.03	.26		.29	vapor barrier, 3000 PSI concrete, float
	Paint	S.F.	.12	.33		.45	finish, and curing.
	Minimum Charge	Job		143		143	
8″							
	Demolish	S.F.		3.58	.54	4.12	Cost includes material and labor to
	Install	S.F.	2.12	.76	.01	2.89	install 8″ concrete slab-on-grade
	Demolish and Install	S.F.	2.12	4.34	.55	7.01	poured by chute, including forms,
	Clean	S.F.	.03	.26		.29	vapor barrier, 3000 PSI concrete, float
	Paint	S.F.	.12	.33		.45	finish, and curing.
	Minimum Charge	Job		143		143	
Lightweight							
	Demolish	S.F.		.68	.14	.82	Includes material and labor to install
	Install	S.F.	1.41	2.13		3.54	lightweight concrete, 3″ - 4″ thick.
	Demolish and Install	S.F.	1.41	2.81	.14	4.36	
	Clean	S.F.	.03	.26		.29	
	Paint	S.F.	.12	.33		.45	
	Minimum Charge	Job		143		143	

Concrete Slab		Unit	Material	Labor	Equip.	Total	Specification
Gunite							
	Demolish	S.F.		.84	.25	1.09	Includes material and labor to install
	Install	S.F.	1.08	.91	.40	2.39	gunite on flat plane per inch of
	Demolish and Install	S.F.	1.08	1.75	.65	3.48	thickness using 1" as base price. No
	Clean	S.F.	.03	.26		.29	forms or reinforcing are included.
	Paint	S.F.	.12	.33		.45	
	Minimum Charge	Job		800		800	
Epoxy Inject							
	Install	L.F.	2.15	2.85		5	Includes labor and material to inject
	Minimum Charge	Job		143		143	epoxy into cracks.
Pressure Grout							
	Install	S.F.	2.44	2.11		4.55	Includes labor, material and
	Minimum Charge	Job		143		143	equipment to install pressure grout under an existing building.
Scrape and Paint							
	Paint	S.F.	.17	.69		.86	Includes labor and materials to scrape
	Minimum Charge	Job		138		138	and paint a concrete floor.
Vapor Barrier							
	Install	S.F.	.02	.08		.10	Cost includes material and labor to
	Minimum Charge	Job		157		157	install 4 mil polyethylene vapor barrier 10' wide sheets with 6" overlaps.

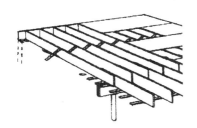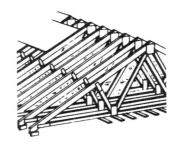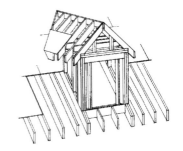

Rough Framing

Beam / Girder		Unit	Material	Labor	Equip.	Total	Specification
Solid Wood Beam							
2" x 8"							
	Demolish	L.F.		.62		.62	Cost includes material and labor to
	Install	L.F.	.88	.97		1.85	install 2" x 8" beam.
	Demolish and Install	L.F.	.88	1.59		2.47	
	Clean	L.F.	.10	.20		.30	
	Paint	L.F.	.12	1.49		1.61	
	Minimum Charge	Job		157		157	
2" x 10"							
	Demolish	L.F.		.77		.77	Cost includes material and labor to
	Install	L.F.	1.19	1.05		2.24	install 2" x 10" beam.
	Demolish and Install	L.F.	1.19	1.82		3.01	
	Clean	L.F.	.11	.25		.36	
	Paint	L.F.	.15	.92		1.07	
	Minimum Charge	Job		157		157	
2" x 12"							
	Demolish	L.F.		.93		.93	Cost includes material and labor to
	Install	L.F.	1.62	1.14		2.76	install 2" x 12" beam.
	Demolish and Install	L.F.	1.62	2.07		3.69	
	Clean	L.F.	.13	.29		.42	
	Paint	L.F.	.18	2.21		2.39	
	Minimum Charge	Job		157		157	
4" x 8"							
	Demolish	L.F.		3.16		3.16	Cost includes material and labor to
	Install	L.F.	3.67	1.40	.67	5.74	install 4" x 8" beam.
	Demolish and Install	L.F.	3.67	4.56	.67	8.90	
	Clean	L.F.	.11	.25		.36	
	Paint	L.F.	.15	.92		1.07	
	Minimum Charge	Job		157		157	
4" x 10"							
	Demolish	L.F.		3.96		3.96	Cost includes material and labor to
	Install	L.F.	4.60	1.47	.71	6.78	install 4" x 10" beam.
	Demolish and Install	L.F.	4.60	5.43	.71	10.74	
	Clean	L.F.	.13	.29		.42	
	Paint	L.F.	.18	1.07		1.25	
	Minimum Charge	Job		157		157	
4" x 12"							
	Demolish	L.F.		4.75		4.75	Cost includes material and labor to
	Install	L.F.	5.50	1.56	.75	7.81	install 4" x 12" beam.
	Demolish and Install	L.F.	5.50	6.31	.75	12.56	
	Clean	L.F.	.15	.33		.48	
	Paint	L.F.	.18	2.21		2.39	
	Minimum Charge	Job		157		157	

27

Rough Frame / Structure

Beam / Girder		Unit	Material	Labor	Equip.	Total	Specification
6″ x 8″							
	Demolish	L.F.		4.75		4.75	Cost includes material and labor to
	Install	L.F.	7.80	2.67	1.28	11.75	install 6″ x 8″ beam.
	Demolish and Install	L.F.	7.80	7.42	1.28	16.50	
	Clean	L.F.	.13	.29		.42	
	Paint	L.F.	.18	1.07		1.25	
	Minimum Charge	Job		157		157	
6″ x 10″							
	Demolish	L.F.		5.95		5.95	Cost includes material and labor to
	Install	L.F.	3.56	1.25		4.81	install 6″ x 10″ triple beam.
	Demolish and Install	L.F.	3.56	7.20		10.76	
	Clean	L.F.	.15	.33		.48	
	Paint	L.F.	.20	1.23		1.43	
	Minimum Charge	Job		157		157	
6″ x 12″							
	Demolish	L.F.		7.05		7.05	Cost includes material and labor to
	Install	L.F.	4.86	1.32		6.18	install 6″ x 12″ triple beam.
	Demolish and Install	L.F.	4.86	8.37		13.23	
	Clean	L.F.	.17	.37		.54	
	Paint	L.F.	.23	1.38		1.61	
	Minimum Charge	Job		157		157	
Steel I-type							
W8 x 31							
	Demolish	L.F.		6.65	3.05	9.70	Includes material and labor to install
	Install	L.F.	23.50	2.10	1.44	27.04	W shaped steel beam / girder.
	Demolish and Install	L.F.	23.50	8.75	4.49	36.74	
	Minimum Charge	Job		385		385	
W8 x 48							
	Demolish	L.F.		6.65	3.05	9.70	Includes material and labor to install
	Install	L.F.	37	2.20	1.51	40.71	W shaped steel beam / girder.
	Demolish and Install	L.F.	37	8.85	4.56	50.41	
	Minimum Charge	Job		385		385	
W8 x 67							
	Demolish	L.F.		6.65	3.05	9.70	Includes material and labor to install
	Install	L.F.	51.50	2.31	1.58	55.39	W shaped steel beam / girder.
	Demolish and Install	L.F.	51.50	8.96	4.63	65.09	
	Minimum Charge	Job		385		385	
W10 x 45							
	Demolish	L.F.		6.65	3.05	9.70	Includes material and labor to install
	Install	L.F.	34.50	2.20	1.51	38.21	W shaped steel beam / girder.
	Demolish and Install	L.F.	34.50	8.85	4.56	47.91	
	Minimum Charge	Job		385		385	
W10 x 68							
	Demolish	L.F.		6.65	3.05	9.70	Includes material and labor to install
	Install	L.F.	52.50	2.31	1.58	56.39	W shaped steel beam / girder.
	Demolish and Install	L.F.	52.50	8.96	4.63	66.09	
	Minimum Charge	Job		385		385	
W12 x 50							
	Demolish	L.F.		6.65	3.05	9.70	Includes material and labor to install
	Install	L.F.	38.50	2.20	1.51	42.21	W shaped steel beam / girder.
	Demolish and Install	L.F.	38.50	8.85	4.56	51.91	
	Minimum Charge	Job		385		385	
W12 x 87							
	Demolish	L.F.		9.75	4.49	14.24	Includes material and labor to install
	Install	L.F.	67	2.31	1.58	70.89	W shaped steel beam / girder.
	Demolish and Install	L.F.	67	12.06	6.07	85.13	
	Minimum Charge	Job		385		385	

Rough Frame / Structure

Beam / Girder

		Unit	Material	Labor	Equip.	Total	Specification
W12 x 120							
	Demolish	L.F.		9.75	4.49	14.24	Includes material and labor to install
	Install	L.F.	92.50	2.37	1.62	96.49	W shaped steel beam / girder.
	Demolish and Install	L.F.	92.50	12.12	6.11	110.73	
	Minimum Charge	Job		385		385	
W14 x 74							
	Demolish	L.F.		9.75	4.49	14.24	Includes material and labor to install
	Install	L.F.	57	2.31	1.58	60.89	W shaped steel beam / girder.
	Demolish and Install	L.F.	57	12.06	6.07	75.13	
	Minimum Charge	Job		385		385	
W14 x 120							
	Demolish	L.F.		9.75	4.49	14.24	Includes material and labor to install
	Install	L.F.	92.50	2.37	1.62	96.49	W shaped steel beam / girder.
	Demolish and Install	L.F.	92.50	12.12	6.11	110.73	
	Minimum Charge	Job		385		385	

Column

		Unit	Material	Labor	Equip.	Total	Specification
Concrete							
Small Diameter							
	Demolish	L.F.		10.90	1.63	12.53	Includes material and labor to install
	Install	L.F.	64	9.80	2.22	76.02	small diameter concrete column.
	Demolish and Install	L.F.	64	20.70	3.85	88.55	
	Minimum Charge	Job		490	111	601	
Large Diameter							
	Demolish	L.F.		32.50	4.89	37.39	Includes material and labor to install
	Install	L.F.	185	11.55	2.61	199.16	large diameter concrete column.
	Demolish and Install	L.F.	185	44.05	7.50	236.55	
	Minimum Charge	Job		490	111	601	

Floor Framing System

		Unit	Material	Labor	Equip.	Total	Specification
12" O.C.							
2" x 6" Joists							
	Demolish	S.F.		.53		.53	Cost includes material and labor to
	Install	S.F.	.67	.25		.92	install 2" x 6" joists, 12" O.C.
	Demolish and Install	S.F.	.67	.78		1.45	including box or band joist. Blocking
	Reinstall	S.F.		.20		.20	or bridging not included.
	Minimum Charge	Job		157		157	
2" x 8" Joists							
	Demolish	S.F.		.54		.54	Cost includes material and labor to
	Install	S.F.	1.03	.29		1.32	install 2" x 8" joists, 12" O.C.
	Demolish and Install	S.F.	1.03	.83		1.86	including box or band joist. Blocking
	Reinstall	S.F.		.23		.23	or bridging not included.
	Minimum Charge	Job		157		157	
2" x 10" Joists							
	Demolish	S.F.		.56		.56	Cost includes material and labor to
	Install	S.F.	1.39	.35		1.74	install 2" x 10" joists, 12" O.C.
	Demolish and Install	S.F.	1.39	.91		2.30	including box or band joist. Blocking
	Reinstall	S.F.		.28		.28	or bridging not included.
	Minimum Charge	Job		157		157	

Rough Frame / Structure

Floor Framing System		Unit	Material	Labor	Equip.	Total	Specification
2″ x 12″ Joists							
	Demolish	S.F.		.58		.58	Cost includes material and labor to
	Install	S.F.	1.89	.36		2.25	install 2″ x 12″ joists, 12″ O.C.
	Demolish and Install	S.F.	1.89	.94		2.83	including box or band joist. Blocking
	Reinstall	S.F.		.29		.29	or bridging not included.
	Minimum Charge	Job		157		157	
Block / Bridge							
	Demolish	Ea.		.80		.80	Includes material and labor to install
	Install	Ea.	1.60	1.76		3.36	set of cross bridging or per block of
	Demolish and Install	Ea.	1.60	2.56		4.16	solid bridging for 2″ x 10″ joists cut to
	Minimum Charge	Job		157		157	size on site.
16″ O.C.							
2″ x 6″ Joists							
	Demolish	S.F.		.39		.39	Cost includes material and labor to
	Install	S.F.	.52	.38		.90	install 2″ x 6″ joists including box or
	Demolish and Install	S.F.	.52	.77		1.29	band joist installed 16″ O.C. Does not
	Reinstall	S.F.		.30		.30	include beams, blocking or bridging.
	Minimum Charge	Job		157		157	
2″ x 8″ Joists							
	Demolish	S.F.		.41		.41	Cost includes material and labor to
	Install	S.F.	.79	.43		1.22	install 2″ x 8″ joists including box or
	Demolish and Install	S.F.	.79	.84		1.63	band joist installed 16″ O.C. Does not
	Reinstall	S.F.		.34		.34	include beams, blocking or bridging.
	Minimum Charge	Job		157		157	
2″ x 10″ Joists							
	Demolish	S.F.		.42		.42	Cost includes material and labor to
	Install	S.F.	1.07	.52		1.59	install 2″ x 10″ joists including box or
	Demolish and Install	S.F.	1.07	.94		2.01	band joist 16″ O.C. Does not include
	Reinstall	S.F.		.42		.42	beams, blocking or bridging.
	Minimum Charge	Job		157		157	
2″ x 12″ Joists							
	Demolish	S.F.		.43		.43	Cost includes material and labor to
	Install	S.F.	1.46	.54		2	install 2″ x 12″ joists including box or
	Demolish and Install	S.F.	1.46	.97		2.43	band joist installed 16″ O.C. Does not
	Reinstall	S.F.		.43		.43	include beams, blocking or bridging.
	Minimum Charge	Job		157		157	
Block / Bridge							
	Demolish	Ea.		.80		.80	Includes material and labor to install
	Install	Ea.	1.60	1.76		3.36	set of cross bridging or per block of
	Demolish and Install	Ea.	1.60	2.56		4.16	solid bridging for 2″ x 10″ joists cut to
	Minimum Charge	Job		157		157	size on site.
24″ O.C.							
2″ x 6″ Joists							
	Demolish	S.F.		.26		.26	Cost includes material and labor to
	Install	S.F.	.36	.13		.49	install 2″ x 6″ joists including box or
	Demolish and Install	S.F.	.36	.39		.75	band joist installed 24″ O.C. Does not
	Reinstall	S.F.		.10		.10	include beams, blocking or bridging.
	Minimum Charge	Job		157		157	
2″ x 8″ Joists							
	Demolish	S.F.		.27		.27	Cost includes material and labor to
	Install	S.F.	.56	.14		.70	install 2″ x 8″ joists including box or
	Demolish and Install	S.F.	.56	.41		.97	band joist installed 24″ O.C. Does not
	Reinstall	S.F.		.11		.11	include beams, blocking or bridging.
	Minimum Charge	Job		157		157	
2″ x 10″ Joists							
	Demolish	S.F.		.28		.28	Cost includes material and labor to
	Install	S.F.	.76	.17		.93	install 2″ x 10″ joists including box or
	Demolish and Install	S.F.	.76	.45		1.21	band joist installed 24″ O.C. Does not
	Reinstall	S.F.		.14		.14	include beams, blocking or bridging.
	Minimum Charge	Job		157		157	

Floor Framing System		Unit	Material	Labor	Equip.	Total	Specification
2″ x 12″ Joists							Cost includes material and labor to
	Demolish	S.F.		.29		.29	install 2″ x 12″ joists including box or
	Install	S.F.	1.03	.18		1.21	band joist installed 24″ O.C. Does not
	Demolish and Install	S.F.	1.03	.47		1.50	include beams, blocking or bridging.
	Reinstall	S.F.		.14		.14	
	Minimum Charge	Job		157		157	
Block / Bridge							Includes material and labor to install
	Demolish	Ea.		.80		.80	set of cross bridging or per block of
	Install	Ea.	1.60	1.76		3.36	solid bridging for 2″ x 10″ joists cut to
	Demolish and Install	Ea.	1.60	2.56		4.16	size on site.
	Minimum Charge	Job		157		157	
Engineered Lumber, Joist							
9-1/2″							Includes material and labor to install
	Demolish	S.F.		.42		.42	engineered lumber truss / joists per
	Install	S.F.	1.57	.31		1.88	S.F. of floor area based on joists 16″
	Demolish and Install	S.F.	1.57	.73		2.30	O.C. Beams, supports and bridging
	Reinstall	S.F.		.25		.25	are not included.
	Minimum Charge	Job		157		157	
11-7/8″							Includes material and labor to install
	Demolish	S.F.		.43		.43	engineered lumber truss / joists per
	Install	S.F.	1.68	.32		2	S.F. of floor area based on joists 16″
	Demolish and Install	S.F.	1.68	.75		2.43	O.C. Beams, supports and bridging
	Reinstall	S.F.		.26		.26	are not included.
	Minimum Charge	Job		157		157	
14″							Includes material and labor to install
	Demolish	S.F.		.45		.45	engineered lumber truss / joists per
	Install	S.F.	1.85	.34		2.19	S.F. of floor area based on joists 16″
	Demolish and Install	S.F.	1.85	.79		2.64	O.C. Beams, supports and bridging
	Reinstall	S.F.		.28		.28	are not included.
	Minimum Charge	Job		157		157	
16″							Includes material and labor to install
	Demolish	S.F.		.47		.47	engineered lumber truss / joists per
	Install	S.F.	2.52	.36		2.88	S.F. of floor area based on joists 16″
	Demolish and Install	S.F.	2.52	.83		3.35	O.C. Beams, supports and bridging
	Reinstall	S.F.		.29		.29	are not included.
	Minimum Charge	Job		157		157	
Block / Bridge							Includes material and labor to install
	Demolish	Pr.		.80		.80	set of steel, one-nail type cross
	Install	Pr.	.96	2.32		3.28	bridging for trusses placed 16″ O.C.
	Demolish and Install	Pr.	.96	3.12		4.08	
	Minimum Charge	Job		157		157	
Block / Bridge							
2″ x 6″							Includes material and labor to install
	Demolish	Ea.		.80		.80	set of cross bridging or per block of
	Install	Ea.	.77	1.41		2.18	solid bridging for 2″ x 6″ joists cut to
	Demolish and Install	Ea.	.77	2.21		2.98	size on site.
	Minimum Charge	Job		157		157	
2″ x 8″							Includes material and labor to install
	Demolish	Ea.		.80		.80	set of cross bridging or per block of
	Install	Ea.	1.18	1.57		2.75	solid bridging for 2″ x 8″ joists cut to
	Demolish and Install	Ea.	1.18	2.37		3.55	size on site.
	Minimum Charge	Job		157		157	

Rough Frame / Structure

Floor Framing System		Unit	Material	Labor	Equip.	Total	Specification
2″ x 10″							
	Demolish	Ea.		.80		.80	Includes material and labor to install
	Install	Ea.	1.60	1.76		3.36	set of cross bridging or per block of
	Demolish and Install	Ea.	1.60	2.56		4.16	solid bridging for 2″ x 10″ joists cut to
	Minimum Charge	Job		157		157	size on site.
2″ x 12″							
	Demolish	Ea.		.80		.80	Includes material and labor to install
	Install	Ea.	2.18	2.08		4.26	set of cross bridging or per block of
	Demolish and Install	Ea.	2.18	2.88		5.06	solid bridging for 2″ x 12″ joists cut to
	Minimum Charge	Job		157		157	size on site.
Ledger Strips							
1″ x 2″							
	Demolish	L.F.		.21		.21	Cost includes material and labor to
	Install	L.F.	.21	.91		1.12	install 1″ x 2″ ledger strip nailed to the
	Demolish and Install	L.F.	.21	1.12		1.33	face of studs, beams or joist.
	Minimum Charge	Job		157		157	
1″ x 3″							
	Demolish	L.F.		.21		.21	Includes material and labor to install
	Install	L.F.	.66	1.05		1.71	up to 1″ x 4″ ledger strip nailed to the
	Demolish and Install	L.F.	.66	1.26		1.92	face of studs, beams or joist.
	Minimum Charge	Job		157		157	
1″ x 4″							
	Demolish	L.F.		.21		.21	Includes material and labor to install
	Install	L.F.	.66	1.05		1.71	up to 1″ x 4″ ledger strip nailed to the
	Demolish and Install	L.F.	.66	1.26		1.92	face of studs, beams or joist.
	Minimum Charge	Job		157		157	
2″ x 2″							
	Demolish	L.F.		.23		.23	Cost includes material and labor to
	Install	L.F.	.26	.95		1.21	install 2″ x 2″ ledger strip nailed to the
	Demolish and Install	L.F.	.26	1.18		1.44	face of studs, beams or joist.
	Minimum Charge	Job		157		157	
2″ x 4″							
	Demolish	L.F.		.26		.26	Cost includes material and labor to
	Install	L.F.	.34	1.25		1.59	install 2″ x 4″ ledger strip nailed to the
	Demolish and Install	L.F.	.34	1.51		1.85	face of studs, beams or joist.
	Reinstall	L.F.		1		1	
	Minimum Charge	Job		157		157	
Ledger Boards							
2″ x 6″							
	Demolish	L.F.		.26		.26	Cost includes material and labor to
	Install	L.F.	1.57	1.96		3.53	install 2″ x 6″ ledger board fastened to
	Demolish and Install	L.F.	1.57	2.22		3.79	a wall, joists or studs.
	Reinstall	L.F.		1.57		1.57	
	Minimum Charge	Job		157		157	
4″ x 6″							
	Demolish	L.F.		.43		.43	Cost includes material and labor to
	Install	L.F.	3.99	1.96		5.95	install 4″ x 6″ ledger board fastened to
	Demolish and Install	L.F.	3.99	2.39		6.38	a wall, joists or studs.
	Reinstall	L.F.		1.57		1.57	
	Minimum Charge	Job		157		157	
4″ x 8″							
	Demolish	L.F.		.57		.57	Cost includes material and labor to
	Install	L.F.	4.91	1.96		6.87	install 4″ x 8″ ledger board fastened to
	Demolish and Install	L.F.	4.91	2.53		7.44	a wall, joists or studs.
	Reinstall	L.F.		1.57		1.57	
	Minimum Charge	Job		157		157	

Rough Frame / Structure

Floor Framing System

Floor Framing System		Unit	Material	Labor	Equip.	Total	Specification
Sill Plate Per L.F.							
2" x 4"							
	Demolish	L.F.		.27		.27	Cost includes material and labor to
	Install	L.F.	.46	1.14		1.60	install 2" x 4" pressure treated lumber,
	Demolish and Install	L.F.	.46	1.41		1.87	drilled and installed on foundation
	Reinstall	L.F.		.91		.91	bolts 48" O.C. Bolts, nuts and washers
	Minimum Charge	Job		157		157	are not included.
2" x 6"							
	Demolish	L.F.		.27		.27	Cost includes material and labor to
	Install	L.F.	.84	1.25		2.09	install 2" x 6" pressure treated lumber,
	Demolish and Install	L.F.	.84	1.52		2.36	drilled and installed on foundation
	Reinstall	L.F.		1		1	bolts 48" O.C. Bolts, nuts and washers
	Minimum Charge	Job		157		157	not included.
2" x 8"							
	Demolish	L.F.		.27		.27	Cost includes material and labor to
	Install	L.F.	1.05	1.39		2.44	install 2" x 8" pressure treated lumber,
	Demolish and Install	L.F.	1.05	1.66		2.71	drilled and installed on foundation
	Reinstall	L.F.		1.12		1.12	bolts 48" O.C. Bolts, nuts and washers
	Minimum Charge	Job		157		157	not included.
Earthquake Strapping							
	Install	Ea.	1.78	1.96		3.74	Includes labor and material to replace
	Minimum Charge	Job		157		157	an earthquake strap.

Subflooring

Subflooring		Unit	Material	Labor	Equip.	Total	Specification
Plywood							
1/2"							
	Demolish	S.F.		.33		.33	Cost includes material and labor to
	Install	SF Flr.	.33	.34		.67	install 1/2" plywood subfloor, CD
	Demolish and Install	SF Flr.	.33	.67		1	standard interior grade, plugged and
	Minimum Charge	Job		157		157	touch sanded.
5/8"							
	Demolish	S.F.		.34		.34	Cost includes material and labor to
	Install	SF Flr.	.38	.37		.75	install 5/8" plywood subfloor, CD
	Demolish and Install	SF Flr.	.38	.71		1.09	standard interior grade, plugged and
	Reinstall	SF Flr.		.30		.30	touch sanded.
	Minimum Charge	Job		157		157	
3/4"							
	Demolish	S.F.		.34		.34	Cost includes material and labor to
	Install	SF Flr.	.48	.40		.88	install 3/4" plywood subfloor, CD
	Demolish and Install	SF Flr.	.48	.74		1.22	standard interior grade, plugged and
	Reinstall	SF Flr.		.32		.32	touch sanded.
	Minimum Charge	Job		157		157	
Particle Board							
1/2"							
	Demolish	S.F.		.33		.33	Cost includes material and labor to
	Install	SF Flr.	.43	.42		.85	install 1/2" particle board subfloor.
	Demolish and Install	SF Flr.	.43	.75		1.18	
	Minimum Charge	Job		157		157	
5/8"							
	Demolish	S.F.		.34		.34	Cost includes material and labor to
	Install	S.F.	.54	.46		1	install 5/8" particle board subfloor.
	Demolish and Install	S.F.	.54	.80		1.34	
	Minimum Charge	Job		157		157	

Rough Frame / Structure

Subflooring

Subflooring		Unit	Material	Labor	Equip.	Total	Specification
3/4"							
	Demolish	S.F.		.34		.34	Cost includes material and labor to
	Install	SF Flr	.59	.50		1.09	install 3/4" particle board subfloor.
	Demolish and Install	SF Flr	.59	.84		1.43	
	Minimum Charge	Job		157		157	
Plank Board							
	Demolish	S.F.		1.07		1.07	Cost includes material and labor to
	Install	S.F.	.90	.35		1.25	install 1" x 6" standard grade plank
	Demolish and Install	S.F.	.90	1.42		2.32	flooring.
	Reinstall	S.F.		.35		.35	
	Minimum Charge	Job		157		157	
Felt Underlay							
	Demolish	S.F.		.06		.06	Includes material and labor to install
	Install	S.F.	.03	.08		.11	#15 felt building paper.
	Demolish and Install	S.F.	.03	.14		.17	
	Minimum Charge	Job		157		157	
Prep (for flooring)							
	Install	S.F.		.27		.27	Preparation of subflooring for installation of new finished flooring. Cost reflects average time to prep area.
	Minimum Charge	Job		142		142	

Wall Framing System

Wall Framing System		Unit	Material	Labor	Equip.	Total	Specification
2" x 4"							
8' High							
	Demolish	L.F.		3.40		3.40	Cost includes material and labor to
	Install	L.F.	3.52	3.69		7.21	install 2" x 4" x 8' high wall system,
	Demolish and Install	L.F.	3.52	7.09		10.61	including studs, treated bottom plate,
	Reinstall	L.F.		2.95		2.95	double top plate and one row of fire
	Minimum Charge	Job		157		157	blocking.
9' High							
	Demolish	L.F.		3.81		3.81	Cost includes material and labor to
	Install	L.F.	3.78	3.69		7.47	install 2" x 4" x 9' wall system
	Demolish and Install	L.F.	3.78	7.50		11.28	including studs, treated bottom plate,
	Reinstall	L.F.		2.95		2.95	double top plate and one row of fire
	Minimum Charge	Job		157		157	blocking.
10' High							
	Demolish	L.F.		4.25		4.25	Cost includes material and labor to
	Install	L.F.	4.04	3.69		7.73	install 2" x 4" x 10' wall system
	Demolish and Install	L.F.	4.04	7.94		11.98	including studs, treated bottom plate,
	Reinstall	L.F.		2.95		2.95	double top plate and one row of fire
	Minimum Charge	Job		157		157	blocking.
12' High							
	Demolish	L.F.		5.10		5.10	Cost includes material and labor to
	Install	L.F.	4.54	4.48		9.02	install 2" x 4" x 12' wall system
	Demolish and Install	L.F.	4.54	9.58		14.12	including studs, treated bottom plate,
	Reinstall	L.F.		3.58		3.58	double top plate and one row of fire
	Minimum Charge	Job		157		157	blocking.

Rough Frame / Structure

Wall Framing System	Unit	Material	Labor	Equip.	Total	Specification
2″ x 6″						
8′ High						
Demolish	L.F.		3.19		3.19	Cost includes material and labor to
Install	L.F.	6	6.80		12.80	install 2″ x 6″ x 8′ wall system
Demolish and Install	L.F.	6	9.99		15.99	including studs, treated bottom plate,
Reinstall	L.F.		5.42		5.42	double top plate and one row of fire
Minimum Charge	Job		157		157	blocking.
9′ High						
Demolish	L.F.		3.59		3.59	Cost includes material and labor to
Install	L.F.	6.45	6.80		13.25	install 2″ x 6″ x 9′ wall system
Demolish and Install	L.F.	6.45	10.39		16.84	including studs, treated bottom plate,
Reinstall	L.F.		5.42		5.42	double top plate and one row of fire
Minimum Charge	Job		157		157	blocking.
10′ High						
Demolish	L.F.		3.92		3.92	Cost includes material and labor to
Install	L.F.	6.90	6.80		13.70	install 2″ x 6″ x 10′ wall system
Demolish and Install	L.F.	6.90	10.72		17.62	including studs, treated bottom plate,
Reinstall	L.F.		5.42		5.42	double top plate and one row of fire
Minimum Charge	Job		157		157	blocking.
12′ High						
Demolish	L.F.		4.64		4.64	Cost includes material and labor to
Install	L.F.	7.70	8.35		16.05	install 2″ x 6″ x 12′ wall system
Demolish and Install	L.F.	7.70	12.99		20.69	including studs, treated bottom plate,
Reinstall	L.F.		6.68		6.68	double top plate and one row of fire
Minimum Charge	Job		157		157	blocking.
Fireblock						
2″ x 4″, 16″ O.C. System						
Demolish	L.F.		.32		.32	Cost includes material and labor to
Install	L.F.	.34	1.05		1.39	install 2″ x 4″ fireblocks in wood frame
Demolish and Install	L.F.	.34	1.37		1.71	walls per L.F. of wall to be blocked.
Reinstall	L.F.		.84		.84	
Minimum Charge	Job		157		157	
2″ x 4″, 24″ O.C. System						
Demolish	L.F.		.32		.32	Cost includes material and labor to
Install	L.F.	.34	1.05		1.39	install 2″ x 4″ fireblocks in wood frame
Demolish and Install	L.F.	.34	1.37		1.71	walls per L.F. of wall to be blocked.
Reinstall	L.F.		.84		.84	
Minimum Charge	Job		157		157	
2″ x 6″, 16″ O.C. System						
Demolish	L.F.		.32		.32	Cost includes material and labor to
Install	L.F.	.57	1.05		1.62	install 2″ x 6″ fireblocks in wood frame
Demolish and Install	L.F.	.57	1.37		1.94	walls per L.F. of wall to be blocked.
Reinstall	L.F.		.84		.84	
Minimum Charge	Job		157		157	
2″ x 6″, 24″ O.C. System						
Demolish	L.F.		.32		.32	Cost includes material and labor to
Install	L.F.	.57	1.05		1.62	install 2″ x 6″ fireblocks in wood frame
Demolish and Install	L.F.	.57	1.37		1.94	walls per L.F. of wall to be blocked.
Reinstall	L.F.		.84		.84	
Minimum Charge	Job		157		157	
Bracing						
1″ x 3″						
Demolish	L.F.		.24		.24	Cost includes material and labor to
Install	L.F.	.23	1.96		2.19	install 1″ x 3″ let-in bracing.
Demolish and Install	L.F.	.23	2.20		2.43	
Minimum Charge	Job		157		157	

Rough Frame / Structure

Wall Framing System	Unit	Material	Labor	Equip.	Total	Specification
1" x 4"						
Demolish	L.F.		.24		.24	Cost includes material and labor to
Install	L.F.	.37	1.96		2.33	install 1" x 4" let-in bracing.
Demolish and Install	L.F.	.37	2.20		2.57	
Minimum Charge	Job		157		157	
1" x 6"						
Demolish	L.F.		.24		.24	Cost includes material and labor to
Install	L.F.	.42	2.09		2.51	install 1" x 6" let-in bracing.
Demolish and Install	L.F.	.42	2.33		2.75	
Minimum Charge	Job		157		157	
2" x 3"						
Demolish	L.F.		.32		.32	Cost includes material and labor to
Install	L.F.	.26	2.09		2.35	install 2" x 3" let-in bracing.
Demolish and Install	L.F.	.26	2.41		2.67	
Reinstall	L.F.		1.67		1.67	
Minimum Charge	Job		157		157	
2" x 4"						
Demolish	L.F.		.32		.32	Cost includes material and labor to
Install	L.F.	.34	2.09		2.43	install 2" x 4" let-in bracing.
Demolish and Install	L.F.	.34	2.41		2.75	
Reinstall	L.F.		1.67		1.67	
Minimum Charge	Job		157		157	
2" x 6"						
Demolish	L.F.		.32		.32	Cost includes material and labor to
Install	L.F.	.57	2.24		2.81	install 2" x 6" let-in bracing.
Demolish and Install	L.F.	.57	2.56		3.13	
Reinstall	L.F.		1.79		1.79	
Minimum Charge	Job		157		157	
2" x 8"						
Demolish	L.F.		.32		.32	Cost includes material and labor to
Install	L.F.	.88	2.24		3.12	install 2" x 8" let-in bracing.
Demolish and Install	L.F.	.88	2.56		3.44	
Reinstall	L.F.		1.79		1.79	
Minimum Charge	Job		157		157	
Earthquake Strapping						
Install	Ea.	1.78	1.96		3.74	Includes labor and material to replace
Minimum Charge	Job		157		157	an earthquake strap.
Hurricane Clips						
Install	Ea.	1.57	2.16		3.73	Includes labor and material to install a
Minimum Charge	Job		157		157	hurricane clip.
Stud						
2" x 4"						
Demolish	L.F.		.26		.26	Cost includes material and labor to
Install	L.F.	.34	.57		.91	install 2" x 4" wall stud per L.F. of stud.
Demolish and Install	L.F.	.34	.83		1.17	
Reinstall	L.F.		.46		.46	
Minimum Charge	Job		157		157	
2" x 6"						
Demolish	L.F.		.32		.32	Cost includes material and labor to
Install	L.F.	.57	.63		1.20	install 2" x 6" stud per L.F. of stud.
Demolish and Install	L.F.	.57	.95		1.52	
Reinstall	L.F.		.50		.50	
Minimum Charge	Job		157		157	

Wall Framing System		Unit	Material	Labor	Equip.	Total	Specification
Plates							
2" x 4"							
	Demolish	L.F.		.23		.23	Cost includes material and labor to
	Install	L.F.	.34	.78		1.12	install 2" x 4" plate per L.F. of plate.
	Demolish and Install	L.F.	.34	1.01		1.35	
	Reinstall	L.F.		.63		.63	
	Minimum Charge	Job		157		157	
2" x 6"							
	Demolish	L.F.		.24		.24	Cost includes material and labor to
	Install	L.F.	.57	.84		1.41	install 2" x 6" plate per L.F. of plate.
	Demolish and Install	L.F.	.57	1.08		1.65	
	Reinstall	L.F.		.67		.67	
	Minimum Charge	Job		157		157	
Headers							
	Demolish	L.F.		.23		.23	Includes material and labor to install
	Install	L.F.	.57	1.74		2.31	header over wall openings and
	Demolish and Install	L.F.	.57	1.97		2.54	around floor, ceiling and roof
	Clean	L.F.	.01	.29		.30	openings or flush beam.
	Paint	L.F.	.09	1.10		1.19	
	Minimum Charge	Job		157		157	
2" x 8"							
	Demolish	L.F.		.24		.24	Includes material and labor to install
	Install	L.F.	.88	1.84		2.72	header over wall openings and
	Demolish and Install	L.F.	.88	2.08		2.96	around floor, ceiling and roof
	Clean	L.F.	.02	.29		.31	openings or flush beam.
	Paint	L.F.	.12	1.49		1.61	
	Minimum Charge	Job		157		157	
2" x 10"							
	Demolish	L.F.		.24		.24	Includes material and labor to install
	Install	L.F.	1.19	1.96		3.15	header over wall openings and
	Demolish and Install	L.F.	1.19	2.20		3.39	around floor, ceiling and roof
	Clean	L.F.	.02	.29		.31	openings or flush beam.
	Paint	L.F.	.14	1.84		1.98	
	Minimum Charge	Job		157		157	
2" x 12"							
	Demolish	L.F.		.26		.26	Includes material and labor to install
	Install	L.F.	1.62	2.09		3.71	header over wall openings and
	Demolish and Install	L.F.	1.62	2.35		3.97	around floor, ceiling and roof
	Clean	L.F.	.03	.29		.32	openings or flush beam.
	Paint	L.F.	.18	2.21		2.39	
	Minimum Charge	Job		157		157	
4" x 8"							
	Demolish	L.F.		.46		.46	Includes material and labor to install
	Install	L.F.	3.67	2.41		6.08	header over wall openings and
	Demolish and Install	L.F.	3.67	2.87		6.54	around floor, ceiling and roof
	Clean	L.F.	.02	.29		.31	openings or flush beam.
	Paint	L.F.	.12	1.49		1.61	
	Minimum Charge	Job		157		157	
4" x 10"							
	Demolish	L.F.		.49		.49	Includes material and labor to install
	Install	L.F.	4.60	2.61		7.21	header over wall openings and
	Demolish and Install	L.F.	4.60	3.10		7.70	around floor, ceiling and roof
	Clean	L.F.	.02	.29		.31	openings or flush beam.
	Paint	L.F.	.14	1.84		1.98	
	Minimum Charge	Job		157		157	

Rough Frame / Structure

Wall Framing System

Wall Framing System		Unit	Material	Labor	Equip.	Total	Specification
4" x 12"							
	Demolish	L.F.		.51		.51	Includes material and labor to install
	Install	L.F.	5.50	3.30		8.80	header over wall openings and
	Demolish and Install	L.F.	5.50	3.81		9.31	around floor, ceiling and roof
	Clean	L.F.	.03	.29		.32	openings or flush beam.
	Paint	L.F.	.18	2.21		2.39	
	Minimum Charge	Job		157		157	
6" x 8"							
	Demolish	L.F.		.46		.46	Includes material and labor to install
	Install	L.F.	7.80	1.74		9.54	header over wall openings and
	Demolish and Install	L.F.	7.80	2.20		10	around floor, ceiling and roof
	Clean	L.F.	.02	.29		.31	openings or flush beam.
	Paint	L.F.	.12	1.49		1.61	
	Minimum Charge	Job		157		157	
6" x 10"							
	Demolish	L.F.		.49		.49	Includes material and labor to install
	Install	L.F.	10.10	3.80		13.90	header over wall openings and
	Demolish and Install	L.F.	10.10	4.29		14.39	around floor, ceiling and roof
	Clean	L.F.	.02	.29		.31	openings or flush beam.
	Paint	L.F.	.14	1.84		1.98	
	Minimum Charge	Job		157		157	
6" x 12"							
	Demolish	L.F.		.51		.51	Includes material and labor to install
	Install	L.F.	11.95	4.48		16.43	header over wall openings and
	Demolish and Install	L.F.	11.95	4.99		16.94	around floor, ceiling and roof
	Clean	L.F.	.03	.29		.32	openings or flush beam.
	Paint	L.F.	.18	.77		.95	
	Minimum Charge	Job		157		157	

Rough-in Opening

Rough-in Opening		Unit	Material	Labor	Equip.	Total	Specification
Door w / 2" x 4" Lumber							
3' Wide							
	Demolish	Ea.		7.95		7.95	Includes material and labor to install
	Install	Ea.	14.65	9.80		24.45	header, double studs each side,
	Demolish and Install	Ea.	14.65	17.75		32.40	cripples, blocking and nails, up to 3'
	Reinstall	Ea.		7.84		7.84	opening in 2" x 4" stud wall 8' high.
	Minimum Charge	Job		157		157	
4' Wide							
	Demolish	Ea.		7.95		7.95	Includes material and labor to install
	Install	Ea.	15.80	9.80		25.60	header, double studs each side,
	Demolish and Install	Ea.	15.80	17.75		33.55	cripples, blocking and nails, up to 4'
	Reinstall	Ea.		7.84		7.84	opening in 2" x 4" stud wall 8' high.
	Minimum Charge	Job		157		157	
5' Wide							
	Demolish	Ea.		7.95		7.95	Includes material and labor to install
	Install	Ea.	20	9.80		29.80	header, double studs each side,
	Demolish and Install	Ea.	20	17.75		37.75	cripples, blocking and nails, up to 5'
	Reinstall	Ea.		7.84		7.84	opening in 2" x 4" stud wall 8' high.
	Minimum Charge	Job		157		157	
6' Wide							
	Demolish	Ea.		7.95		7.95	Includes material and labor to install
	Install	Ea.	22	9.80		31.80	header, double studs each side,
	Demolish and Install	Ea.	22	17.75		39.75	cripples, blocking and nails, up to 6'
	Reinstall	Ea.		7.84		7.84	opening in 2" x 4" stud wall 8' high.
	Minimum Charge	Job		157		157	

Rough Frame / Structure

Rough-in Opening		Unit	Material	Labor	Equip.	Total	Specification
8' Wide							
	Demolish	Ea.		8.50		8.50	Includes material and labor to install
	Install	Ea.	30.50	10.45		40.95	header, double studs each side,
	Demolish and Install	Ea.	30.50	18.95		49.45	cripples, blocking and nails, up to 8'
	Reinstall	Ea.		8.36		8.36	opening in 2" x 4" stud wall 8' high.
	Minimum Charge	Job		157		157	
10' Wide							
	Demolish	Ea.		8.50		8.50	Includes material and labor to install
	Install	Ea.	44	10.45		54.45	header, double studs each side,
	Demolish and Install	Ea.	44	18.95		62.95	cripples, blocking and nails, up to 10'
	Reinstall	Ea.		8.36		8.36	opening in 2" x 4" stud wall 8' high.
	Minimum Charge	Job		157		157	
12' Wide							
	Demolish	Ea.		8.50		8.50	Includes material and labor to install
	Install	Ea.	59	10.45		69.45	header, double studs each side,
	Demolish and Install	Ea.	59	18.95		77.95	cripples, blocking and nails, up to 12'
	Reinstall	Ea.		8.36		8.36	opening in 2" x 4" stud wall 8' high.
	Minimum Charge	Job		157		157	
Door w / 2" x 6" Lumber							
3' Wide							
	Demolish	Ea.		7.95		7.95	Includes material and labor to install
	Install	Ea.	22	9.80		31.80	header, double studs each side,
	Demolish and Install	Ea.	22	17.75		39.75	cripples, blocking and nails, up to 3'
	Reinstall	Ea.		7.84		7.84	opening in 2" x 6" stud wall 8' high.
	Minimum Charge	Job		157		157	
4' Wide							
	Demolish	Ea.		7.95		7.95	Includes material and labor to install
	Install	Ea.	23	9.80		32.80	header, double studs each side,
	Demolish and Install	Ea.	23	17.75		40.75	cripples, blocking and nails, up to 4'
	Reinstall	Ea.		7.84		7.84	opening in 2" x 6" stud wall 8' high.
	Minimum Charge	Job		157		157	
5' Wide							
	Demolish	Ea.		7.95		7.95	Includes material and labor to install
	Install	Ea.	27.50	9.80		37.30	header, double studs each side,
	Demolish and Install	Ea.	27.50	17.75		45.25	cripples, blocking and nails up to 5'
	Reinstall	Ea.		7.84		7.84	opening in 2" x 6" stud wall 8' high.
	Minimum Charge	Job		157		157	
6' Wide							
	Demolish	Ea.		7.95		7.95	Includes material and labor to install
	Install	Ea.	29	9.80		38.80	header, double studs each side,
	Demolish and Install	Ea.	29	17.75		46.75	cripples, blocking and nails up to 6'
	Reinstall	Ea.		7.84		7.84	opening in 2" x 6" stud wall 8' high.
	Minimum Charge	Job		157		157	
8' Wide							
	Demolish	Ea.		8.50		8.50	Includes material and labor to install
	Install	Ea.	38	10.45		48.45	header, double studs each side,
	Demolish and Install	Ea.	38	18.95		56.95	cripples, blocking and nails up to 8'
	Reinstall	Ea.		8.36		8.36	opening in 2" x 6" stud wall 8' high.
	Minimum Charge	Job		157		157	
10' Wide							
	Demolish	Ea.		8.50		8.50	Includes material and labor to install
	Install	Ea.	51.50	10.45		61.95	header, double studs each side,
	Demolish and Install	Ea.	51.50	18.95		70.45	cripples, blocking and nails up to 10'
	Reinstall	Ea.		8.36		8.36	opening in 2" x 6" stud wall 8' high.
	Minimum Charge	Job		157		157	
12' Wide							
	Demolish	Ea.		8.50		8.50	Includes material and labor to install
	Install	Ea.	66	10.45		76.45	header, double studs each side,
	Demolish and Install	Ea.	66	18.95		84.95	cripples, blocking and nails, up to 12'
	Reinstall	Ea.		8.36		8.36	opening in 2" x 6" stud wall 8' high.
	Minimum Charge	Job		157		157	

Rough-in Opening	Unit	Material	Labor	Equip.	Total	Specification
Window w / 2″ x 4″ Lumber						
2′ Wide						
Demolish	Ea.		10.65		10.65	Includes material and labor to install
Install	Ea.	15.15	13.05		28.20	header, double studs each side of
Demolish and Install	Ea.	15.15	23.70		38.85	opening, cripples, blocking, nails and
Reinstall	Ea.		10.45		10.45	sub-sills for opening up to 2′ in a 2″ x
Minimum Charge	Job		157		157	4″ stud wall 8′ high.
3′ Wide						
Demolish	Ea.		10.65		10.65	Includes material and labor to install
Install	Ea.	18.05	13.05		31.10	header, double studs each side of
Demolish and Install	Ea.	18.05	23.70		41.75	opening, cripples, blocking, nails and
Reinstall	Ea.		10.45		10.45	sub-sills for opening up to 3′ in a 2″ x
Minimum Charge	Job		157		157	4″ stud wall 8′ high.
4′ Wide						
Demolish	Ea.		10.65		10.65	Includes material and labor to install
Install	Ea.	19.85	13.05		32.90	header, double studs each side of
Demolish and Install	Ea.	19.85	23.70		43.55	opening, cripples, blocking, nails and
Reinstall	Ea.		10.45		10.45	sub-sills for opening up to 4′ in a 2″ x
Minimum Charge	Job		157		157	4″ stud wall 8′ high.
5′ Wide						
Demolish	Ea.		10.65		10.65	Includes material and labor to install
Install	Ea.	24	13.05		37.05	header, double studs each side of
Demolish and Install	Ea.	24	23.70		47.70	opening, cripples, blocking, nails and
Reinstall	Ea.		10.45		10.45	sub-sills for opening up to 5′ in a 2″ x
Minimum Charge	Job		157		157	4″ stud wall 8′ high.
6′ Wide						
Demolish	Ea.		10.65		10.65	Includes material and labor to install
Install	Ea.	27	13.05		40.05	header, double studs each side of
Demolish and Install	Ea.	27	23.70		50.70	opening, cripples, blocking, nails and
Reinstall	Ea.		10.45		10.45	sub-sills for opening up to 6′ in a 2″ x
Minimum Charge	Job		157		157	4″ stud wall 8′ high.
7′ Wide						
Demolish	Ea.		10.65		10.65	Includes material and labor to install
Install	Ea.	34	13.05		47.05	header, double studs each side of
Demolish and Install	Ea.	34	23.70		57.70	opening, cripples, blocking, nails and
Reinstall	Ea.		10.45		10.45	sub-sills for opening up to 7′ in a 2″ x
Minimum Charge	Job		157		157	4″ stud wall 8′ high.
8′ Wide						
Demolish	Ea.		11.60		11.60	Includes material and labor to install
Install	Ea.	37.50	14.25		51.75	header, double studs each side of
Demolish and Install	Ea.	37.50	25.85		63.35	opening, cripples, blocking, nails and
Reinstall	Ea.		11.40		11.40	sub-sills for opening up to 8′ in a 2″ x
Minimum Charge	Job		157		157	4″ stud wall 8′ high.
10′ Wide						
Demolish	Ea.		11.60		11.60	Includes material and labor to install
Install	Ea.	52.50	14.25		66.75	header, double studs each side of
Demolish and Install	Ea.	52.50	25.85		78.35	opening, cripples, blocking, nails and
Reinstall	Ea.		11.40		11.40	sub-sills for opening up to 10′ in a 2″
Minimum Charge	Job		157		157	x 4″ stud wall 8′ high.

Rough-in Opening		Unit	Material	Labor	Equip.	Total	Specification
12' Wide							
	Demolish	Ea.		11.60		11.60	Includes material and labor to install
	Install	Ea.	69	14.25		83.25	header, double studs each side of
	Demolish and Install	Ea.	69	25.85		94.85	opening, cripples, blocking, nails and
	Reinstall	Ea.		11.40		11.40	sub-sills for opening up to 12' in a 2"
	Minimum Charge	Job		157		157	x 4" stud wall 8' high.
Window w / 2" x 6" Lumber							
2' Wide							
	Demolish	Ea.		10.65		10.65	Includes material and labor to install
	Install	Ea.	24	13.05		37.05	header, double studs each side of
	Demolish and Install	Ea.	24	23.70		47.70	opening, cripples, blocking, nails and
	Reinstall	Ea.		10.45		10.45	sub-sills for opening up to 2' in a 2" x
	Minimum Charge	Job		157		157	6" stud wall 8' high.
3' Wide							
	Demolish	Ea.		10.65		10.65	Includes material and labor to install
	Install	Ea.	28	13.05		41.05	header, double studs each side of
	Demolish and Install	Ea.	28	23.70		51.70	opening, cripples, blocking, nails and
	Reinstall	Ea.		10.45		10.45	sub-sills for opening up to 3' in a 2" x
	Minimum Charge	Job		157		157	6" stud wall 8' high.
4' Wide							
	Demolish	Ea.		10.65		10.65	Includes material and labor to install
	Install	Ea.	29.50	13.05		42.55	header, double studs each side of
	Demolish and Install	Ea.	29.50	23.70		53.20	opening, cripples, blocking, nails and
	Reinstall	Ea.		10.45		10.45	sub-sills for opening up to 4' in a 2" x
	Minimum Charge	Job		157		157	6" stud wall 8' high.
5' Wide							
	Demolish	Ea.		10.65		10.65	Includes material and labor to install
	Install	Ea.	34.50	13.05		47.55	header, double studs each side of
	Demolish and Install	Ea.	34.50	23.70		58.20	opening, cripples, blocking, nails and
	Reinstall	Ea.		10.45		10.45	sub-sills for opening up to 5' in a 2" x
	Minimum Charge	Job		157		157	6" stud wall 8' high.
6' Wide							
	Demolish	Ea.		10.65		10.65	Includes material and labor to install
	Install	Ea.	38	13.05		51.05	header, double studs each side of
	Demolish and Install	Ea.	38	23.70		61.70	opening, cripples, blocking, nails and
	Reinstall	Ea.		10.45		10.45	sub-sills for opening up to 6' in a 2" x
	Minimum Charge	Job		157		157	6" stud wall 8' high.
7' Wide							
	Demolish	Ea.		10.65		10.65	Includes material and labor to install
	Install	Ea.	45.50	13.05		58.55	header, double studs each side of
	Demolish and Install	Ea.	45.50	23.70		69.20	opening, cripples, blocking, nails and
	Reinstall	Ea.		10.45		10.45	sub-sills for opening up to 7' in a 2" x
	Minimum Charge	Job		157		157	6" stud wall 8' high.
8' Wide							
	Demolish	Ea.		11.60		11.60	Includes material and labor to install
	Install	Ea.	50	14.25		64.25	header, double studs each side of
	Demolish and Install	Ea.	50	25.85		75.85	opening, cripples, blocking, nails and
	Reinstall	Ea.		11.40		11.40	sub-sills for opening up to 8' in a 2" x
	Minimum Charge	Job		157		157	6" stud wall 8' high.

Rough Frame / Structure

Rough-in Opening

Rough-in Opening	Unit	Material	Labor	Equip.	Total	Specification
10' Wide						
Demolish	Ea.		11.60		11.60	Includes material and labor to install
Install	Ea.	65.50	14.25		79.75	header, double studs each side of
Demolish and Install	Ea.	65.50	25.85		91.35	opening, cripples, blocking, nails and
Reinstall	Ea.		11.40		11.40	sub-sills for opening up to 10' in a 2"
Minimum Charge	Job		157		157	x 6" stud wall 8' high.
12' Wide						
Demolish	Ea.		11.60		11.60	Includes material and labor to install
Install	Ea.	83	14.25		97.25	header, double studs each side of
Demolish and Install	Ea.	83	25.85		108.85	opening, cripples, blocking, nails and
Reinstall	Ea.		11.40		11.40	sub-sills for opening up to 12' in a 2"
Minimum Charge	Job		157		157	x 6" stud wall 8' high.

Glue-Laminated Beams

Glue-Laminated Beams	Unit	Material	Labor	Equip.	Total	Specification
3-1/2" x 6"						
Demolish	L.F.		3.16		3.16	Includes material and labor to install
Install	L.F.	4.52	2.33	1.12	7.97	glue-laminated wood beam.
Demolish and Install	L.F.	4.52	5.49	1.12	11.13	
Reinstall	L.F.		1.87	.90	2.77	
Clean	L.F.	.10	.29		.39	
Paint	L.F.	.28	.81		1.09	
Minimum Charge	Job		157		157	
3-1/2" x 9"						
Demolish	L.F.		3.16		3.16	Includes material and labor to install
Install	L.F.	6.05	2.33	1.12	9.50	glue-laminated wood beam.
Demolish and Install	L.F.	6.05	5.49	1.12	12.66	
Reinstall	L.F.		1.87	.90	2.77	
Clean	L.F.	.12	.30		.42	
Paint	L.F.	.36	1.05		1.41	
Minimum Charge	Job		157		157	
3-1/2" x 12"						
Demolish	L.F.		4.75		4.75	Includes material and labor to install
Install	L.F.	8.10	2.33	1.12	11.55	glue-laminated wood beam.
Demolish and Install	L.F.	8.10	7.08	1.12	16.30	
Reinstall	L.F.		1.87	.90	2.77	
Clean	L.F.	.15	.31		.46	
Paint	L.F.	.44	1.31		1.75	
Minimum Charge	Job		157		157	
3-1/2" x 15"						
Demolish	L.F.		4.75		4.75	Includes material and labor to install
Install	L.F.	10.05	2.42	1.16	13.63	glue-laminated wood beam.
Demolish and Install	L.F.	10.05	7.17	1.16	18.38	
Reinstall	L.F.		1.93	.93	2.86	
Clean	L.F.	.18	.32		.50	
Paint	L.F.	.52	1.53		2.05	
Minimum Charge	Job		157		157	
3-1/2" x 18"						
Demolish	L.F.		5.95		5.95	Includes material and labor to install
Install	L.F.	13.55	2.42	1.16	17.13	glue-laminated wood beam.
Demolish and Install	L.F.	13.55	8.37	1.16	23.08	
Reinstall	L.F.		1.93	.93	2.86	
Clean	L.F.	.21	.33		.54	
Paint	L.F.	.62	1.78		2.40	
Minimum Charge	Job		157		157	

Glue-Laminated Beams		Unit	Material	Labor	Equip.	Total	Specification
5-1/8" x 6"							
	Demolish	L.F.		4.75		4.75	Includes material and labor to install
	Install	L.F.	6.60	2.33	1.12	10.05	glue-laminated wood beam.
	Demolish and Install	L.F.	6.60	7.08	1.12	14.80	
	Reinstall	L.F.		1.87	.90	2.77	
	Clean	L.F.	.11	.29		.40	
	Paint	L.F.	.30	.89		1.19	
	Minimum Charge	Job		157		157	
5-1/8" x 9"							
	Demolish	L.F.		5.95		5.95	Includes material and labor to install
	Install	L.F.	9.90	2.33	1.12	13.35	glue-laminated wood beam.
	Demolish and Install	L.F.	9.90	8.28	1.12	19.30	
	Reinstall	L.F.		1.87	.90	2.77	
	Clean	L.F.	.13	.30		.43	
	Paint	L.F.	.39	1.14		1.53	
	Minimum Charge	Job		157		157	
5-1/8" x 12"							
	Demolish	L.F.		7.05		7.05	Includes material and labor to install
	Install	L.F.	13.25	2.33	1.12	16.70	glue-laminated wood beam.
	Demolish and Install	L.F.	13.25	9.38	1.12	23.75	
	Reinstall	L.F.		1.87	.90	2.77	
	Clean	L.F.	.17	.31		.48	
	Paint	L.F.	.47	1.38		1.85	
	Minimum Charge	Job		157		157	
5-1/8" x 18"							
	Demolish	L.F.		11.85		11.85	Includes material and labor to install
	Install	L.F.	19.85	2.42	1.16	23.43	glue-laminated wood beam.
	Demolish and Install	L.F.	19.85	14.27	1.16	35.28	
	Reinstall	L.F.		1.93	.93	2.86	
	Clean	L.F.	.22	.33		.55	
	Paint	L.F.	.55	1.86		2.41	
	Minimum Charge	Job		157		157	
6-3/4" x 12"							
	Demolish	L.F.		7.05		7.05	Includes material and labor to install
	Install	L.F.	17.45	2.42	1.16	21.03	glue-laminated wood beam.
	Demolish and Install	L.F.	17.45	9.47	1.16	28.08	
	Reinstall	L.F.		1.93	.93	2.86	
	Clean	L.F.	.19	.32		.51	
	Paint	L.F.	.52	1.53		2.05	
	Minimum Charge	Job		157		157	
6-3/4" x 15"							
	Demolish	L.F.		11.85		11.85	Includes material and labor to install
	Install	L.F.	22	2.42	1.16	25.58	glue-laminated wood beam.
	Demolish and Install	L.F.	22	14.27	1.16	37.43	
	Reinstall	L.F.		1.93	.93	2.86	
	Clean	L.F.	.21	.33		.54	
	Paint	L.F.	.62	1.78		2.40	
	Minimum Charge	Job		157		157	
6-3/4" x 18"							
	Demolish	L.F.		11.85		11.85	Includes material and labor to install
	Install	L.F.	26.50	2.50	1.20	30.20	glue-laminated wood beam.
	Demolish and Install	L.F.	26.50	14.35	1.20	42.05	
	Reinstall	L.F.		2	.96	2.96	
	Clean	L.F.	.24	.34		.58	
	Paint	L.F.	.69	2.03		2.72	
	Minimum Charge	Job		157		157	

Rough Frame / Structure

Glue-Laminated Beams

Glue-Laminated Beams		Unit	Material	Labor	Equip.	Total	Specification
Hardware							
5-1/4" Glue-Lam Seat							
	Install	Ea.	51.50	1.74		53.24	Includes labor and material to install
	Minimum Charge	Job		157		157	beam hangers.
6-3/4" Glue-Lam Seat							
	Install	Ea.	53.50	1.74		55.24	Includes labor and material to install
	Minimum Charge	Job		157		157	beam hangers.
8-3/4" Glue-Lam Seat							
	Install	Ea.	61	1.74		62.74	Includes labor and material to install
	Minimum Charge	Job		157		157	beam hangers.
Earthquake Strapping							
	Install	Ea.	1.78	1.96		3.74	Includes labor and material to replace
	Minimum Charge	Job		157		157	an earthquake strap.
Hurricane Clips							
	Install	Ea.	1.57	2.16		3.73	Includes labor and material to install a
	Minimum Charge	Job		157		157	hurricane clip.

Metal Stud Framing

Metal Stud Framing		Unit	Material	Labor	Equip.	Total	Specification
16" O.C. System							
4", 16 Ga.							
	Demolish	S.F.		.95		.95	Includes material and labor to install
	Install	S.F.	.84	.97		1.81	load bearing cold rolled metal stud
	Demolish and Install	S.F.	.84	1.92		2.76	walls, to 10' high, including studs, top
	Reinstall	S.F.		.77		.77	and bottom track and screws.
	Minimum Charge	Job		157		157	
6", 16 Ga.							
	Demolish	S.F.		.95		.95	Includes material and labor to install
	Install	S.F.	1.05	.98		2.03	load bearing cold rolled metal stud
	Demolish and Install	S.F.	1.05	1.93		2.98	walls, to 10' high, including studs, top
	Reinstall	S.F.		.78		.78	and bottom track and screws.
	Minimum Charge	Job		157		157	
4", 25 Ga.							
	Demolish	S.F.		.66		.66	Includes material and labor to install
	Install	S.F.	.26	.66		.92	cold rolled metal stud walls, to 10'
	Demolish and Install	S.F.	.26	1.32		1.58	high, including studs, top and bottom
	Reinstall	S.F.		.53		.53	track and screws.
	Minimum Charge	Job		157		157	
6", 25 Ga.							
	Demolish	S.F.		.66		.66	Includes material and labor to install
	Install	S.F.	.37	.67		1.04	cold rolled metal stud walls, to 10'
	Demolish and Install	S.F.	.37	1.33		1.70	high, including studs, top and bottom
	Reinstall	S.F.		.53		.53	track and screws.
	Minimum Charge	Job		157		157	
24" O.C. System							
4", 25 Ga.							
	Demolish	S.F.		.56		.56	Includes material and labor to install
	Install	S.F.	.20	.42		.62	cold rolled metal stud walls, to 10'
	Demolish and Install	S.F.	.20	.98		1.18	high, including studs, top and bottom
	Reinstall	S.F.		.34		.34	track and screws.
	Minimum Charge	Job		157		157	

Rough Frame / Structure

Metal Stud Framing

		Unit	Material	Labor	Equip.	Total	Specification
6", 25 Ga.							
	Demolish	S.F.		.56		.56	Includes material and labor to install
	Install	S.F.	.28	.43		.71	cold rolled metal stud walls, to 10'
	Demolish and Install	S.F.	.28	.99		1.27	high, including studs, top and bottom
	Reinstall	S.F.		.35		.35	track and screws.
	Minimum Charge	Job		157		157	
4", 16 Ga.							
	Demolish	L.F.		6.35		6.35	Includes material and labor to install
	Install	S.F.	.61	.70		1.31	load bearing cold rolled metal stud
	Demolish and Install	S.F.	.61	7.05		7.66	walls, to 10' high, including studs, top
	Reinstall	S.F.		.56		.56	and bottom track and screws.
	Minimum Charge	Job		157		157	
6", 16 Ga.							
	Demolish	L.F.		6.35		6.35	Includes material and labor to install
	Install	S.F.	.76	.71		1.47	load bearing cold rolled metal stud
	Demolish and Install	S.F.	.76	7.06		7.82	walls, to 10' high, including studs, top
	Reinstall	S.F.		.57		.57	and bottom track and screws.
	Minimum Charge	Job		157		157	

Metal Joist

		Unit	Material	Labor	Equip.	Total	Specification
Bar Joist							
18K9							
	Demolish	L.F.		.75	.35	1.10	Includes material, labor and
	Install	L.F.	4.71	1.71	.86	7.28	equipment to install open web joist.
	Demolish and Install	L.F.	4.71	2.46	1.21	8.38	
	Reinstall	L.F.		1.37	.69	2.06	
	Minimum Charge	Job		1400	675	2075	
16K6							
	Demolish	L.F.		.75	.35	1.10	Includes material, labor and
	Install	L.F.	3.23	1.16	.06	4.45	equipment to install open web joist.
	Demolish and Install	L.F.	3.23	1.91	.41	5.55	
	Reinstall	L.F.		.93	.05	.98	
	Minimum Charge	Job		192		192	

Exterior Sheathing

		Unit	Material	Labor	Equip.	Total	Specification
CDX Plywood							
5/16"							
	Demolish	S.F.		.22		.22	Cost includes material and labor to
	Install	S.F.	.46	.39		.85	install 5/16" CDX plywood sheathing.
	Demolish and Install	S.F.	.46	.61		1.07	
	Minimum Charge	Job		157		157	
3/8"							
	Demolish	S.F.		.22		.22	Cost includes material and labor to
	Install	S.F.	.48	.52		1	install 3/8" CDX plywood sheathing.
	Demolish and Install	S.F.	.48	.74		1.22	
	Minimum Charge	Job		157		157	
1/2"							
	Demolish	S.F.		.23		.23	Cost includes material and labor to
	Install	S.F.	.43	.56		.99	install 1/2" CDX plywood sheathing.
	Demolish and Install	S.F.	.43	.79		1.22	
	Minimum Charge	Job		157		157	

Rough Frame / Structure

Exterior Sheathing		Unit	Material	Labor	Equip.	Total	Specification
5/8″							
	Demolish	S.F.		.23		.23	Cost includes material and labor to
	Install	S.F.	.50	.60		1.10	install 5/8″ CDX plywood sheathing.
	Demolish and Install	S.F.	.50	.83		1.33	
	Minimum Charge	Job		157		157	
3/4″							
	Demolish	S.F.		.24		.24	Cost includes material and labor to
	Install	S.F.	.63	.64		1.27	install 3/4″ CDX plywood sheathing.
	Demolish and Install	S.F.	.63	.88		1.51	
	Minimum Charge	Job		157		157	
OSB							
1/2″							
	Demolish	S.F.		.23		.23	Cost includes material and labor to
	Install	S.F.	.31	.24		.55	install 4′ x 8′ x 1/2″ OSB sheathing.
	Demolish and Install	S.F.	.31	.47		.78	
	Minimum Charge	Job		157		157	
5/8″							
	Demolish	S.F.		.23		.23	Cost includes material and labor to
	Install	S.F.	.46	.25		.71	install 4′ x 8′ x 5/8″ OSB sheathing.
	Demolish and Install	S.F.	.46	.48		.94	
	Minimum Charge	Job		157		157	
Vapor Barrier							
Black Paper							
	Install	S.F.	.03	.08		.11	Includes material and labor to install
	Minimum Charge	Job		157		157	#15 felt building paper.

Plywood Sheathing		Unit	Material	Labor	Equip.	Total	Specification
Finish Plywood							
5/16″							
	Demolish	S.F.		.22		.22	Includes material and labor to install
	Install	S.F.	.58	.20		.78	exterior 5/16″ AC plywood on walls.
	Demolish and Install	S.F.	.58	.42		1	
	Minimum Charge	Job		157		157	
3/8″							
	Demolish	S.F.		.22		.22	Includes material and labor to install
	Install	S.F.	.58	.26		.84	exterior 3/8″ AC plywood on walls.
	Demolish and Install	S.F.	.58	.48		1.06	
	Minimum Charge	Job		157		157	
1/2″							
	Demolish	S.F.		.23		.23	Includes material and labor to install
	Install	S.F.	.68	.28		.96	exterior 1/2″ AC plywood on walls.
	Demolish and Install	S.F.	.68	.51		1.19	
	Minimum Charge	Job		157		157	
5/8″							
	Demolish	S.F.		.23		.23	Includes material and labor to install
	Install	S.F.	.79	.28		1.07	exterior 5/8″ AC plywood on walls.
	Demolish and Install	S.F.	.79	.51		1.30	
	Reinstall	S.F.		.28		.28	
	Minimum Charge	Job		157		157	
3/4″							
	Demolish	S.F.		.24		.24	Includes material and labor to install
	Install	S.F.	.91	.28		1.19	exterior 3/4″ AC plywood on walls.
	Demolish and Install	S.F.	.91	.52		1.43	
	Reinstall	S.F.		.28		.28	
	Minimum Charge	Job		157		157	

Rough Frame / Structure

Stairs

Stairs	Unit	Material	Labor	Equip.	Total	Specification
Job-built						
Treads and Risers						
Demolish	Ea.		32		32	Includes material and labor to install
Install	Ea.	430	39		469	three 2" x 12" stringers, treads and
Demolish and Install	Ea.	430	71		501	risers of 3/4" CDX plywood, installed
Clean	Flight	2.88	19.55		22.43	in a straight or "L" shaped run
Paint	Ea.	.08	55		55.08	including carpet.
Minimum Charge	Job		157		157	
Landing						
Demolish	S.F.		.26		.26	Includes material and labor to install
Install	S.F.	44	2.18		46.18	landing framing, 3/4" CDX plywood
Demolish and Install	S.F.	44	2.44		46.44	surface and carpet.
Clean	S.F.		.29		.29	
Paint	S.F.	.26	.68		.94	
Minimum Charge	Job		157		157	

Ceiling Framing

Ceiling Framing	Unit	Material	Labor	Equip.	Total	Specification
16" O.C. System						
2" x 6" Joists						
Demolish	S.F.		.39		.39	Cost includes material and labor to
Install	S.F.	.44	.52		.96	install 2" x 6" joists including end and
Demolish and Install	S.F.	.44	.91		1.35	header joist installed 16" O.C. Does
Reinstall	S.F.		.41		.41	not include beams, ledger strips,
Minimum Charge	Job		157		157	blocking or bridging.
2" x 8" Joists						
Demolish	S.F.		.41		.41	Cost includes material and labor to
Install	S.F.	.67	.63		1.30	install 2" x 8" joists including end and
Demolish and Install	S.F.	.67	1.04		1.71	header joist installed 16" O.C. Does
Reinstall	S.F.		.51		.51	not include beams, ledger strips,
Minimum Charge	Job		157		157	blocking or bridging.
2" x 10" Joists						
Demolish	S.F.		.42		.42	Cost includes material and labor to
Install	S.F.	.90	.75		1.65	install 2" x 10" joists including end
Demolish and Install	S.F.	.90	1.17		2.07	and header joist installed 16" O.C.
Reinstall	S.F.		.60		.60	Does not include beams, ledger strips,
Minimum Charge	Job		157		157	blocking or bridging.
2" x 12" Joists						
Demolish	S.F.		.43		.43	Cost includes material and labor to
Install	S.F.	1.23	.92		2.15	install 2" x 12" joists including end
Demolish and Install	S.F.	1.23	1.35		2.58	and header joist installed 16" O.C.
Reinstall	S.F.		.73		.73	Does not include beams, ledger strips,
Minimum Charge	Job		157		157	blocking or bridging.
Block / Bridge						
Demolish	Ea.		.80		.80	Includes material and labor to install
Install	Ea.	1.60	1.76		3.36	set of cross bridging or per block of
Demolish and Install	Ea.	1.60	2.56		4.16	solid bridging for 2" x 10" joists cut to
Minimum Charge	Job		157		157	size on site.

Rough Frame / Structure

Ceiling Framing	Unit	Material	Labor	Equip.	Total	Specification
24" O.C. System						
2" x 6" Joists						
Demolish	S.F.		.26		.26	Cost includes material and labor to
Install	S.F.	.30	.34		.64	install 2" x 6" joists including end and
Demolish and Install	S.F.	.30	.60		.90	header joist installed 24" O.C. Does
Reinstall	S.F.		.28		.28	not include beams, ledger strips,
Minimum Charge	Job		157		157	blocking or bridging.
2" x 8" Joists						
Demolish	S.F.		.27		.27	Cost includes material and labor to
Install	S.F.	.45	.42		.87	install 2" x 8" joists including end and
Demolish and Install	S.F.	.45	.69		1.14	header joist installed 24" O.C. Does
Reinstall	S.F.		.34		.34	not include beams, ledger strips,
Minimum Charge	Job		157		157	blocking or bridging.
2" x 10" Joists						
Demolish	S.F.		.28		.28	Cost includes material and labor to
Install	S.F.	.61	.50		1.11	install 2" x 10" joists including end
Demolish and Install	S.F.	.61	.78		1.39	and header joist installed 24" O.C.
Reinstall	S.F.		.40		.40	Does not include beams, ledger strips,
Minimum Charge	Job		157		157	blocking or bridging.
2" x 12" Joists						
Demolish	S.F.		.29		.29	Cost includes material and labor to
Install	S.F.	.83	.61		1.44	install 2" x 12" joists including end
Demolish and Install	S.F.	.83	.90		1.73	and header joist installed 24" O.C.
Reinstall	S.F.		.49		.49	Does not include beams, ledger strips,
Minimum Charge	Job		157		157	blocking or bridging.
Block / Bridge						
Demolish	Ea.		.80		.80	Includes material and labor to install
Install	Ea.	1.60	1.76		3.36	set of cross bridging or per block of
Demolish and Install	Ea.	1.60	2.56		4.16	solid bridging for 2" x 10" joists cut to
Minimum Charge	Job		157		157	size on site.
Ledger Strips						
1" x 2"						
Demolish	L.F.		.21		.21	Cost includes material and labor to
Install	L.F.	.21	.91		1.12	install 1" x 2" ledger strip nailed to the
Demolish and Install	L.F.	.21	1.12		1.33	face of studs, beams or joist.
Minimum Charge	Job		157		157	
1" x 3"						
Demolish	L.F.		.21		.21	Includes material and labor to install
Install	L.F.	.66	1.05		1.71	up to 1" x 4" ledger strip nailed to the
Demolish and Install	L.F.	.66	1.26		1.92	face of studs, beams or joist.
Minimum Charge	Job		157		157	
1" x 4"						
Demolish	L.F.		.21		.21	Includes material and labor to install
Install	L.F.	.66	1.05		1.71	up to 1" x 4" ledger strip nailed to the
Demolish and Install	L.F.	.66	1.26		1.92	face of studs, beams or joist.
Minimum Charge	Job		157		157	
2" x 2"						
Demolish	L.F.		.23		.23	Cost includes material and labor to
Install	L.F.	.26	.95		1.21	install 2" x 2" ledger strip nailed to the
Demolish and Install	L.F.	.26	1.18		1.44	face of studs, beams or joist.
Minimum Charge	Job		157		157	
2" x 4"						
Demolish	L.F.		.26		.26	Cost includes material and labor to
Install	L.F.	.34	1.25		1.59	install 2" x 4" ledger strip nailed to the
Demolish and Install	L.F.	.34	1.51		1.85	face of studs, beams or joist.
Reinstall	L.F.		1		1	
Minimum Charge	Job		157		157	

Rough Frame / Structure

Ceiling Framing

Ceiling Framing		Unit	Material	Labor	Equip.	Total	Specification
Ledger Boards							
	Demolish	L.F.		.26		.26	Cost includes material and labor to
	Install	L.F.	1.51	1.74		3.25	install 2" x 4" ledger board fastened to
	Demolish and Install	L.F.	1.51	2		3.51	a wall, joists or studs.
	Reinstall	L.F.		1.39		1.39	
	Minimum Charge	Job		157		157	
4" x 6"							
	Demolish	L.F.		.43		.43	Cost includes material and labor to
	Install	L.F.	3.99	1.96		5.95	install 4" x 6" ledger board fastened to
	Demolish and Install	L.F.	3.99	2.39		6.38	a wall, joists or studs.
	Reinstall	L.F.		1.57		1.57	
	Minimum Charge	Job		157		157	
4" x 8"							
	Demolish	L.F.		.57		.57	Cost includes material and labor to
	Install	L.F.	4.91	1.96		6.87	install 4" x 8" ledger board fastened to
	Demolish and Install	L.F.	4.91	2.53		7.44	a wall, joists or studs.
	Reinstall	L.F.		1.57		1.57	
	Minimum Charge	Job		157		157	
Earthquake Strapping							
	Install	Ea.	1.78	1.96		3.74	Includes labor and material to replace
	Minimum Charge	Job		157		157	an earthquake strap.
Hurricane Clips							
	Install	Ea.	1.57	2.16		3.73	Includes labor and material to install a
	Minimum Charge	Job		157		157	hurricane clip.

Roof Framing

Roof Framing		Unit	Material	Labor	Equip.	Total	Specification
16" O.C. System							
2" x 4" Rafters							
	Demolish	S.F.		.58		.58	Cost includes material and labor to
	Install	S.F.	.25	.28		.53	install 2" x 4", 16" O.C. rafter framing
	Demolish and Install	S.F.	.25	.86		1.11	for flat, shed or gable roofs, 25' span,
	Reinstall	S.F.		.22		.22	up to 5/12 slope, per S.F. of roof.
	Minimum Charge	Job		157		157	
2" x 6" Rafters							
	Demolish	S.F.		.61		.61	Cost includes material and labor to
	Install	S.F.	.43	.29		.72	install 2" x 6", 16" O.C. rafter framing
	Demolish and Install	S.F.	.43	.90		1.33	for flat, shed or gable roofs, 25' span,
	Reinstall	S.F.		.24		.24	up to 5/12 slope, per S.F. of roof.
	Minimum Charge	Job		157		157	
2" x 8" Rafters							
	Demolish	S.F.		.62		.62	Cost includes material and labor to
	Install	S.F.	.66	.31		.97	install 2" x 8", 16" O.C. rafter framing
	Demolish and Install	S.F.	.66	.93		1.59	for flat, shed or gable roofs, 25' span,
	Reinstall	S.F.		.25		.25	up to 5/12 slope, per S.F. of roof.
	Minimum Charge	Job		157		157	
2" x 10" Rafters							
	Demolish	S.F.		.62		.62	Cost includes material and labor to
	Install	S.F.	.89	.48		1.37	install 2" x 10", 16" O.C. rafter
	Demolish and Install	S.F.	.89	1.10		1.99	framing for flat, shed or gable roofs,
	Reinstall	S.F.		.38		.38	25' span, up to 5/12 slope, per S.F.
	Minimum Charge	Job		157		157	of roof.

Rough Frame / Structure

Roof Framing		Unit	Material	Labor	Equip.	Total	Specification
2″ x 12″ Rafters							
	Demolish	S.F.		.63		.63	Cost includes material and labor to
	Install	S.F.	1.22	.52		1.74	install 2″ x 12″, 16″ O.C. rafter
	Demolish and Install	S.F.	1.22	1.15		2.37	framing for flat, shed or gable roofs,
	Reinstall	S.F.		.41		.41	25′ span, up to 5/12 slope, per S.F.
	Minimum Charge	Job		157		157	of roof.
24″ O.C. System							
2″ x 4″ Rafters							
	Demolish	S.F.		.44		.44	Cost includes material and labor to
	Install	S.F.	.17	.21		.38	install 2″ x 4″, 24″ O.C. rafter framing
	Demolish and Install	S.F.	.17	.65		.82	for flat, shed or gable roofs, 25′ span,
	Reinstall	S.F.		.17		.17	up to 5/12 slope, per S.F. of roof.
	Minimum Charge	Job		157		157	
2″ x 6″ Rafters							
	Demolish	S.F.		.46		.46	Cost includes material and labor to
	Install	S.F.	.29	.22		.51	install 2″ x 6″, 24″ O.C. rafter framing
	Demolish and Install	S.F.	.29	.68		.97	for flat, shed or gable roofs, 25′ span,
	Reinstall	S.F.		.18		.18	up to 5/12 slope, per S.F. of roof.
	Minimum Charge	Job		157		157	
2″ x 8″ Rafters							
	Demolish	S.F.		.47		.47	Cost includes material and labor to
	Install	S.F.	.44	.24		.68	install 2″ x 8″, 24″ O.C. rafter framing
	Demolish and Install	S.F.	.44	.71		1.15	for flat, shed or gable roofs, 25′ span,
	Reinstall	S.F.		.19		.19	up to 5/12 slope, per S.F. of roof.
	Minimum Charge	Job		157		157	
2″ x 10″ Rafters							
	Demolish	S.F.		.47		.47	Cost includes material and labor to
	Install	S.F.	.59	.36		.95	install 2″ x 10″, 24″ O.C. rafter
	Demolish and Install	S.F.	.59	.83		1.42	framing for flat, shed or gable roofs,
	Reinstall	S.F.		.29		.29	25′ span, up to 5/12 slope, per S.F.
	Minimum Charge	Job		157		157	of roof.
2″ x 12″ Rafters							
	Demolish	S.F.		.47		.47	Cost includes material and labor to
	Install	S.F.	.81	.39		1.20	install 2″ x 12″, 24″ O.C. rafter
	Demolish and Install	S.F.	.81	.86		1.67	framing for flat, shed or gable roofs,
	Reinstall	S.F.		.31		.31	25′ span, up to 5/12 slope, per S.F.
	Minimum Charge	Job		157		157	of roof.
Rafter							
2″ x 4″							
	Demolish	L.F.		.59		.59	Cost includes material and labor to
	Install	L.F.	.34	.91		1.25	install 2″ x 4″ rafters for flat, shed or
	Demolish and Install	L.F.	.34	1.50		1.84	gable roofs, up to 5/12 slope, 25′
	Reinstall	L.F.		.73		.73	span, per L.F.
	Minimum Charge	Job		157		157	
2″ x 6″							
	Demolish	L.F.		.60		.60	Cost includes material and labor to
	Install	L.F.	.57	.78		1.35	install 2″ x 6″ rafters for flat, shed or
	Demolish and Install	L.F.	.57	1.38		1.95	gable roofs, up to 5/12 slope, 25′
	Reinstall	L.F.		.63		.63	span, per L.F.
	Minimum Charge	Job		157		157	
2″ x 8″							
	Demolish	L.F.		.61		.61	Cost includes material and labor to
	Install	L.F.	.88	.84		1.72	install 2″ x 8″ rafters for flat, shed or
	Demolish and Install	L.F.	.88	1.45		2.33	gable roofs, up to 5/12 slope, 25′
	Reinstall	L.F.		.67		.67	span, per L.F.
	Minimum Charge	Job		157		157	

Rough Frame / Structure

Roof Framing

		Unit	Material	Labor	Equip.	Total	Specification
2" x 10"							Cost includes material and labor to install 2" x 10" rafters for flat, shed or gable roofs, up to 5/12 slope, 25' span, per L.F.
	Demolish	L.F.		.62		.62	
	Install	L.F.	1.19	1.27		2.46	
	Demolish and Install	L.F.	1.19	1.89		3.08	
	Reinstall	L.F.		1.01		1.01	
	Minimum Charge	Job		157		157	
2" x 12"							Cost includes material and labor to install 2" x 12" rafters for flat, shed or gable roofs, up to 5/12 slope, 25' span, per L.F.
	Demolish	L.F.		.63		.63	
	Install	L.F.	1.62	1.38		3	
	Demolish and Install	L.F.	1.62	2.01		3.63	
	Reinstall	L.F.		1.10		1.10	
	Minimum Charge	Job		157		157	
2" x 4" Valley / Jack							Includes material and labor to install up to 2" x 6" valley/jack rafters for flat, shed or gable roofs, up to 5/12 slope, 25' span, per L.F.
	Demolish	L.F.		.59		.59	
	Install	L.F.	.57	1.32		1.89	
	Demolish and Install	L.F.	.57	1.91		2.48	
	Reinstall	L.F.		1.06		1.06	
	Minimum Charge	Job		157		157	
2" x 6" Valley / Jack							Includes material and labor to install up to 2" x 6" valley/jack rafters for flat, shed or gable roofs, up to 5/12 slope, 25' span, per L.F.
	Demolish	L.F.		.60		.60	
	Install	L.F.	.57	1.32		1.89	
	Demolish and Install	L.F.	.57	1.92		2.49	
	Reinstall	L.F.		1.06		1.06	
	Minimum Charge	Job		157		157	
Ridgeboard							
2" x 4"							Cost includes material and labor to install 2" x 4" ridgeboard for flat, shed or gable roofs, up to 5/12 slope, 25' span, per L.F.
	Demolish	L.F.		.57		.57	
	Install	L.F.	.34	.57		.91	
	Demolish and Install	L.F.	.34	1.14		1.48	
	Reinstall	L.F.		.46		.46	
	Minimum Charge	Job		157		157	
2" x 6"							Cost includes material and labor to install 2" x 6" ridgeboard for flat, shed or gable roofs, up to 5/12 slope, 25' span, per L.F.
	Demolish	L.F.		.58		.58	
	Install	L.F.	.57	1.25		1.82	
	Demolish and Install	L.F.	.57	1.83		2.40	
	Reinstall	L.F.		1		1	
	Minimum Charge	Job		157		157	
2" x 8"							Cost includes material and labor to install 2" x 8" ridgeboard for flat, shed or gable roofs, up to 5/12 slope, 25' span, per L.F.
	Demolish	L.F.		.60		.60	
	Install	L.F.	.88	1.39		2.27	
	Demolish and Install	L.F.	.88	1.99		2.87	
	Reinstall	L.F.		1.12		1.12	
	Minimum Charge	Job		157		157	
2" x 10"							Cost includes material and labor to install 2" x 10" ridgeboard for flat, shed or gable roofs, up to 5/12 slope, 25' span, per L.F.
	Demolish	L.F.		.62		.62	
	Install	L.F.	1.19	1.57		2.76	
	Demolish and Install	L.F.	1.19	2.19		3.38	
	Reinstall	L.F.		1.25		1.25	
	Minimum Charge	Job		157		157	
2" x 12"							Cost includes material and labor to install 2" x 12" ridgeboard for flat, shed or gable roofs, up to 5/12 slope, 25' span, per L.F.
	Demolish	L.F.		.64		.64	
	Install	L.F.	1.62	.90		2.52	
	Demolish and Install	L.F.	1.62	1.54		3.16	
	Reinstall	L.F.		.72		.72	
	Minimum Charge	Job		157		157	

Rough Frame / Structure

Roof Framing		Unit	Material	Labor	Equip.	Total	Specification
Collar Beam							
1″ x 6″							
	Demolish	L.F.		.45		.45	Cost includes material and labor to
	Install	L.F.	.42	.39		.81	install 1″ x 6″ collar beam or tie for
	Demolish and Install	L.F.	.42	.84		1.26	roof framing.
	Reinstall	L.F.		.31		.31	
	Minimum Charge	Job		157		157	
2″ x 6″							
	Demolish	L.F.		.53		.53	Cost includes material and labor to
	Install	L.F.	.57	.39		.96	install 2″ x 6″ collar beam or tie for
	Demolish and Install	L.F.	.57	.92		1.49	roof framing.
	Reinstall	L.F.		.31		.31	
	Minimum Charge	Job		157		157	
Purlins							
	Demolish	L.F.		.51		.51	Cost includes material and labor to
	Install	L.F.	.57	.35		.92	install 2″ x 6″ purlins below roof
	Demolish and Install	L.F.	.57	.86		1.43	rafters.
	Reinstall	L.F.		.28		.28	
	Minimum Charge	Job		157		157	
2″ x 8″							
	Demolish	L.F.		.52		.52	Cost includes material and labor to
	Install	L.F.	.88	.35		1.23	install 2″ x 8″ purlins below roof
	Demolish and Install	L.F.	.88	.87		1.75	rafters.
	Reinstall	L.F.		.28		.28	
	Minimum Charge	Job		157		157	
2″ x 10″							
	Demolish	L.F.		.52		.52	Cost includes material and labor to
	Install	L.F.	1.19	.36		1.55	install 2″ x 10″ purlins below roof
	Demolish and Install	L.F.	1.19	.88		2.07	rafters.
	Reinstall	L.F.		.29		.29	
	Minimum Charge	Job		157		157	
2″ x 12″							
	Demolish	L.F.		.53		.53	Cost includes material and labor to
	Install	L.F.	1.62	.36		1.98	install 2″ x 12″ purlins below roof
	Demolish and Install	L.F.	1.62	.89		2.51	rafters.
	Reinstall	L.F.		.29		.29	
	Minimum Charge	Job		157		157	
4″ x 6″							
	Demolish	L.F.		.53		.53	Cost includes material and labor to
	Install	L.F.	2.76	.36		3.12	install 4″ x 6″ purlins below roof
	Demolish and Install	L.F.	2.76	.89		3.65	rafters.
	Reinstall	L.F.		.29		.29	
	Minimum Charge	Job		157		157	
4″ x 8″							
	Demolish	L.F.		.53		.53	Cost includes material and labor to
	Install	L.F.	3.67	.37		4.04	install 4″ x 8″ purlins below roof
	Demolish and Install	L.F.	3.67	.90		4.57	rafters.
	Reinstall	L.F.		.30		.30	
	Minimum Charge	Job		157		157	
Ledger Board							
2″ x 6″							
	Demolish	L.F.		.26		.26	Cost includes material and labor to
	Install	L.F.	.57	1.41		1.98	install 2″ x 6″ ledger board nailed to
	Demolish and Install	L.F.	.57	1.67		2.24	the face of a rafter.
	Reinstall	L.F.		1.13		1.13	
	Minimum Charge	Job		157		157	

Roof Framing		Unit	Material	Labor	Equip.	Total	Specification
4″ x 6″							
	Demolish	L.F.		.43		.43	Cost includes material and labor to
	Install	L.F.	2.76	2.08		4.84	install 4″ x 6″ ledger board nailed to
	Demolish and Install	L.F.	2.76	2.51		5.27	the face of a rafter.
	Reinstall	L.F.		1.66		1.66	
	Minimum Charge	Job		157		157	
4″ x 8″							
	Demolish	L.F.		.57		.57	Cost includes material and labor to
	Install	L.F.	3.67	2.39		6.06	install 4″ x 8″ ledger board nailed to
	Demolish and Install	L.F.	3.67	2.96		6.63	the face of a rafter.
	Reinstall	L.F.		1.92		1.92	
	Minimum Charge	Job		157		157	
Outriggers							
2″ x 4″							
	Demolish	L.F.		.18		.18	Cost includes material and labor to
	Install	L.F.	.34	1.25		1.59	install 2″ x 4″ outrigger rafters for flat,
	Demolish and Install	L.F.	.34	1.43		1.77	shed or gabled roofs, up to 5/12
	Reinstall	L.F.		1		1	slope, per L.F.
	Minimum Charge	Job		157		157	
2″ x 6″							
	Demolish	L.F.		.26		.26	Cost includes material and labor to
	Install	L.F.	.57	1.25		1.82	install 2″ x 6″ outrigger rafters for flat,
	Demolish and Install	L.F.	.57	1.51		2.08	shed or gabled roofs, up to 5/12
	Reinstall	L.F.		1		1	slope, per L.F.
	Minimum Charge	Job		157		157	
Lookout Rafter							
2″ x 4″							
	Demolish	L.F.		.59		.59	Cost includes material and labor to
	Install	L.F.	.34	.52		.86	install 2″ x 4″ lookout rafters for flat,
	Demolish and Install	L.F.	.34	1.11		1.45	shed or gabled roofs, up to 5/12
	Reinstall	L.F.		.42		.42	slope, per L.F.
	Minimum Charge	Job		157		157	
2″ x 6″							
	Demolish	L.F.		.60		.60	Cost includes material and labor to
	Install	L.F.	.57	.53		1.10	install 2″ x 6″ lookout rafters for flat,
	Demolish and Install	L.F.	.57	1.13		1.70	shed or gabled roofs, up to 5/12
	Reinstall	L.F.		.43		.43	slope, per L.F.
	Minimum Charge	Job		157		157	
Fly Rafter							
2″ x 4″							
	Demolish	L.F.		.59		.59	Cost includes material and labor to
	Install	L.F.	.34	.52		.86	install 2″ x 4″ fly rafters for flat, shed
	Demolish and Install	L.F.	.34	1.11		1.45	or gabled roofs, up to 5/12 slope,
	Reinstall	L.F.		.42		.42	per L.F.
	Minimum Charge	Job		157		157	
2″ x 6″							
	Demolish	L.F.		.60		.60	Cost includes material and labor to
	Install	L.F.	.57	.53		1.10	install 2″ x 6″ fly rafters for flat, shed
	Demolish and Install	L.F.	.57	1.13		1.70	or gabled roofs, up to 5/12 slope,
	Reinstall	L.F.		.43		.43	per L.F.
	Minimum Charge	Job		157		157	

Rough Frame / Structure

Residential Trusses

	Unit	Material	Labor	Equip.	Total	Specification
W or Fink Truss						
24' Span						
Demolish	Ea.		7.75		7.75	Includes material, labor and
Install	Ea.	51	23.50	11.20	85.70	equipment to install wood gang-nailed
Demolish and Install	Ea.	51	31.25	11.20	93.45	residential truss, up to 24' span.
Reinstall	Ea.		23.37	11.22	34.59	
Minimum Charge	Job		157		157	
28' Span						
Demolish	Ea.		8.25		8.25	Includes material, labor and
Install	Ea.	62	26.50	12.70	101.20	equipment to install wood gang-nailed
Demolish and Install	Ea.	62	34.75	12.70	109.45	residential truss, up to 28' span.
Reinstall	Ea.		26.46	12.70	39.16	
Minimum Charge	Job		157		157	
32' Span						
Demolish	Ea.		9.10		9.10	Includes material, labor and
Install	Ea.	87.50	28	13.45	128.95	equipment to install wood gang-nailed
Demolish and Install	Ea.	87.50	37.10	13.45	138.05	residential truss, up to 32' span.
Reinstall	Ea.		28.05	13.46	41.51	
Minimum Charge	Job		157		157	
36' Span						
Demolish	Ea.		9.80		9.80	Includes material, labor and
Install	Ea.	109	30.50	14.65	154.15	equipment to install wood gang-nailed
Demolish and Install	Ea.	109	40.30	14.65	163.95	residential truss, up to 36' span.
Reinstall	Ea.		30.49	14.63	45.12	
Minimum Charge	Job		157		157	
40' Span						
Demolish	Ea.		10.65		10.65	Includes material, labor and
Install	Ea.	161	32.50	15.65	209.15	equipment to install wood gang-nailed
Demolish and Install	Ea.	161	43.15	15.65	219.80	residential truss, up to 40' span.
Reinstall	Ea.		32.61	15.66	48.27	
Minimum Charge	Job		157		157	
Gable End						
24' Span						
Demolish	Ea.		7.75		7.75	Includes material, labor and
Install	Ea.	72.50	25	12	109.50	equipment to install wood gang-nailed
Demolish and Install	Ea.	72.50	32.75	12	117.25	residential gable end truss, up to 24'
Reinstall	Ea.		25.04	12.02	37.06	span.
Clean	Ea.	1.20	30.50		31.70	
Paint	Ea.	3.48	37		40.48	
Minimum Charge	Job		157		157	
28' Span						
Demolish	Ea.		8.25		8.25	Includes material, labor and
Install	Ea.	88	30	14.30	132.30	equipment to install wood gang-nailed
Demolish and Install	Ea.	88	38.25	14.30	140.55	residential gable end truss, up to 30'
Reinstall	Ea.		29.84	14.32	44.16	span.
Clean	Ea.	1.25	42		43.25	
Paint	Ea.	3.63	50		53.63	
Minimum Charge	Job		157		157	
32' Span						
Demolish	Ea.		9.10		9.10	Includes material, labor and
Install	Ea.	98.50	30.50	14.65	143.65	equipment to install wood gang-nailed
Demolish and Install	Ea.	98.50	39.60	14.65	152.75	residential gable end truss, up to 32'
Reinstall	Ea.		30.49	14.63	45.12	span.
Clean	Ea.	1.31	54.50		55.81	
Paint	Ea.	3.80	65.50		69.30	
Minimum Charge	Job		157		157	

Rough Frame / Structure

Residential Trusses

	Unit	Material	Labor	Equip.	Total	Specification
36' Span						Includes material, labor and
Demolish	Ea.		9.80		9.80	equipment to install wood gang-nailed
Install	Ea.	117	35	16.85	168.85	residential gable end truss, up to 36'
Demolish and Install	Ea.	117	44.80	16.85	178.65	span.
Reinstall	Ea.		35.06	16.83	51.89	
Clean	Ea.	1.38	69		70.38	
Paint	Ea.	3.98	83		86.98	
Minimum Charge	Job		157		157	
40' Span						Includes material, labor and
Demolish	Ea.		10.65		10.65	equipment to install wood gang-nailed
Install	Ea.	130	37	17.70	184.70	residential gable end truss, up to 40'
Demolish and Install	Ea.	130	47.65	17.70	195.35	span.
Reinstall	Ea.		36.91	17.72	54.63	
Clean	Ea.	4.18	85		89.18	
Paint	Ea.	4.18	102		106.18	
Minimum Charge	Job		157		157	
Truss Hardware						
5-1/4" Glue-Lam Seat						Includes labor and material to install
Install	Ea.	51.50	1.74		53.24	beam hangers.
Minimum Charge	Job		157		157	
6-3/4" Glue-Lam Seat						Includes labor and material to install
Install	Ea.	53.50	1.74		55.24	beam hangers.
Minimum Charge	Job		157		157	
8-3/4" Glue-Lam Seat						Includes labor and material to install
Install	Ea.	61	1.74		62.74	beam hangers.
Minimum Charge	Job		157		157	
2" x 4" Joist Hanger						Includes labor and material to install
Install	Ea.	.63	1.79		2.42	joist and beam hangers.
Minimum Charge	Job		157		157	
2" x 12" Joist Hanger						Cost includes labor and material to
Install	Ea.	.66	1.90		2.56	install 2" x 12" joist hanger.
Minimum Charge	Job		157		157	

Commercial Trusses

	Unit	Material	Labor	Equip.	Total	Specification
Engineered Lumber, Truss / Joist						
9-1/2"						Includes material and labor to install
Demolish	S.F.		.42		.42	engineered lumber truss / joists per
Install	S.F.	1.57	.31		1.88	S.F. of floor area based on joists 16"
Demolish and Install	S.F.	1.57	.73		2.30	O.C. Beams, supports and bridging
Reinstall	S.F.		.25		.25	are not included.
Minimum Charge	Job		157		157	
11-7/8"						Includes material and labor to install
Demolish	S.F.		.43		.43	engineered lumber truss / joists per
Install	S.F.	1.68	.32		2	S.F. of floor area based on joists 16"
Demolish and Install	S.F.	1.68	.75		2.43	O.C. Beams, supports and bridging
Reinstall	S.F.		.26		.26	are not included.
Minimum Charge	Job		157		157	
14"						Includes material and labor to install
Demolish	S.F.		.45		.45	engineered lumber truss / joists per
Install	S.F.	1.85	.34		2.19	S.F. of floor area based on joists 16"
Demolish and Install	S.F.	1.85	.79		2.64	O.C. Beams, supports and bridging
Reinstall	S.F.		.28		.28	are not included.
Minimum Charge	Job		157		157	

Rough Frame / Structure

Commercial Trusses

	Unit	Material	Labor	Equip.	Total	Specification
16"						Includes material and labor to install
Demolish	S.F.		.47		.47	engineered lumber truss / joists per
Install	S.F.	2.52	.36		2.88	S.F. of floor area based on joists 16"
Demolish and Install	S.F.	2.52	.83		3.35	O.C. Beams, supports and bridging
Reinstall	S.F.		.29		.29	are not included.
Minimum Charge	Job		157		157	
Bowstring Truss						
100' Clear Span						
Demolish	SF Flr.		.42	.31	.73	Includes material, labor and
Install	SF Flr.	5.95	.35	.25	6.55	equipment to install bow string truss
Demolish and Install	SF Flr.	5.95	.77	.56	7.28	system up to 100' clear span per S.F.
Minimum Charge	Job		157		157	of floor area.
120' Clear Span						
Demolish	SF Flr.		.47	.35	.82	Includes material, labor and
Install	SF Flr.	6.40	.39	.28	7.07	equipment to install bow string truss
Demolish and Install	SF Flr.	6.40	.86	.63	7.89	system up to 120' clear span per S.F.
Minimum Charge	Job		157		157	of floor area.

Roof Sheathing

	Unit	Material	Labor	Equip.	Total	Specification
Plywood						
3/8"						
Demolish	S.F.		.33		.33	Cost includes material and labor to
Install	S.F.	.48	.41		.89	install 3/8" CDX plywood roof
Demolish and Install	S.F.	.48	.74		1.22	sheathing.
Reinstall	S.F.		.41		.41	
Minimum Charge	Job		157		157	
1/2"						
Demolish	S.F.		.33		.33	Cost includes material and labor to
Install	S.F.	.43	.45		.88	install 1/2" CDX plywood roof
Demolish and Install	S.F.	.43	.78		1.21	sheathing.
Minimum Charge	Job		157		157	
5/8"						
Demolish	S.F.		.33		.33	Cost includes material and labor to
Install	S.F.	.50	.48		.98	install 5/8" CDX plywood roof
Demolish and Install	S.F.	.50	.81		1.31	sheathing.
Minimum Charge	Job		157		157	
3/4"						
Demolish	S.F.		.33		.33	Cost includes material and labor to
Install	S.F.	.63	.52		1.15	install 3/4" CDX plywood roof
Demolish and Install	S.F.	.63	.85		1.48	sheathing.
Minimum Charge	Job		157		157	
OSB						
1/2"						
Demolish	S.F.		.33		.33	Cost includes material and labor to
Install	S.F.	.31	.24		.55	install 4' x 8' x 1/2" OSB sheathing.
Demolish and Install	S.F.	.31	.57		.88	
Minimum Charge	Job		157		157	
5/8"						
Demolish	S.F.		.33		.33	Cost includes material and labor to
Install	S.F.	.46	.25		.71	install 4' x 8' x 5/8" OSB sheathing.
Demolish and Install	S.F.	.46	.58		1.04	
Minimum Charge	Job		157		157	

Rough Frame / Structure

Roof Sheathing

Roof Sheathing		Unit	Material	Labor	Equip.	Total	Specification
Plank							Cost includes material and labor to
	Demolish	S.F.		.33		.33	install 1" x 6" or 1" x 8" utility T&G
	Install	S.F.	.88	.87		1.75	board sheathing laid diagonal.
	Demolish and Install	S.F.	.88	1.20		2.08	
	Reinstall	S.F.		.87		.87	
	Minimum Charge	Job		157		157	
Skip-type							
1" x 4"							
	Demolish	S.F.		.51		.51	Includes material and labor to install
	Install	S.F.	.46	.26		.72	sheathing board material 1" x 4", 7"
	Demolish and Install	S.F.	.46	.77		1.23	O.C.
	Minimum Charge	Job		157		157	
1" x 6"							
	Demolish	S.F.		.51		.51	Includes material and labor to install
	Install	S.F.	.42	.22		.64	sheathing board material 1" x 6", 7"
	Demolish and Install	S.F.	.42	.73		1.15	or 9" O.C.
	Minimum Charge	Job		157		157	

Wood Deck

Wood Deck		Unit	Material	Labor	Equip.	Total	Specification
Treated							Includes material and labor to install
	Demolish	S.F.		1.06		1.06	unfinished deck using 2" x 4" treated
	Install	S.F.	4.16	7.85		12.01	decking, with 2" x 6" double beams
	Demolish and Install	S.F.	4.16	8.91		13.07	24" O.C., 4" x 4" posts 5' O.C. set in
	Reinstall	S.F.		6.27		6.27	concrete.
	Clean	S.F.	.06	.24		.30	
	Paint	S.F.	.06	.21		.27	
	Minimum Charge	Job		157		157	
Redwood							Includes material and labor to install
	Demolish	S.F.		1.06		1.06	unfinished deck using 2" x 4" select
	Install	S.F.	16.35	7.85		24.20	redwood decking, with 2" x 6" double
	Demolish and Install	S.F.	16.35	8.91		25.26	beams 24" O.C., 4" x 4" posts 5'
	Reinstall	S.F.		6.27		6.27	O.C. set in concrete.
	Clean	S.F.	.06	.24		.30	
	Paint	S.F.	.06	.21		.27	
	Minimum Charge	Job		157		157	
Cedar							Includes material and labor to install
	Demolish	S.F.		1.06		1.06	unfinished deck using 2" x 4" select
	Install	S.F.	7.10	7.85		14.95	cedar decking, with 2" x 6" double
	Demolish and Install	S.F.	7.10	8.91		16.01	beams 24" O.C., 4" x 4" posts 5'
	Reinstall	S.F.		6.27		6.27	O.C. set in concrete.
	Clean	S.F.	.06	.24		.30	
	Paint	S.F.	.06	.21		.27	
	Minimum Charge	Job		157		157	
Stairs							Includes material and labor to install
	Demolish	Ea.		300		300	wood steps 5-7 risers included.
	Install	Ea.	143	545		688	
	Demolish and Install	Ea.	143	845		988	
	Reinstall	Ea.		434.30		434.30	
	Clean	Flight	2.88	19.55		22.43	
	Paint	Flight	10.10	37		47.10	
	Minimum Charge	Job		157		157	

Rough Frame / Structure

Wood Deck

Wood Deck	Unit	Material	Labor	Equip.	Total	Specification
Railing						
Demolish	L.F.		3.92		3.92	Includes material and labor to install
Install	L.F.	15.05	14.25		29.30	porch rail with balusters.
Demolish and Install	L.F.	15.05	18.17		33.22	
Clean	L.F.	.57	1.44		2.01	
Paint	L.F.	.48	2.30		2.78	
Minimum Charge	Job		157		157	
Bench Seating						
Demolish	L.F.		4.25		4.25	Includes material and labor to install
Install	L.F.	2.15	15.70		17.85	pressure treated bench seating.
Demolish and Install	L.F.	2.15	19.95		22.10	
Reinstall	L.F.		12.54		12.54	
Clean	L.F.	.11	1.15		1.26	
Paint	L.F.	.50	2.21		2.71	
Minimum Charge	Job		157		157	
Privacy Wall						
Demolish	SF Flr.		1.02		1.02	Includes material, labor and
Install	SF Flr.	.44	.54		.98	equipment to install privacy wall.
Demolish and Install	SF Flr.	.44	1.56		2	
Reinstall	SF Flr.		.43		.43	
Clean	SF Flr.	.06	.24		.30	
Paint	S.F.	.06	.21		.27	
Minimum Charge	Job		157		157	

Porch

Porch	Unit	Material	Labor	Equip.	Total	Specification
Treated						
Demolish	S.F.		2.13		2.13	Includes material and labor to install
Install	S.F.	4.94	15.70		20.64	floor and roof frame and decking
Demolish and Install	S.F.	4.94	17.83		22.77	using 2" x 4" treated decking, with 2"
Reinstall	S.F.		12.54		12.54	x 6" double beams 24" O.C., 4" x 4"
Clean	S.F.	.11	.47		.58	posts 5' O.C. set in concrete.
Paint	S.F.	.17	.55		.72	
Minimum Charge	Job		157		157	
Redwood						
Demolish	S.F.		2.13		2.13	Includes material and labor to install
Install	S.F.	17.40	15.70		33.10	floor and roof frame and decking
Demolish and Install	S.F.	17.40	17.83		35.23	using 2" x 4" redwood decking, with
Reinstall	S.F.		12.54		12.54	2" x 6" double beams 24" O.C., 4" x
Clean	S.F.	.11	.47		.58	4" posts 5' O.C. set in concrete.
Paint	S.F.	.17	.55		.72	
Minimum Charge	Job		157		157	
Cedar						
Demolish	S.F.		2.13		2.13	Includes material and labor to install
Install	S.F.	8.15	15.70		23.85	floor and roof frame and cedar
Demolish and Install	S.F.	8.15	17.83		25.98	decking using 2" x 4" select cedar,
Reinstall	S.F.		12.54		12.54	with 2" x 6" double beams 24" O.C.,
Clean	S.F.	.11	.47		.58	4" x 4" posts 5' O.C. set in concrete.
Paint	S.F.	.17	.55		.72	
Minimum Charge	Job		157		157	
Stairs						
Demolish	Ea.		300		300	Includes material and labor to install
Install	Ea.	143	545		688	wood steps 5-7 risers included.
Demolish and Install	Ea.	143	845		988	
Reinstall	Ea.		434.30		434.30	
Clean	Flight	2.88	19.55		22.43	
Paint	Flight	10.10	37		47.10	
Minimum Charge	Job		157		157	

Rough Frame / Structure

Porch		Unit	Material	Labor	Equip.	Total	Specification
Railing							
	Demolish	L.F.		3.92		3.92	Includes material and labor to install
	Install	L.F.	15.05	14.25		29.30	porch rail with balusters.
	Demolish and Install	L.F.	15.05	18.17		33.22	
	Clean	L.F.	.57	1.79		2.36	
	Paint	L.F.	.48	2.30		2.78	
	Minimum Charge	Job		157		157	
Bench Seating							
	Demolish	L.F.		4.25		4.25	Includes material and labor to install
	Install	L.F.	2.15	15.70		17.85	pressure treated bench seating.
	Demolish and Install	L.F.	2.15	19.95		22.10	
	Reinstall	L.F.		12.54		12.54	
	Clean	L.F.	.11	1.15		1.26	
	Paint	L.F.	.50	2.21		2.71	
	Minimum Charge	Job		157		157	
Privacy Wall							
	Demolish	SF Flr.		1.02		1.02	Includes material, labor and
	Install	SF Flr.	.44	.54		.98	equipment to install privacy wall.
	Demolish and Install	SF Flr.	.44	1.56		2	
	Reinstall	SF Flr.		.43		.43	
	Clean	SF Flr.	.06	.24		.30	
	Paint	S.F.	.06	.21		.27	
	Minimum Charge	Job		157		157	
Wood Board Ceiling							
	Demolish	S.F.		.43		.43	Includes material and labor to install
	Install	S.F.	1.77	1.39		3.16	wood board on ceiling.
	Demolish and Install	S.F.	1.77	1.82		3.59	
	Clean	S.F.	.02	.13		.15	
	Paint	S.F.	.15	.35		.50	
	Minimum Charge	Job		157		157	
Re-screen							
	Install	S.F.	.53	1.25		1.78	Includes labor and material to
	Minimum Charge	Job		157		157	re-screen wood frame.

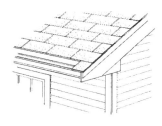
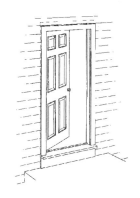

Exterior Trim

Exterior Trim	Unit	Material	Labor	Equip.	Total	Specification
Stock Lumber						
1″ x 2″						
Demolish	L.F.		.21		.21	Cost includes material and labor to
Install	L.F.	.35	.95		1.30	install 1″ x 2″ pine trim.
Demolish and Install	L.F.	.35	1.16		1.51	
Clean	L.F.	.01	.15		.16	
Paint	L.F.	.02	.43		.45	
Minimum Charge	Job		157		157	
1″ x 3″						
Demolish	L.F.		.21		.21	Cost includes material and labor to
Install	L.F.	.23	1.05		1.28	install 1″ x 3″ pine trim.
Demolish and Install	L.F.	.23	1.26		1.49	
Clean	L.F.	.01	.15		.16	
Paint	L.F.	.02	.43		.45	
Minimum Charge	Job		157		157	
1″ x 4″						
Demolish	L.F.		.21		.21	Cost includes material and labor to
Install	L.F.	.73	1.25		1.98	install 1″ x 4″ pine trim.
Demolish and Install	L.F.	.73	1.46		2.19	
Clean	L.F.	.01	.15		.16	
Paint	L.F.	.02	.43		.45	
Minimum Charge	Job		157		157	
1″ x 6″						
Demolish	L.F.		.21		.21	Cost includes material and labor to
Install	L.F.	1.27	1.25		2.52	install 1″ x 6″ pine trim.
Demolish and Install	L.F.	1.27	1.46		2.73	
Clean	L.F.	.01	.15		.16	
Paint	L.F.	.07	.43		.50	
Minimum Charge	Job		157		157	
1″ x 8″						
Demolish	L.F.		.23		.23	Cost includes material and labor to
Install	L.F.	1.56	1.57		3.13	install 1″ x 8″ pine trim.
Demolish and Install	L.F.	1.56	1.80		3.36	
Reinstall	L.F.		1.25		1.25	
Clean	L.F.	.02	.16		.18	
Paint	L.F.	.07	.43		.50	
Minimum Charge	Job		157		157	

Exterior Trim

Exterior Trim		Unit	Material	Labor	Equip.	Total	Specification
1" x 10"							
	Demolish	L.F.		.35		.35	Cost includes material and labor to
	Install	L.F.	1.73	1.74		3.47	install 1" x 10" pine trim.
	Demolish and Install	L.F.	1.73	2.09		3.82	
	Minimum Charge	Job		157		157	
1" x 12"							
	Demolish	L.F.		.35		.35	Cost includes material and labor to
	Install	L.F.	1.89	1.74		3.63	install 1" x 12" pine trim.
	Demolish and Install	L.F.	1.89	2.09		3.98	
	Minimum Charge	Job		157		157	

Fascia		Unit	Material	Labor	Equip.	Total	Specification
Aluminum							
	Demolish	L.F.		.44		.44	Includes material and labor to install
	Install	S.F.	1.11	1.49		2.60	aluminum fascia.
	Demolish and Install	S.F.	1.11	1.93		3.04	
	Reinstall	S.F.		1.19		1.19	
	Clean	L.F.		.29		.29	
	Minimum Charge	Job		157		157	
Vinyl							
	Demolish	L.F.		.44		.44	Cost includes material and labor to
	Install	L.F.	2.34	1.79		4.13	install 6" vinyl fascia with 12" vinyl
	Demolish and Install	L.F.	2.34	2.23		4.57	soffit and J-channel.
	Reinstall	L.F.		1.43		1.43	
	Clean	L.F.		.29		.29	
	Minimum Charge	Job		157		157	
Plywood							
	Demolish	L.F.		.44		.44	Cost includes material and labor to
	Install	L.F.	1.10	.70		1.80	install 12" wide plywood fascia.
	Demolish and Install	L.F.	1.10	1.14		2.24	
	Reinstall	L.F.		.56		.56	
	Clean	L.F.	.03	.19		.22	
	Paint	L.F.	.14	.69		.83	
	Minimum Charge	Job		157		157	
Cedar							
	Demolish	L.F.		.44		.44	Cost includes material and labor to
	Install	L.F.	1.72	.70		2.42	install 12" wide cedar fascia.
	Demolish and Install	L.F.	1.72	1.14		2.86	
	Reinstall	L.F.		.56		.56	
	Clean	L.F.	.03	.19		.22	
	Paint	L.F.	.14	.69		.83	
	Minimum Charge	Job		157		157	
Redwood							
1" x 6"							
	Demolish	L.F.		.34		.34	Cost includes material and labor to
	Install	L.F.	2.34	1.25		3.59	install 1" x 6" redwood fascia board.
	Demolish and Install	L.F.	2.34	1.59		3.93	
	Reinstall	L.F.		1		1	
	Minimum Charge	Job		157		157	
1" x 8"							
	Demolish	L.F.		.38		.38	Cost includes material and labor to
	Install	L.F.	3.20	1.39		4.59	install 1" x 8" redwood fascia board.
	Demolish and Install	L.F.	3.20	1.77		4.97	
	Reinstall	L.F.		1.12		1.12	
	Minimum Charge	Job		157		157	

Exterior Trim

Fascia		Unit	Material	Labor	Equip.	Total	Specification
2″ x 6″							
	Demolish	L.F.		.34		.34	Cost includes material and labor to
	Install	L.F.	2.27	1.39		3.66	install 2″ x 6″ redwood fascia board.
	Demolish and Install	L.F.	2.27	1.73		4	
	Reinstall	L.F.		1.12		1.12	
	Minimum Charge	Job		157		157	
2″ x 8″							
	Demolish	L.F.		.38		.38	Cost includes material and labor to
	Install	L.F.	3.07	1.57		4.64	install 2″ x 8″ redwood fascia board.
	Demolish and Install	L.F.	3.07	1.95		5.02	
	Reinstall	L.F.		1.25		1.25	
	Minimum Charge	Job		157		157	

Hem-fir Std & Better

		Unit	Material	Labor	Equip.	Total	Specification
2″ x 6″							
	Demolish	L.F.		.34		.34	Cost includes material and labor to
	Install	L.F.	.57	2.51		3.08	install 2″ x 6″ hem-fir exterior trim
	Demolish and Install	L.F.	.57	2.85		3.42	board.
	Reinstall	L.F.		2.01		2.01	
	Clean	L.F.	.01	.19		.20	
	Paint	L.F.	.14	.69		.83	
	Minimum Charge	Job		157		157	
2″ x 8″							
	Demolish	L.F.		.38		.38	Cost includes material and labor to
	Install	L.F.	.88	2.79		3.67	install 2″ x 8″ hem-fir exterior trim
	Demolish and Install	L.F.	.88	3.17		4.05	board.
	Reinstall	L.F.		2.23		2.23	
	Clean	L.F.	.02	.19		.21	
	Paint	L.F.	.14	.69		.83	
	Minimum Charge	Job		157		157	
2″ x 10″							
	Demolish	L.F.		.44		.44	Cost includes material and labor to
	Install	L.F.	1.19	3.48		4.67	install 2″ x 10″ hem-fir exterior trim
	Demolish and Install	L.F.	1.19	3.92		5.11	board.
	Reinstall	L.F.		2.79		2.79	
	Clean	L.F.	.03	.21		.24	
	Paint	L.F.	.14	.69		.83	
	Minimum Charge	Job		157		157	

Barge Rafter

		Unit	Material	Labor	Equip.	Total	Specification
2″ x 6″							
	Demolish	L.F.		.68		.68	Cost includes material and labor to
	Install	L.F.	.57	.35		.92	install 2″ x 6″ barge rafters.
	Demolish and Install	L.F.	.57	1.03		1.60	
	Reinstall	L.F.		.28		.28	
	Clean	L.F.	.01	.19		.20	
	Paint	L.F.	.14	.69		.83	
	Minimum Charge	Job		157		157	
2″ x 8″							
	Demolish	L.F.		.69		.69	Cost includes material and labor to
	Install	L.F.	.88	.37		1.25	install 2″ x 8″ barge rafters.
	Demolish and Install	L.F.	.88	1.06		1.94	
	Reinstall	L.F.		.30		.30	
	Clean	L.F.	.02	.19		.21	
	Paint	L.F.	.14	.69		.83	
	Minimum Charge	Job		157		157	

Fascia		Unit	Material	Labor	Equip.	Total	Specification
2″ x 10″							
	Demolish	L.F.		.70		.70	Cost includes material and labor to
	Install	L.F.	1.19	.38		1.57	install 2″ x 10″ barge rafters.
	Demolish and Install	L.F.	1.19	1.08		2.27	
	Reinstall	L.F.		.30		.30	
	Clean	L.F.	.03	.21		.24	
	Paint	L.F.	.14	.69		.83	
	Minimum Charge	Job		157		157	

Soffit		Unit	Material	Labor	Equip.	Total	Specification
Aluminum							
12″ w / Fascia							
	Demolish	L.F.		.44		.44	Cost includes material and labor to
	Install	L.F.	2.70	1.79		4.49	install 12″ aluminum soffit with 6″
	Demolish and Install	L.F.	2.70	2.23		4.93	fascia and J-channel.
	Reinstall	L.F.		1.43		1.43	
	Clean	L.F.		.29		.29	
	Paint	L.F.	.18	.69		.87	
	Minimum Charge	Job		157		157	
18″ w / Fascia							
	Demolish	L.F.		.44		.44	Cost includes material and labor to
	Install	L.F.	3.21	2.09		5.30	install 18″ aluminum soffit with 6″
	Demolish and Install	L.F.	3.21	2.53		5.74	fascia and J-channel.
	Reinstall	L.F.		1.67		1.67	
	Clean	L.F.		.29		.29	
	Paint	L.F.	.18	.69		.87	
	Minimum Charge	Job		157		157	
24″ w / Fascia							
	Demolish	L.F.		.44		.44	Cost includes material and labor to
	Install	L.F.	3.76	2.32		6.08	install 24″ aluminum soffit with 6″
	Demolish and Install	L.F.	3.76	2.76		6.52	fascia and J-channel.
	Reinstall	L.F.		1.86		1.86	
	Clean	L.F.		.42		.42	
	Paint	L.F.	.18	.69		.87	
	Minimum Charge	Job		157		157	
Vinyl							
	Demolish	L.F.		.44		.44	Cost includes material and labor to
	Install	L.F.	2.34	1.79		4.13	install 6″ vinyl fascia with 12″ vinyl
	Demolish and Install	L.F.	2.34	2.23		4.57	soffit and J-channel.
	Reinstall	L.F.		1.43		1.43	
	Clean	L.F.		.29		.29	
	Paint	L.F.	.18	.69		.87	
	Minimum Charge	Job		157		157	
Plywood							
	Demolish	L.F.		.49		.49	Cost includes material and labor to
	Install	L.F.	1.10	.90		2	install 12″ wide plywood soffit.
	Demolish and Install	L.F.	1.10	1.39		2.49	
	Reinstall	L.F.		.72		.72	
	Clean	L.F.	.03	.19		.22	
	Paint	L.F.	.14	.69		.83	
	Minimum Charge	Job		157		157	

Exterior Trim

Soffit	Unit	Material	Labor	Equip.	Total	Specification
Redwood						
Demolish	L.F.		.49		.49	Cost includes material and labor to
Install	L.F.	4.69	.90		5.59	install 12″ wide redwood soffit.
Demolish and Install	L.F.	4.69	1.39		6.08	
Reinstall	L.F.		.72		.72	
Clean	L.F.	.03	.19		.22	
Paint	L.F.	.14	.69		.83	
Minimum Charge	Job		157		157	
Cedar						
Demolish	L.F.		.49		.49	Cost includes material and labor to
Install	L.F.	1.72	.90		2.62	install 12″ wide cedar soffit.
Demolish and Install	L.F.	1.72	1.39		3.11	
Reinstall	L.F.		.72		.72	
Clean	L.F.	.03	.19		.22	
Paint	L.F.	.14	.69		.83	
Minimum Charge	Job		157		157	
Douglas Fir						
Demolish	L.F.		.49		.49	Cost includes material and labor to
Install	L.F.	1.89	.90		2.79	install 12″ wide Douglas fir soffit.
Demolish and Install	L.F.	1.89	1.39		3.28	
Reinstall	L.F.		.72		.72	
Clean	L.F.	.03	.19		.22	
Paint	L.F.	.14	.69		.83	
Minimum Charge	Job		157		157	

Roofing and Flashing

Tile Roofing

Metal Flashing	Unit	Material	Labor	Equip.	Total	Specification
Galvanized						
6″						
Demolish	L.F.		.17		.17	Includes material and labor to install
Install	L.F.	.41	.72		1.13	flat galvanized steel flashing, 6″ wide.
Demolish and Install	L.F.	.41	.89		1.30	
Paint	L.F.	.10	.35		.45	
Minimum Charge	Job		144		144	
14″						
Demolish	L.F.		.17		.17	Includes material and labor to install
Install	L.F.	.95	.72		1.67	flat galvanized steel flashing, 14″
Demolish and Install	L.F.	.95	.89		1.84	wide.
Paint	L.F.	.10	.35		.45	
Minimum Charge	Job		144		144	
20″						
Demolish	L.F.		.17		.17	Includes material and labor to install
Install	L.F.	1.34	.72		2.06	flat galvanized steel flashing, 20″
Demolish and Install	L.F.	1.34	.89		2.23	wide.
Paint	L.F.	.10	.35		.45	
Minimum Charge	Job		144		144	
Aluminum						
6″						
Demolish	L.F.		.17		.17	Includes material and labor to install
Install	L.F.	.25	.72		.97	flat aluminum flashing, 6″ wide.
Demolish and Install	L.F.	.25	.89		1.14	
Paint	L.F.	.10	.35		.45	
Minimum Charge	Job		144		144	
14″						
Demolish	L.F.		.17		.17	Includes material and labor to install
Install	L.F.	.48	.72		1.20	flat aluminum flashing, 14″ wide.
Demolish and Install	L.F.	.48	.89		1.37	
Paint	L.F.	.10	.35		.45	
Minimum Charge	Job		144		144	
20″						
Demolish	L.F.		.17		.17	Includes material and labor to install
Install	L.F.	.64	.72		1.36	flat aluminum flashing, 20″ wide.
Demolish and Install	L.F.	.64	.89		1.53	
Paint	L.F.	.10	.35		.45	
Minimum Charge	Job		144		144	

Roofing

Metal Flashing		Unit	Material	Labor	Equip.	Total	Specification
Copper							
6"							
	Demolish	L.F.		.17		.17	Includes material and labor to install
	Install	L.F.	1.72	.96		2.68	flat copper flashing, 6" wide.
	Demolish and Install	L.F.	1.72	1.13		2.85	
	Minimum Charge	Job		171		171	
14"							
	Demolish	L.F.		.17		.17	Includes material and labor to install
	Install	L.F.	3.98	.96		4.94	flat copper flashing, 14" wide.
	Demolish and Install	L.F.	3.98	1.13		5.11	
	Minimum Charge	Job		171		171	
18"							
	Demolish	L.F.		.17		.17	Includes material and labor to install
	Install	L.F.	5.70	.96		6.66	flat copper flashing, 18" wide.
	Demolish and Install	L.F.	5.70	1.13		6.83	
	Minimum Charge	Job		171		171	
Aluminum Valley							
	Demolish	L.F.		.17		.17	Includes material and labor to install
	Install	L.F.	2.40	1.84		4.24	aluminum valley flashing, .024" thick.
	Demolish and Install	L.F.	2.40	2.01		4.41	
	Paint	L.F.	.10	.35		.45	
	Minimum Charge	Job		157		157	
Gravel Stop							
	Demolish	L.F.		.15		.15	Includes material and labor to install
	Install	L.F.	3.80	2.36		6.16	aluminum gravel stop, .050" thick, 4"
	Demolish and Install	L.F.	3.80	2.51		6.31	high, mill finish.
	Reinstall	L.F.		1.89		1.89	
	Paint	L.F.	.15	.35		.50	
	Minimum Charge	Job		157		157	
Drip Edge							
	Demolish	L.F.		.15		.15	Includes material and labor to install
	Install	L.F.	.28	.78		1.06	aluminum drip edge, .016" thick, 5"
	Demolish and Install	L.F.	.28	.93		1.21	wide, mill finish.
	Reinstall	L.F.		.63		.63	
	Paint	L.F.	.15	.35		.50	
	Minimum Charge	Job		157		157	
End Wall							
	Demolish	L.F.		.78		.78	Includes material and labor to install
	Install	L.F.	1.42	1.84		3.26	aluminum end wall flashing, .024"
	Demolish and Install	L.F.	1.42	2.62		4.04	thick.
	Minimum Charge	Job		157		157	
Side Wall							
	Demolish	L.F.		.15		.15	Includes material and labor to install
	Install	L.F.	1.42	1.84		3.26	aluminum side wall flashing, .024"
	Demolish and Install	L.F.	1.42	1.99		3.41	thick.
	Minimum Charge	Job		157		157	
Coping and Wall Cap							
	Demolish	L.F.		1.06		1.06	Includes material and labor to install
	Install	L.F.	1.53	1.99		3.52	aluminum flashing, mill finish, 0.40
	Demolish and Install	L.F.	1.53	3.05		4.58	thick, 12" wide.
	Clean	L.F.	.08	.72		.80	
	Minimum Charge	Job		144		144	

Roofing

Roof Shingles

Roof Shingles	Unit	Material	Labor	Equip.	Total	Specification
Underlayment						
15#						
Demolish	Sq.		8.50		8.50	Includes material and labor to install
Install	Sq.	2.95	4.50		7.45	#15 felt underlayment.
Demolish and Install	Sq.	2.95	13		15.95	
Minimum Charge	Job		144		144	
30#						
Demolish	Sq.		8.50		8.50	Includes material and labor to install
Install	Sq.	7.80	4.97		12.77	#30 felt underlayment.
Demolish and Install	Sq.	7.80	13.47		21.27	
Minimum Charge	Job		144		144	
Self Adhering						
Demolish	Sq.		21.50		21.50	Includes material and labor to install
Install	Sq.	42	13.10		55.10	self adhering ice barrier roofing
Demolish and Install	Sq.	42	34.60		76.60	underlayment.
Minimum Charge	Job		144		144	

Rolled Roofing

Rolled Roofing	Unit	Material	Labor	Equip.	Total	Specification
90 lb						
Demolish	Sq.		21.50		21.50	Cost includes material and labor to
Install	Sq.	16.90	19.20		36.10	install 90 lb mineral surface rolled
Demolish and Install	Sq.	16.90	40.70		57.60	roofing.
Minimum Charge	Job		144		144	
110 lb						
Demolish	Sq.		21.50		21.50	Cost includes material and labor to
Install	Sq.	44.50	19.20		63.70	install 110 lb mineral surface rolled
Demolish and Install	Sq.	44.50	40.70		85.20	roofing.
Minimum Charge	Job		144		144	
140 lb						
Demolish	Sq.		21.50		21.50	Cost includes material and labor to
Install	Sq.	44.50	29		73.50	install 140 lb mineral surface rolled
Demolish and Install	Sq.	44.50	50.50		95	roofing.
Minimum Charge	Job		144		144	

Composition Shingles

Composition Shingles	Unit	Material	Labor	Equip.	Total	Specification
Fiberglass (3-tab)						
225 lb, 20 Year						
Demolish	Sq.		23		23	Cost includes material and labor to
Install	Sq.	27	56		83	install 225 lb, 20 year fiberglass
Demolish and Install	Sq.	27	79		106	shingles.
Minimum Charge	Job		144		144	
300 lb, 25 Year						
Demolish	Sq.		23		23	Cost includes material and labor to
Install	Sq.	28.50	56		84.50	install 300 lb, 25 year fiberglass
Demolish and Install	Sq.	28.50	79		107.50	shingles.
Minimum Charge	Job		144		144	
Emergency Repairs						
Install	Ea.	72	144		216	Includes labor and material for
Minimum Charge	Job		144		144	emergency repairs.

Roofing

Composition Shingles	Unit	Material	Labor	Equip.	Total	Specification
Architectural						
Laminated, 25 Year						
Demolish	Sq.		23		23	Cost includes material and labor to
Install	Sq.	51	56		107	install 25 year fiberglass roof shingles.
Demolish and Install	Sq.	51	79		130	
Minimum Charge	Job		144		144	
Laminated, 30 Year						
Demolish	Sq.		23		23	Cost includes material and labor to
Install	Sq.	56	56		112	install 30 year fiberglass roof shingles.
Demolish and Install	Sq.	56	79		135	
Minimum Charge	Job		144		144	
Laminated, 40 Year						
Demolish	Sq.		23		23	Cost includes material and labor to
Install	Sq.	69	56		125	install 40 year fiberglass roof shingles.
Demolish and Install	Sq.	69	79		148	
Minimum Charge	Job		144		144	
Laminated Shake, 40 Year						
Demolish	Sq.		23		23	Cost includes material and labor to
Install	Sq.	85	75		160	install 40 year fiberglass roof shakes.
Demolish and Install	Sq.	85	98		183	
Minimum Charge	Job		144		144	
Emergency Repairs						
Install	Ea.	72	144		216	Includes labor and material for
Minimum Charge	Job		144		144	emergency repairs.
Roof Jack						
Demolish	Ea.		4.64		4.64	Includes material and labor to install
Install	Ea.	13.55	31		44.55	residential roof jack, w/bird screen,
Demolish and Install	Ea.	13.55	35.64		49.19	backdraft damper, 3" & 4" dia. round
Paint	Ea.	.06	3.45		3.51	duct.
Minimum Charge	Job		172		172	
Ridge Vent						
Demolish	L.F.		.82		.82	Includes material and labor to install a
Install	L.F.	2.53	2.21		4.74	mill finish aluminum ridge vent strip.
Demolish and Install	L.F.	2.53	3.03		5.56	
Minimum Charge	Job		157		157	
Shingle Molding						
1" x 2"						
Demolish	L.F.		.16		.16	Cost includes material and labor to
Install	L.F.	.35	.95		1.30	install 1" x 2" pine trim.
Demolish and Install	L.F.	.35	1.11		1.46	
Paint	L.F.	.02	.43		.45	
Minimum Charge	Job		157		157	
1" x 3"						
Demolish	L.F.		.16		.16	Cost includes material and labor to
Install	L.F.	.43	1.08		1.51	install 1" x 3" pine trim.
Demolish and Install	L.F.	.43	1.24		1.67	
Paint	L.F.	.02	.43		.45	
Minimum Charge	Job		157		157	
1" x 4"						
Demolish	L.F.		.16		.16	Cost includes material and labor to
Install	L.F.	.73	1.25		1.98	install 1" x 4" pine trim.
Demolish and Install	L.F.	.73	1.41		2.14	
Paint	L.F.	.02	.43		.45	
Minimum Charge	Job		157		157	

Roofing

Composition Shingles

Composition Shingles		Unit	Material	Labor	Equip.	Total	Specification
Add for Steep Pitch	Install	Sq.		18		18	Includes labor and material added for 2nd story or steep roofs.
Add for Additional Story	Install	Sq.		18		18	Includes labor and material added for 2nd story or steep roofs.

Wood Shingles

Wood Shingles		Unit	Material	Labor	Equip.	Total	Specification
Cedar Wood Shingles							
16" Red Label (#2)							
	Demolish	Sq.		56.50		56.50	Includes material and labor to install
	Install	Sq.	127	125		252	wood shingles/shakes.
	Demolish and Install	Sq.	127	181.50		308.50	
	Minimum Charge	Job		157		157	
16" Blue Label (#1)							
	Demolish	Sq.		56.50		56.50	Includes material and labor to install
	Install	Sq.	179	96.50		275.50	wood shingles/shakes.
	Demolish and Install	Sq.	179	153		332	
	Minimum Charge	Job		157		157	
18" Red Label (#2)							
	Demolish	Sq.		56.50		56.50	Includes material and labor to install
	Install	Sq.	219	80.50		299.50	wood shingles/shakes.
	Demolish and Install	Sq.	219	137		356	
	Minimum Charge	Job		157		157	
18" Blue Label (#1)							
	Demolish	Sq.		56.50		56.50	Includes material and labor to install
	Install	Sq.	176	88		264	wood shingles/shakes.
	Demolish and Install	Sq.	176	144.50		320.50	
	Minimum Charge	Job		157		157	
Replace Shingle							
	Install	Ea.	13.20	7.85		21.05	Includes labor and material to replace individual wood shingle/shake. Cost is per shingle.
	Minimum Charge	Job		157		157	
Cedar Shake							
18" Medium Handsplit							
	Demolish	Sq.		56.50		56.50	Includes material and labor to install
	Install	Sq.	107	121		228	wood shingles/shakes.
	Demolish and Install	Sq.	107	177.50		284.50	
	Minimum Charge	Job		157		157	
18" Heavy Handsplit							
	Demolish	Sq.		56.50		56.50	Includes material and labor to install
	Install	Sq.	107	134		241	wood shingles/shakes.
	Demolish and Install	Sq.	107	190.50		297.50	
	Minimum Charge	Job		157		157	
24" Heavy Handsplit							
	Demolish	Sq.		56.50		56.50	Includes material and labor to install
	Install	Sq.	152	107		259	wood shingles/shakes.
	Demolish and Install	Sq.	152	163.50		315.50	
	Minimum Charge	Job		157		157	
24" Medium Handsplit							
	Demolish	Sq.		56.50		56.50	Includes material and labor to install
	Install	Sq.	152	96.50		248.50	wood shingles/shakes.
	Demolish and Install	Sq.	152	153		305	
	Minimum Charge	Job		157		157	

Roofing

Wood Shingles		Unit	Material	Labor	Equip.	Total	Specification
24" Straight Split							
	Demolish	Sq.		56.50		56.50	Includes material and labor to install
	Install	Sq.	163	143		306	wood shingles/shakes.
	Demolish and Install	Sq.	163	199.50		362.50	
	Minimum Charge	Job		157		157	
24" Tapersplit							
	Demolish	Sq.		56.50		56.50	Includes material and labor to install
	Install	Sq.	163	143		306	wood shingles/shakes.
	Demolish and Install	Sq.	163	199.50		362.50	
	Minimum Charge	Job		157		157	
18" Straight Split							
	Demolish	Sq.		56.50		56.50	Includes material and labor to install
	Install	Sq.	121	143		264	wood shingles/shakes.
	Demolish and Install	Sq.	121	199.50		320.50	
	Minimum Charge	Job		157		157	
Replace Shingle							
	Install	Ea.	13.20	7.85		21.05	Includes labor and material to replace
	Minimum Charge	Job		157		157	individual wood shingle/shake. Cost is per shingle.

Slate Tile Roofing		Unit	Material	Labor	Equip.	Total	Specification
Unfading Green							
	Demolish	Sq.		51		51	Includes material and labor to install
	Install	Sq.	380	165		545	clear Vermont slate tile.
	Demolish and Install	Sq.	380	216		596	
	Reinstall	Sq.		132.21		132.21	
	Minimum Charge	Job		145		145	
Unfading Purple							
	Demolish	Sq.		51		51	Includes material and labor to install
	Install	Sq.	405	165		570	clear Vermont slate tile.
	Demolish and Install	Sq.	405	216		621	
	Reinstall	Sq.		132.21		132.21	
	Minimum Charge	Job		145		145	
Replace Slate Tiles							
	Install	Ea.	5.95	15.20		21.15	Includes labor and material to replace
	Minimum Charge	Job		145		145	individual slate tiles. Cost is per tile.
Variegated Purple							
	Demolish	Sq.		51		51	Includes material and labor to install
	Install	Sq.	395	165		560	clear Vermont slate tile.
	Demolish and Install	Sq.	395	216		611	
	Reinstall	Sq.		132.21		132.21	
	Minimum Charge	Job		145		145	
Unfading Grey / Black							
	Demolish	Sq.		51		51	Includes material and labor to install
	Install	Sq.	360	165		525	clear Vermont slate tile.
	Demolish and Install	Sq.	360	216		576	
	Reinstall	Sq.		132.21		132.21	
	Minimum Charge	Job		145		145	
Unfading Red							
	Demolish	Sq.		51		51	Includes material and labor to install
	Install	Sq.	1200	165		1365	clear Vermont slate tile.
	Demolish and Install	Sq.	1200	216		1416	
	Reinstall	Sq.		132.21		132.21	
	Minimum Charge	Job		145		145	

Roofing

Slate Tile Roofing

Slate Tile Roofing		Unit	Material	Labor	Equip.	Total	Specification
Weathering Black							
	Demolish	Sq.		51		51	Includes material and labor to install
	Install	Sq.	495	165		660	clear Pennsylvania slate tile.
	Demolish and Install	Sq.	495	216		711	
	Reinstall	Sq.		132.21		132.21	
	Minimum Charge	Job		145		145	
Weathering Green							
	Demolish	Sq.		51		51	Includes material and labor to install
	Install	Sq.	365	165		530	clear Vermont slate tile.
	Demolish and Install	Sq.	365	216		581	
	Reinstall	Sq.		132.21		132.21	
	Minimum Charge	Job		145		145	

Flat Clay Tile Roofing

Flat Clay Tile Roofing		Unit	Material	Labor	Equip.	Total	Specification
Glazed							
	Demolish	Sq.		56.50		56.50	Includes material and labor to install
	Install	Sq.	495	116		611	tile roofing. Cost based on terra cotta
	Demolish and Install	Sq.	495	172.50		667.50	red flat clay tile and includes ridge
	Reinstall	Sq.		92.54		92.54	and rake tiles.
	Minimum Charge	Job		145		145	
Terra Cotta Red							
	Demolish	Sq.		56.50		56.50	Includes material and labor to install
	Install	Sq.	495	116		611	tile roofing. Cost based on terra cotta
	Demolish and Install	Sq.	495	172.50		667.50	red flat clay tile and includes ridge
	Reinstall	Sq.		92.54		92.54	and rake tiles.
	Minimum Charge	Job		145		145	

Mission Tile Roofing

Mission Tile Roofing		Unit	Material	Labor	Equip.	Total	Specification
Glazed Red							
	Demolish	Sq.		56.50		56.50	Includes material and labor to install
	Install	Sq.	715	193		908	tile roofing. Cost based on glazed red
	Demolish and Install	Sq.	715	249.50		964.50	tile and includes birdstop, ridge and
	Reinstall	Sq.		154.24		154.24	rake tiles.
	Minimum Charge	Job		145		145	
Unglazed Red							
	Demolish	Sq.		56.50		56.50	Includes material and labor to install
	Install	Sq.	715	193		908	tile roofing. Cost based on unglazed
	Demolish and Install	Sq.	715	249.50		964.50	red tile and includes birdstop, ridge
	Reinstall	Sq.		154.24		154.24	and rake tiles.
	Minimum Charge	Job		145		145	
Glazed Blue							
	Demolish	Sq.		56.50		56.50	Includes material and labor to install
	Install	Sq.	715	193		908	tile roofing. Cost based on glazed
	Demolish and Install	Sq.	715	249.50		964.50	blue tile and includes birdstop, ridge
	Reinstall	Sq.		154.24		154.24	and rake tiles.
	Minimum Charge	Job		145		145	
Glazed White							
	Demolish	Sq.		56.50		56.50	Includes material and labor to install
	Install	Sq.	715	193		908	tile roofing. Cost based on glazed
	Demolish and Install	Sq.	715	249.50		964.50	white tile and includes birdstop, ridge
	Reinstall	Sq.		154.24		154.24	and rake tiles.
	Minimum Charge	Job		145		145	

Mission Tile Roofing		Unit	Material	Labor	Equip.	Total	Specification
Unglazed White							
	Demolish	Sq.		56.50		56.50	Includes material and labor to install
	Install	Sq.	715	193		908	tile roofing. Cost based on unglazed
	Demolish and Install	Sq.	715	249.50		964.50	white tile and includes birdstop, ridge
	Reinstall	Sq.		154.24		154.24	and rake tiles.
	Minimum Charge	Job		145		145	
Color Blend							
	Demolish	Sq.		56.50		56.50	Includes material and labor to install
	Install	Sq.	715	193		908	tile roofing. Cost based on color blend
	Demolish and Install	Sq.	715	249.50		964.50	tile and includes birdstop, ridge and
	Reinstall	Sq.		154.24		154.24	rake tiles.
	Minimum Charge	Job		145		145	
Accessories / Extras							
Add for Cap Furring Strips							
	Install	Sq.	78	181		259	Includes labor and material to install vertically placed wood furring strips under cap tiles.
Add for Tile Adhesive							
	Install	Sq.	13.35	10.55		23.90	Includes material and labor for adhesive material placed between tiles.
Add for Wire Attachment							
	Install	Sq.	13.20	126		139.20	Includes labor and material to install single wire anchors for attaching tile.
Add for Braided Runners							
	Install	Sq.	23.50	121		144.50	Includes labor and material to install braided wire runners for secure attachment in high wind areas.
Add for Stainless Steel Nails							
	Install	Sq.	18.20			18.20	Includes material only for stainless steel nails.
Add for Brass Nails							
	Install	Sq.	24			24	Includes material only for brass nails.
Add for Hurricane Clips							
	Install	Sq.	15.95	170		185.95	Includes material and labor for Hurricane or wind clips for high wind areas.
Add for Copper Nails							
	Install	Sq.	24			24	Includes material only for copper nails.
Add for Vertical Furring Strips							
	Install	Sq.	15.40	33.50		48.90	Includes labor and material to install vertically placed wood furring strips under tiles.
Add for Horizontal Furring Strips							
	Install	Sq.	9.90	15.05		24.95	Includes labor and material to install horizontally placed wood furring strips under tiles.

Spanish Tile Roofing		Unit	Material	Labor	Equip.	Total	Specification
Glazed Red							Includes material and labor to install
	Demolish	Sq.		56.50		56.50	tile roofing. Cost based on glazed red
	Install	Sq.	495	121		616	tile and includes birdstop, ridge and
	Demolish and Install	Sq.	495	177.50		672.50	rake tiles.
	Reinstall	Sq.		96.40		96.40	
	Minimum Charge	Job		145		145	
Glazed White							Includes material and labor to install
	Demolish	Sq.		56.50		56.50	tile roofing. Cost based on glazed
	Install	Sq.	495	121		616	white tile and includes birdstop, ridge
	Demolish and Install	Sq.	495	177.50		672.50	and rake tiles.
	Reinstall	Sq.		96.40		96.40	
	Minimum Charge	Job		145		145	
Glazed Blue							Includes material and labor to install
	Demolish	Sq.		56.50		56.50	tile roofing. Cost based on glazed
	Install	Sq.	495	121		616	blue tile and includes birdstop, ridge
	Demolish and Install	Sq.	495	177.50		672.50	and rake tiles.
	Reinstall	Sq.		96.40		96.40	
	Minimum Charge	Job		145		145	
Unglazed Red							Includes material and labor to install
	Demolish	Sq.		56.50		56.50	tile roofing. Cost based on unglazed
	Install	Sq.	495	121		616	red tile and includes birdstop, ridge
	Demolish and Install	Sq.	495	177.50		672.50	and rake tiles.
	Reinstall	Sq.		96.40		96.40	
	Minimum Charge	Job		145		145	
Unglazed White							Includes material and labor to install
	Demolish	Sq.		56.50		56.50	tile roofing. Cost based on unglazed
	Install	Sq.	495	121		616	white tile and includes birdstop, ridge
	Demolish and Install	Sq.	495	177.50		672.50	and rake tiles.
	Reinstall	Sq.		96.40		96.40	
	Minimum Charge	Job		145		145	
Color Blend							Includes material and labor to install
	Demolish	Sq.		56.50		56.50	tile roofing. Cost based on color blend
	Install	Sq.	495	121		616	tile and includes birdstop, ridge and
	Demolish and Install	Sq.	495	177.50		672.50	rake tiles.
	Reinstall	Sq.		96.40		96.40	
	Minimum Charge	Job		145		145	
Accessories / Extras							
Add for Vertical Furring Strips							Includes labor and material to install
	Install	Sq.	15.40	33.50		48.90	vertically placed wood furring strips under tiles.
Add for Wire Attachment							Includes labor and material to install
	Install	Sq.	11	96.50		107.50	single wire ties.
Add for Braided Runners							Includes labor and material to install
	Install	Sq.	19.45	82.50		101.95	braided wire runners for secure attachment in high wind areas.
Add for Stainless Steel Nails							Includes material only for stainless
	Install	Sq.	15.40			15.40	steel nails.
Add for Copper Nails							Includes material only for copper nails.
	Install	Sq.	20.50			20.50	

Roofing

Spanish Tile Roofing		Unit	Material	Labor	Equip.	Total	Specification
Add for Brass Nails	Install	Sq.	20.50			20.50	Includes material only for brass nails.
Add for Hurricane Clips	Install	Sq.	12.20	99.50		111.70	Includes material and labor to install hurricane or wind clips for high wind areas.
Add for Mortar Set Tiles	Install	Sq.	38	145		183	Includes material and labor to install setting tiles by using mortar material placed between tiles.
Add for Tile Adhesive	Install	Sq.	12.40	9.40		21.80	Includes material and labor for adhesive placed between tiles.
Add for Cap Furring Strips	Install	Sq.	78	181		259	Includes labor and material to install vertically placed wood furring strips under cap tiles.
Add for Horizontal Furring Strips	Install	Sq.	9.90	15.05		24.95	Includes labor and material to install horizontally placed wood furring strips under tiles.

Concrete Tile Roofing		Unit	Material	Labor	Equip.	Total	Specification
Corrugated Red / Brown	Demolish	Sq.		56.50		56.50	Includes material and labor to install tile roofing. Cost includes birdstop, booster, ridge and rake tiles.
	Install	Sq.	121	214		335	
	Demolish and Install	Sq.	121	270.50		391.50	
	Reinstall	Sq.		171.38		171.38	
	Minimum Charge	Job		145		145	
Corrugated Gray	Demolish	Sq.		56.50		56.50	Includes material and labor to install tile roofing. Cost includes birdstop, booster, ridge and rake tiles.
	Install	Sq.	121	214		335	
	Demolish and Install	Sq.	121	270.50		391.50	
	Reinstall	Sq.		171.38		171.38	
	Minimum Charge	Job		145		145	
Corrugated Bright Red	Demolish	Sq.		56.50		56.50	Includes material and labor to install tile roofing. Cost includes birdstop, booster, ridge and rake tiles.
	Install	Sq.	121	214		335	
	Demolish and Install	Sq.	121	270.50		391.50	
	Reinstall	Sq.		171.38		171.38	
	Minimum Charge	Job		145		145	
Corrugated Black	Demolish	Sq.		56.50		56.50	Includes material and labor to install tile roofing. Cost includes birdstop, booster, ridge and rake tiles.
	Install	Sq.	121	214		335	
	Demolish and Install	Sq.	121	270.50		391.50	
	Reinstall	Sq.		171.38		171.38	
	Minimum Charge	Job		145		145	
Corrugated Green	Demolish	Sq.		56.50		56.50	Includes material and labor to install tile roofing. Cost includes birdstop, booster, ridge and rake tiles.
	Install	Sq.	121	214		335	
	Demolish and Install	Sq.	121	270.50		391.50	
	Reinstall	Sq.		171.38		171.38	
	Minimum Charge	Job		145		145	

Roofing

Concrete Tile Roofing		Unit	Material	Labor	Equip.	Total	Specification
Corrugated Blue							
	Demolish	Sq.		56.50		56.50	Includes material and labor to install
	Install	Sq.	121	214		335	tile roofing. Cost includes birdstop,
	Demolish and Install	Sq.	121	270.50		391.50	booster, ridge and rake tiles.
	Reinstall	Sq.		171.38		171.38	
	Minimum Charge	Job		145		145	
Flat Natural Gray							
	Demolish	Sq.		56.50		56.50	Includes material and labor to install
	Install	Sq.	105	214		319	tile roofing. Cost includes birdstop,
	Demolish and Install	Sq.	105	270.50		375.50	booster, ridge and rake tiles.
	Reinstall	Sq.		171.38		171.38	
	Minimum Charge	Job		145		145	
Flat Red / Brown							
	Demolish	Sq.		56.50		56.50	Includes material and labor to install
	Install	Sq.	105	214		319	tile roofing. Cost includes birdstop,
	Demolish and Install	Sq.	105	270.50		375.50	booster, ridge and rake tiles.
	Reinstall	Sq.		171.38		171.38	
	Minimum Charge	Job		145		145	
Flat Bright Red							
	Demolish	Sq.		56.50		56.50	Includes material and labor to install
	Install	Sq.	105	214		319	tile roofing. Cost includes birdstop,
	Demolish and Install	Sq.	105	270.50		375.50	booster, ridge and rake tiles.
	Reinstall	Sq.		171.38		171.38	
	Minimum Charge	Job		145		145	
Flat Green							
	Demolish	Sq.		56.50		56.50	Includes material and labor to install
	Install	Sq.	105	214		319	tile roofing. Cost includes birdstop,
	Demolish and Install	Sq.	105	270.50		375.50	booster, ridge and rake tiles.
	Minimum Charge	Job		145		145	
Flat Blue							
	Demolish	Sq.		56.50		56.50	Includes material and labor to install
	Install	Sq.	105	214		319	tile roofing. Cost includes birdstop,
	Demolish and Install	Sq.	105	270.50		375.50	booster, ridge and rake tiles.
	Reinstall	Sq.		171.38		171.38	
	Minimum Charge	Job		145		145	
Flat Black							
	Demolish	Sq.		56.50		56.50	Includes material and labor to install
	Install	Sq.	105	214		319	tile roofing. Cost includes birdstop,
	Demolish and Install	Sq.	105	270.50		375.50	booster, ridge and rake tiles.
	Reinstall	Sq.		171.38		171.38	
	Minimum Charge	Job		145		145	

Aluminum Sheet		Unit	Material	Labor	Equip.	Total	Specification
Corrugated							
.019" Thick Natural							
	Demolish	S.F.		.61		.61	Includes material and labor to install
	Install	S.F.	.88	1.43		2.31	aluminum sheet metal roofing.
	Demolish and Install	S.F.	.88	2.04		2.92	
	Reinstall	S.F.		1.14		1.14	
	Minimum Charge	Job		345		345	
.016" Thick Natural							
	Demolish	S.F.		.61		.61	Includes material and labor to install
	Install	S.F.	.68	1.43		2.11	aluminum sheet metal roofing.
	Demolish and Install	S.F.	.68	2.04		2.72	
	Reinstall	S.F.		1.14		1.14	
	Minimum Charge	Job		345		345	

Roofing

Aluminum Sheet

Aluminum Sheet	Unit	Material	Labor	Equip.	Total	Specification
.016″ Colored Finish						
Demolish	S.F.		.61		.61	Includes material and labor to install
Install	S.F.	.98	1.43		2.41	aluminum sheet metal roofing.
Demolish and Install	S.F.	.98	2.04		3.02	
Reinstall	S.F.		1.14		1.14	
Minimum Charge	Job		345		345	
.019″ Colored Finish						
Demolish	S.F.		.61		.61	Includes material and labor to install
Install	S.F.	1.09	1.43		2.52	aluminum sheet metal roofing.
Demolish and Install	S.F.	1.09	2.04		3.13	
Reinstall	S.F.		1.14		1.14	
Minimum Charge	Job		345		345	

Ribbed

	Unit	Material	Labor	Equip.	Total	Specification
.019″ Thick Natural						
Demolish	S.F.		.61		.61	Includes material and labor to install
Install	S.F.	.88	1.43		2.31	aluminum sheet metal roofing.
Demolish and Install	S.F.	.88	2.04		2.92	
Reinstall	S.F.		1.14		1.14	
Minimum Charge	Job		345		345	
.016″ Thick Natural						
Demolish	S.F.		.61		.61	Includes material and labor to install
Install	S.F.	.68	1.43		2.11	aluminum sheet metal roofing.
Demolish and Install	S.F.	.68	2.04		2.72	
Reinstall	S.F.		1.14		1.14	
Minimum Charge	Job		345		345	
.050″ Colored Finish						
Demolish	S.F.		.61		.61	Includes material and labor to install
Install	S.F.	3.01	1.43		4.44	aluminum sheet metal roofing.
Demolish and Install	S.F.	3.01	2.04		5.05	
Reinstall	S.F.		1.14		1.14	
Minimum Charge	Job		345		345	
.032″ Thick Natural						
Demolish	S.F.		.61		.61	Includes material and labor to install
Install	S.F.	1.67	1.43		3.10	aluminum sheet metal roofing.
Demolish and Install	S.F.	1.67	2.04		3.71	
Reinstall	S.F.		1.14		1.14	
Minimum Charge	Job		345		345	
.040″ Thick Natural						
Demolish	S.F.		.61		.61	Includes material and labor to install
Install	S.F.	2.04	1.43		3.47	aluminum sheet metal roofing.
Demolish and Install	S.F.	2.04	2.04		4.08	
Reinstall	S.F.		1.14		1.14	
Minimum Charge	Job		345		345	
.050″ Thick Natural						
Demolish	S.F.		.61		.61	Includes material and labor to install
Install	S.F.	2.50	1.43		3.93	aluminum sheet metal roofing.
Demolish and Install	S.F.	2.50	2.04		4.54	
Reinstall	S.F.		1.14		1.14	
Minimum Charge	Job		345		345	
.016″ Colored Finish						
Demolish	S.F.		.61		.61	Includes material and labor to install
Install	S.F.	.98	1.43		2.41	aluminum sheet metal roofing.
Demolish and Install	S.F.	.98	2.04		3.02	
Reinstall	S.F.		1.14		1.14	
Minimum Charge	Job		345		345	

Roofing

Aluminum Sheet

Aluminum Sheet	Unit	Material	Labor	Equip.	Total	Specification
.019" Colored Finish						
Demolish	S.F.		.61		.61	Includes material and labor to install
Install	S.F.	1.09	1.43		2.52	aluminum sheet metal roofing.
Demolish and Install	S.F.	1.09	2.04		3.13	
Reinstall	S.F.		1.14		1.14	
Minimum Charge	Job		345		345	
.032" Colored Finish						
Demolish	S.F.		.61		.61	Includes material and labor to install
Install	S.F.	2.10	1.43		3.53	aluminum sheet metal roofing.
Demolish and Install	S.F.	2.10	2.04		4.14	
Reinstall	S.F.		1.14		1.14	
Minimum Charge	Job		345		345	
.040" Colored Finish						
Demolish	S.F.		.61		.61	Includes material and labor to install
Install	S.F.	2.51	1.43		3.94	aluminum sheet metal roofing.
Demolish and Install	S.F.	2.51	2.04		4.55	
Reinstall	S.F.		1.14		1.14	
Minimum Charge	Job		345		345	

Fiberglass Sheet

Fiberglass Sheet	Unit	Material	Labor	Equip.	Total	Specification
12 Ounce Corrugated						
Demolish	S.F.		1.09		1.09	Cost includes material and labor to
Install	S.F.	3.55	1.15		4.70	install 12 ounce corrugated fiberglass
Demolish and Install	S.F.	3.55	2.24		5.79	sheet roofing.
Reinstall	S.F.		.92		.92	
Minimum Charge	Job		144		144	
8 Ounce Corrugated						
Demolish	S.F.		1.09		1.09	Cost includes material and labor to
Install	S.F.	2.46	1.15		3.61	install 8 ounce corrugated fiberglass
Demolish and Install	S.F.	2.46	2.24		4.70	roofing.
Reinstall	S.F.		.92		.92	
Minimum Charge	Job		144		144	

Galvanized Steel

Galvanized Steel	Unit	Material	Labor	Equip.	Total	Specification
Corrugated						
24 Gauge						
Demolish	S.F.		.61		.61	Includes material and labor to install
Install	S.F.	1.16	1.21		2.37	galvanized steel sheet metal roofing.
Demolish and Install	S.F.	1.16	1.82		2.98	Cost based on 24 gauge corrugated
Reinstall	S.F.		.96		.96	or ribbed steel roofing.
Minimum Charge	Job		144		144	
26 Gauge						
Demolish	S.F.		.61		.61	Includes material and labor to install
Install	S.F.	1.07	1.15		2.22	galvanized steel sheet metal roofing.
Demolish and Install	S.F.	1.07	1.76		2.83	Cost based on 26 gauge corrugated
Reinstall	S.F.		.92		.92	or ribbed steel roofing.
Minimum Charge	Job		144		144	
28 Gauge						
Demolish	S.F.		.61		.61	Includes material and labor to install
Install	S.F.	.89	1.09		1.98	galvanized steel sheet metal roofing.
Demolish and Install	S.F.	.89	1.70		2.59	Cost based on 28 gauge corrugated
Reinstall	S.F.		.87		.87	or ribbed steel roofing.
Minimum Charge	Job		144		144	

Roofing

Galvanized Steel

Galvanized Steel		Unit	Material	Labor	Equip.	Total	Specification
30 Gauge							
	Demolish	S.F.		.61		.61	Includes material and labor to install
	Install	S.F.	.85	1.04		1.89	galvanized steel sheet metal roofing.
	Demolish and Install	S.F.	.85	1.65		2.50	Cost based on 30 gauge corrugated
	Reinstall	S.F.		.83		.83	or ribbed steel roofing.
	Minimum Charge	Job		144		144	

Ribbed

		Unit	Material	Labor	Equip.	Total	Specification
24 Gauge							
	Demolish	S.F.		.61		.61	Includes material and labor to install
	Install	S.F.	1.16	1.21		2.37	galvanized steel sheet metal roofing.
	Demolish and Install	S.F.	1.16	1.82		2.98	Cost based on 24 gauge corrugated
	Reinstall	S.F.		.96		.96	or ribbed steel roofing.
	Minimum Charge	Job		144		144	
26 Gauge							
	Demolish	S.F.		.61		.61	Includes material and labor to install
	Install	S.F.	1.07	1.15		2.22	galvanized steel sheet metal roofing.
	Demolish and Install	S.F.	1.07	1.76		2.83	Cost based on 26 gauge corrugated
	Reinstall	S.F.		.92		.92	or ribbed steel roofing.
	Minimum Charge	Job		144		144	
28 Gauge							
	Demolish	S.F.		.61		.61	Includes material and labor to install
	Install	S.F.	.89	1.09		1.98	galvanized steel sheet metal roofing.
	Demolish and Install	S.F.	.89	1.70		2.59	Cost based on 28 gauge corrugated
	Reinstall	S.F.		.87		.87	or ribbed steel roofing.
	Minimum Charge	Job		144		144	
30 Gauge							
	Demolish	S.F.		.61		.61	Includes material and labor to install
	Install	S.F.	.85	1.04		1.89	galvanized steel sheet metal roofing.
	Demolish and Install	S.F.	.85	1.65		2.50	Cost based on 30 gauge corrugated
	Reinstall	S.F.		.83		.83	or ribbed steel roofing.
	Minimum Charge	Job		144		144	

Standing Seam

Standing Seam		Unit	Material	Labor	Equip.	Total	Specification
Copper							
16 Ounce							
	Demolish	Sq.		60.50		60.50	Includes material and labor to install
	Install	Sq.	430	264		694	copper standing seam metal roofing.
	Demolish and Install	Sq.	430	324.50		754.50	Cost based on 16 ounce copper
	Reinstall	Sq.		210.95		210.95	roofing.
	Minimum Charge	Job		171		171	
18 Ounce							
	Demolish	Sq.		60.50		60.50	Includes material and labor to install
	Install	Sq.	480	286		766	copper standing seam metal roofing.
	Demolish and Install	Sq.	480	346.50		826.50	Cost based on 18 ounce copper
	Reinstall	Sq.		228.53		228.53	roofing.
	Minimum Charge	Job		171		171	
20 Ounce							
	Demolish	Sq.		60.50		60.50	Includes material and labor to install
	Install	Sq.	510	310		820	copper standing seam metal roofing.
	Demolish and Install	Sq.	510	370.50		880.50	Cost based on 20 ounce copper
	Reinstall	Sq.		249.31		249.31	roofing.

Roofing

Standing Seam

Standing Seam		Unit	Material	Labor	Equip.	Total	Specification
Stainless Steel							
26 Gauge							Cost includes material and labor to
	Demolish	Sq.		60.50		60.50	install 26 gauge stainless steel
	Install	Sq.	420	298		718	standing seam metal roofing.
	Demolish and Install	Sq.	420	358.50		778.50	
	Reinstall	Sq.		238.47		238.47	
	Minimum Charge	Job		171		171	
28 Gauge							Includes material and labor to install
	Demolish	Sq.		60.50		60.50	standing seam metal roofing. Cost
	Install	Sq.	335	286		621	based on 28 gauge stainless steel
	Demolish and Install	Sq.	335	346.50		681.50	roofing.
	Reinstall	Sq.		228.53		228.53	
	Minimum Charge	Job		171		171	

Elastomeric Roofing

Elastomeric Roofing		Unit	Material	Labor	Equip.	Total	Specification
Neoprene							
1/16"							Includes material and labor to install
	Demolish	Sq.		21		21	neoprene membrane, fully adhered,
	Install	Sq.	156	138	24	318	1/16" thick.
	Demolish and Install	Sq.	156	159	24	339	
	Minimum Charge	Job		660	76	736	
GRM Membrane							Includes material and labor to install
	Demolish	Sq.		21		21	coal tar based elastomeric roofing
	Install	Sq.	239	219	25.50	483.50	membrane, polyester fiber reinforced.
	Demolish and Install	Sq.	239	240	25.50	504.50	
	Minimum Charge	Job		660	76	736	
Cant Strips							
3"							Includes material and labor to install
	Demolish	L.F.		.65		.65	cants 3" x 3", treated timber, cut
	Install	L.F.	.44	.89		1.33	diagonally.
	Demolish and Install	L.F.	.44	1.54		1.98	
	Minimum Charge	Job		144		144	
4"							Includes material and labor to install
	Demolish	L.F.		.65		.65	cants 4" x 4", treated timber, cut
	Install	L.F.	1.46	.89		2.35	diagonally.
	Demolish and Install	L.F.	1.46	1.54		3	
	Minimum Charge	Job		144		144	
Polyvinyl Chloride (PVC)							
45 mil (loose laid)							Cost includes material and labor to
	Demolish	Sq.		21		21	install 45 mil PVC elastomeric roofing,
	Install	Sq.	90	26	2.99	118.99	loose-laid with stone.
	Demolish and Install	Sq.	90	47	2.99	139.99	
	Minimum Charge	Job		660	76	736	
48 mil (loose laid)							Cost includes material and labor to
	Demolish	Sq.		21		21	install 48 mil PVC elastomeric roofing,
	Install	Sq.	105	26	2.99	133.99	loose-laid with stone.
	Demolish and Install	Sq.	105	47	2.99	154.99	
	Minimum Charge	Job		660	76	736	
60 mil (loose laid)							Cost includes material and labor to
	Demolish	Sq.		21		21	install 60 mil PVC elastomeric roofing,
	Install	Sq.	96.50	26	2.99	125.49	loose-laid with stone.
	Demolish and Install	Sq.	96.50	47	2.99	146.49	
	Minimum Charge	Job		660	76	736	

Roofing

Elastomeric Roofing	Unit	Material	Labor	Equip.	Total	Specification
45 mil (attached)						
Demolish	Sq.		31		31	Cost includes material and labor to
Install	Sq.	78.50	50.50	5.85	134.85	install 45 mil PVC elastomeric roofing
Demolish and Install	Sq.	78.50	81.50	5.85	165.85	fully attached to roof deck.
Minimum Charge	Job		660	76	736	
48 mil (attached)						
Demolish	Sq.		31		31	Cost includes material and labor to
Install	Sq.	127	50.50	5.85	183.35	install 48 mil PVC elastomeric roofing
Demolish and Install	Sq.	127	81.50	5.85	214.35	fully attached to roof deck.
Minimum Charge	Job		660	76	736	
60 mil (attached)						
Demolish	Sq.		31		31	Cost includes material and labor to
Install	Sq.	119	50.50	5.85	175.35	install 60 mil PVC elastomeric roofing
Demolish and Install	Sq.	119	81.50	5.85	206.35	fully attached to roof deck.
Minimum Charge	Job		660	76	736	
CSPE Type						
45 mil (loose laid w / stone)						
Demolish	Sq.		21		21	Cost includes material and labor to
Install	Sq.	143	26	2.99	171.99	install 45 mil CSPE (chlorosulfonated
Demolish and Install	Sq.	143	47	2.99	192.99	polyethylene-hypalon) roofing material
Minimum Charge	Job		660	76	736	loose laid on deck and ballasted with stone.
45 mil (attached at seams)						
Demolish	Sq.		50		50	Cost includes material and labor to
Install	Sq.	138	37.50	4.35	179.85	install 45 mil CSPE (chlorosulfonated
Demolish and Install	Sq.	138	87.50	4.35	229.85	polyethylene-hypalon) roofing material
Minimum Charge	Job		660	76	736	attached to deck at seams with batten strips.
45 mil (fully attached)						
Demolish	Sq.		31		31	Cost includes material and labor to
Install	Sq.	135	50.50	5.85	191.35	install 45 mil CSPE (chlorosulfonated
Demolish and Install	Sq.	135	81.50	5.85	222.35	polyethylene-hypalon) roofing material
Minimum Charge	Job		660	76	736	fully attached to deck.
EPDM Type						
45 mil (attached at seams)						
Demolish	Sq.		50		50	Cost includes material and labor to
Install	Sq.	52.50	37.50	4.35	94.35	install 45 mil EPDM (ethylene
Demolish and Install	Sq.	52.50	87.50	4.35	144.35	propylene diene monomer) roofing
Minimum Charge	Job		660	76	736	material attached to deck at seams.
45 mil (fully attached)						
Demolish	Sq.		31		31	Cost includes material and labor to
Install	Sq.	73.50	50.50	5.85	129.85	install 45 mil EPDM (ethylene
Demolish and Install	Sq.	73.50	81.50	5.85	160.85	propylene diene monomer) roofing
Minimum Charge	Job		660	76	736	material fully attached to deck.
45 mil (loose laid w / stone)						
Demolish	Sq.		21		21	Cost includes material and labor to
Install	Sq.	61	26	2.99	89.99	install 45 mil EPDM (ethylene
Demolish and Install	Sq.	61	47	2.99	110.99	propylene diene monomer) roofing
Minimum Charge	Job		660	76	736	material loose laid on deck and ballasted with stones (approx. 1/2 ton of stone per square).
60 mil (attached at seams)						
Demolish	Sq.		50		50	Cost includes material and labor to
Install	Sq.	64.50	37.50	4.35	106.35	install 60 mil EPDM (ethylene
Demolish and Install	Sq.	64.50	87.50	4.35	156.35	propylene diene monomer) roofing
Minimum Charge	Job		660	76	736	material attached to deck at seams.

Roofing

Elastomeric Roofing

Elastomeric Roofing	Unit	Material	Labor	Equip.	Total	Specification
60 mil (fully attached)						Cost includes material and labor to
Demolish	Sq.		31		31	install 60 mil EPDM (ethylene
Install	Sq.	86.50	50.50	5.85	142.85	propylene diene monomer) roofing
Demolish and Install	Sq.	86.50	81.50	5.85	173.85	material fully attached to deck.
Minimum Charge	Job		660	76	736	
60 mil (loose laid w / stone)						Cost includes material and labor to
Demolish	Sq.		21		21	install 60 mil EPDM (ethylene
Install	Sq.	75	26	2.99	103.99	propylene diene monomer) roofing
Demolish and Install	Sq.	75	47	2.99	124.99	material loose laid on deck and
Minimum Charge	Job		660	76	736	ballasted with stones (approx. 1/2 ton of stone per square).
Silicone-urethane Foam						
Prep Roof for Foam System						Includes labor and material to removal
Install	Sq.	5.25	115		120.25	of any gravel and prepare existing
Minimum Charge	Job		660	76	736	built-up roof for application of silicone-urethane foam system.

Modified Bitumen Roofing

Modified Bitumen Roofing	Unit	Material	Labor	Equip.	Total	Specification
Fully Attached						
120 mil						Cost includes material and labor to
Demolish	Sq.		50		50	install 120 mil modified bitumen
Install	Sq.	78	6.40	1.12	85.52	roofing, fully attached to deck by
Demolish and Install	Sq.	78	56.40	1.12	135.52	torch.
Minimum Charge	Job		144		144	
150 mil						Cost includes material and labor to
Demolish	Sq.		50		50	install 150 mil modified bitumen
Install	Sq.	112	15.75	2.75	130.50	roofing, fully attached to deck by
Demolish and Install	Sq.	112	65.75	2.75	180.50	torch.
Minimum Charge	Job		945	165	1110	
160 mil						Cost includes material and labor to
Demolish	Sq.		50		50	install 160 mil modified bitumen
Install	Sq.	89.50	16.45	2.87	108.82	roofing, fully attached to deck by
Demolish and Install	Sq.	89.50	66.45	2.87	158.82	torch.
Minimum Charge	Job		945	165	1110	

Built-up Roofing

Built-up Roofing	Unit	Material	Labor	Equip.	Total	Specification
Emergency Repairs						Minimum charge for work requiring
Install	Ea.	142	1350	236	1728	hot asphalt.
Minimum Charge	Job		945	165	1110	
Asphalt						
3-ply Asphalt						Cost includes material and labor to
Demolish	Sq.		52		52	install 3-ply asphalt built-up roof
Install	Sq.	39	79	13.75	131.75	including one 30 lb and two 15 lb
Demolish and Install	Sq.	39	131	13.75	183.75	layers of felt.
Minimum Charge	Job		945	165	1110	

Roofing

Built-up Roofing

Built-up Roofing		Unit	Material	Labor	Equip.	Total	Specification
4-ply Asphalt							Cost includes material and labor to install 4-ply asphalt built-up roof including one 30 lb and three 15 lb layers of felt.
	Demolish	Sq.		62		62	
	Install	Sq.	51	86	15	152	
	Demolish and Install	Sq.	51	148	15	214	
	Minimum Charge	Job		945	165	1110	
5-ply Asphalt							Cost includes material and labor to install 5-ply asphalt built-up roof including one 30 lb and four 15 lb layers of felt.
	Demolish	Sq.		81.50		81.50	
	Install	Sq.	73	94.50	16.50	184	
	Demolish and Install	Sq.	73	176	16.50	265.50	
	Minimum Charge	Job		945	165	1110	
Components / Accessories							
Flood Coat							Includes labor and material to install flood coat.
	Install	Sq.	13.55	44.50		58.05	
R / R Gravel and Flood Coat							Includes labor and material to install flood coat and gravel.
	Install	Sq.	17.60	144		161.60	
Aluminum UV Coat							Includes labor and material to install aluminum-based UV coating.
	Install	Sq.	14.15	165	19.05	198.20	
Gravel Coat							Includes labor and material to install gravel coat.
	Demolish	Sq.		26		26	
	Install	Sq.	4.06	24		28.06	
	Demolish and Install	Sq.	4.06	50		54.06	
Mineral Surface Cap Sheet							Cost includes labor and material to install 50 # mineral surfaced cap sheet.
	Install	Sq.	18.15	4.97		23.12	
Fibrated Aluminum UV Coating							Includes labor and material to install fibered aluminum-based UV coating.
	Install	Sq.	16.10	44.50		60.60	
Impregnated Walk							
1"							Cost includes labor and material to install 1" layer of asphalt coating.
	Install	S.F.	2.37	1.54		3.91	
1/2"							Cost includes labor and material to install 1/2" layer of asphalt coating.
	Install	S.F.	1.60	1.21		2.81	
3/4"							Cost includes labor and material to install 3/4" layer of asphalt coating.
	Install	S.F.	1.98	1.28		3.26	

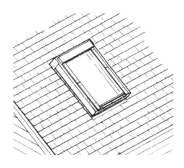

| Doors | Roof Window | Double Hung Window |

Exterior Type Door		Unit	Material	Labor	Equip.	Total	Specification
Entry							
Flush (1-3/4")							
	Demolish	Ea.		29		29	Cost includes material and labor to
	Install	Ea.	210	16.50		226.50	install 3' x 6' 8" pre-hung 1-3/4"
	Demolish and Install	Ea.	210	45.50		255.50	door, including casing and stop,
	Reinstall	Ea.		16.51		16.51	hinges, jamb, aluminum sill,
	Clean	Ea.	2.88	9.55		12.43	weatherstripped.
	Paint	Ea.	7.60	39.50		47.10	
	Minimum Charge	Job		157		157	
Raised Panel							
	Demolish	Ea.		29		29	Cost includes material and labor to
	Install	Ea.	370	17.40		387.40	install 3' x 6' 8" pre-hung 1-3/4"
	Demolish and Install	Ea.	370	46.40		416.40	door, jamb, including casing and
	Reinstall	Ea.		17.42		17.42	stop, hinges, aluminum sill,
	Clean	Ea.	2.88	9.55		12.43	weatherstripping.
	Paint	Ea.	7.60	39.50		47.10	
	Minimum Charge	Job		157		157	
Entry w / Lights							
Good Quality							
	Demolish	Ea.		29		29	Cost includes material and labor to
	Install	Ea.	340	18.45		358.45	install 3' x 6' 8" pre-hung 1-3/4"
	Demolish and Install	Ea.	340	47.45		387.45	door, solid wood lower panel, with up
	Reinstall	Ea.		18.45		18.45	to 9-light glass panels, casing, sill,
	Clean	Ea.	2.88	9.55		12.43	weatherstripped.
	Paint	Ea.	5.60	18.40		24	
	Minimum Charge	Job		157		157	
Better Quality							
	Demolish	Ea.		29		29	Cost includes material and labor to
	Install	Ea.	375	18.45		393.45	install 3' x 6' 8" pre-hung 1-3/4" Ash
	Demolish and Install	Ea.	375	47.45		422.45	door, solid wood, raised panel door
	Reinstall	Ea.		18.45		18.45	without glass panels.
	Clean	Ea.	2.88	9.55		12.43	
	Paint	Ea.	7.60	39.50		47.10	
	Minimum Charge	Job		157		157	
Premium Quality							
	Demolish	Ea.		29		29	Cost includes material and labor to
	Install	Ea.	580	18.45		598.45	install 3' x 6' 8" pre-hung 1-3/4" Oak
	Demolish and Install	Ea.	580	47.45		627.45	door, solid wood, raised panel door
	Reinstall	Ea.		18.45		18.45	without glass panels.
	Clean	Ea.	2.88	9.55		12.43	
	Paint	Ea.	7.60	39.50		47.10	
	Minimum Charge	Job		157		157	

Exterior Doors and Windows

Exterior Type Door		Unit	Material	Labor	Equip.	Total	Specification
Custom Quality							
	Demolish	Ea.		29		29	Cost includes material and labor to
	Install	Ea.	1200	18.45		1218.45	install 3' x 6' 8" pre-hung 1-3/4"
	Demolish and Install	Ea.	1200	47.45		1247.45	hand-made, solid wood, raised panel
	Reinstall	Ea.		18.45		18.45	door with custom stained glass.
	Clean	Ea.	.20	9.55		9.75	
	Paint	Ea.	2.10	92		94.10	
	Minimum Charge	Job		157		157	
Miami Style							
	Demolish	Ea.		29		29	Cost includes material and labor to
	Install	Ea.	200	15.70		215.70	install 3' x 6' 8" pre-hung 1-3/4"
	Demolish and Install	Ea.	200	44.70		244.70	lauan door of solid wood.
	Reinstall	Ea.		15.68		15.68	
	Clean	Ea.	.20	9.55		9.75	
	Paint	Ea.	7.60	39.50		47.10	
	Minimum Charge	Job		157		157	
French Style							
1-Light							
	Demolish	Ea.		29		29	Cost includes material and labor to
	Install	Ea.	340	22.50		362.50	install 3' x 6' 8" pre-hung 1-3/4"
	Demolish and Install	Ea.	340	51.50		391.50	french style door with 1 light.
	Reinstall	Ea.		22.40		22.40	
	Clean	Ea.	1.80	14.35		16.15	
	Paint	Ea.	8.95	69		77.95	
	Minimum Charge	Job		157		157	
5-Light							
	Demolish	Ea.		29		29	Cost includes material and labor to
	Install	Ea.	340	22.50		362.50	install 3' x 6' 8" pre-hung 1-3/4"
	Demolish and Install	Ea.	340	51.50		391.50	french style door with 5 lights.
	Reinstall	Ea.		22.40		22.40	
	Clean	Ea.	.40	19.15		19.55	
	Paint	Ea.	8.95	69		77.95	
	Minimum Charge	Job		157		157	
9-Light							
	Demolish	Ea.		29		29	Cost includes material and labor to
	Install	Ea.	325	22.50		347.50	install 3' x 6' 8" pre-hung 1-3/4"
	Demolish and Install	Ea.	325	51.50		376.50	french style door with 9 lights.
	Reinstall	Ea.		22.40		22.40	
	Clean	Ea.	.40	19.15		19.55	
	Paint	Ea.	8.95	69		77.95	
	Minimum Charge	Job		157		157	
10-Light							
	Demolish	Ea.		32		32	Cost includes material and labor to
	Install	Ea.	550	39		589	install 3' x 6' 8" pre-hung 1-3/4"
	Demolish and Install	Ea.	550	71		621	french style door with 10 lights.
	Reinstall	Ea.		39.20		39.20	
	Clean	Ea.	.40	19.15		19.55	
	Paint	Ea.	8.95	69		77.95	
	Minimum Charge	Job		157		157	
15-Light							
	Demolish	Ea.		32		32	Cost includes material and labor to
	Install	Ea.	615	45		660	install 3' x 6' 8" pre-hung 1-3/4"
	Demolish and Install	Ea.	615	77		692	french style door with 15 lights.
	Reinstall	Ea.		44.80		44.80	
	Clean	Ea.	.40	19.15		19.55	
	Paint	Ea.	8.95	69		77.95	
	Minimum Charge	Job		157		157	

Exterior Type Door		Unit	Material	Labor	Equip.	Total	Specification
Jamb							
Wood							
	Demolish	Ea.		17.40		17.40	Includes material and labor to install
	Install	Ea.	80	8.50		88.50	flat pine jamb with square cut heads
	Demolish and Install	Ea.	80	25.90		105.90	and rabbeted sides for 6' 8" high and
	Clean	Ea.	.09	5.75		5.84	3-9/16" door including trim sets for
	Paint	Ea.	5.60	18.40		24	both sides.
	Minimum Charge	Job		157		157	
Metal 6' 8" High							
	Demolish	Ea.		17.40		17.40	Cost includes material and labor to
	Install	Ea.	78	19.60		97.60	install 18 gauge hollow metal door
	Demolish and Install	Ea.	78	37		115	frame to fit a 4-1/2" jamb 6' 8" high
	Reinstall	Ea.		19.60		19.60	and up to 3' 6" opening.
	Clean	Ea.	.61	4.78		5.39	
	Paint	Ea.	1.05	5.50		6.55	
	Minimum Charge	Job		157		157	
Metal 7' High							
	Demolish	Ea.		17.40		17.40	Cost includes material and labor to
	Install	Ea.	81	21		102	install 18 gauge hollow metal door
	Demolish and Install	Ea.	81	38.40		119.40	frame to fit a 4-1/2" jamb 7' high and
	Reinstall	Ea.		20.91		20.91	up to 3' 6" opening.
	Clean	Ea.	.61	4.78		5.39	
	Paint	Ea.	1.05	5.50		6.55	
	Minimum Charge	Job		157		157	
Metal 8' High							
	Demolish	Ea.		17.40		17.40	Cost includes material and labor to
	Install	Ea.	97.50	26		123.50	install 18 gauge hollow metal door
	Demolish and Install	Ea.	97.50	43.40		140.90	frame to fit a 4-1/2" jamb 8' high and
	Reinstall	Ea.		26.13		26.13	up to 3' 6" opening.
	Clean	Ea.	.85	5.75		6.60	
	Paint	Ea.	1.18	6.15		7.33	
	Minimum Charge	Job		157		157	
Metal 9' High							
	Demolish	Ea.		17.40		17.40	Cost includes material and labor to
	Install	Ea.	111	31.50		142.50	install 18 gauge hollow metal door
	Demolish and Install	Ea.	111	48.90		159.90	frame to fit a 4-1/2" jamb 9' high and
	Reinstall	Ea.		31.36		31.36	up to 3' 6" opening.
	Clean	Ea.	.85	5.75		6.60	
	Paint	Ea.	1.18	6.15		7.33	
	Minimum Charge	Job		157		157	
Casing Trim							
Single Width							
	Demolish	Ea.		5.55		5.55	Cost includes material and labor to
	Install	Opng.	15.90	53		68.90	install 11/16" x 2-1/2" pine ranch
	Demolish and Install	Opng.	15.90	58.55		74.45	style casing for one side of a standard
	Reinstall	Opng.		42.52		42.52	door opening.
	Clean	Opng.	.41	2.30		2.71	
	Paint	Ea.	3.08	5.50		8.58	
	Minimum Charge	Job		157		157	
Double Width							
	Demolish	Ea.		6.55		6.55	Cost includes material and labor to
	Install	Opng.	18.70	62.50		81.20	install 11/16" x 2-1/2" pine ranch
	Demolish and Install	Opng.	18.70	69.05		87.75	style door casing for one side of a
	Reinstall	Opng.		50.18		50.18	double door opening.
	Clean	Opng.	.48	4.59		5.07	
	Paint	Ea.	3.08	6.25		9.33	
	Minimum Charge	Job		157		157	

Exterior Doors and Windows

Residential Metal Door

Residential Metal Door	Unit	Material	Labor	Equip.	Total	Specification
Pre-hung, Steel-clad						
2' 8" x 6' 8"						Cost includes material and labor to
Demolish	Ea.		29		29	install 2' 6" x 6' 8" pre-hung 1-3/4"
Install	Ea.	211	39		250	24 gauge steel clad door with 4-1/2"
Demolish and Install	Ea.	211	68		279	wood frame, trims, hinges, aluminum
Reinstall	Ea.		39.20		39.20	sill, weatherstripped.
Clean	Ea.	2.88	9.55		12.43	
Paint	Ea.	8.95	39.50		48.45	
Minimum Charge	Job		157		157	
3' x 6' 8"						Cost includes material and labor to
Demolish	Ea.		29		29	install 3' x 6' 8" pre-hung 1-3/4" 24
Install	Ea.	215	42		257	gauge steel clad door with 4-1/2"
Demolish and Install	Ea.	215	71		286	wood frame, trims, hinges, aluminum
Reinstall	Ea.		41.81		41.81	sill, weatherstripped.
Clean	Ea.	2.88	9.55		12.43	
Paint	Ea.	8.95	39.50		48.45	
Minimum Charge	Job		157		157	

Metal Door

Metal Door	Unit	Material	Labor	Equip.	Total	Specification
2' 8" x 6' 8", 18 Gauge						Cost includes material and labor to
Demolish	Ea.		6.20		6.20	install 18 gauge 2' 8" x 6' 8" steel
Install	Ea.	217	19.60		236.60	door only, 1-3/4" thick, hinges.
Demolish and Install	Ea.	217	25.80		242.80	
Reinstall	Ea.		19.60		19.60	
Clean	Ea.	1.20	7.20		8.40	
Paint	Ea.	3.80	23		26.80	
Minimum Charge	Job		157		157	
3' x 6' 8", 18 Gauge						Cost includes material and labor to
Demolish	Ea.		6.20		6.20	install 18 gauge 3' x 6' 8" steel door
Install	Ea.	211	21		232	only, 1-3/4" thick, hinges.
Demolish and Install	Ea.	211	27.20		238.20	
Reinstall	Ea.		20.91		20.91	
Clean	Ea.	1.20	7.20		8.40	
Paint	Ea.	3.80	23		26.80	
Minimum Charge	Job		157		157	
2' 8" x 7', 18 Gauge						Cost includes material and labor to
Demolish	Ea.		6.20		6.20	install 18 gauge 2' 8" x 7' steel door
Install	Ea.	234	21		255	only, 1-3/4" thick, hinges.
Demolish and Install	Ea.	234	27.20		261.20	
Reinstall	Ea.		20.91		20.91	
Clean	Ea.	1.20	7.20		8.40	
Paint	Ea.	3.80	23		26.80	
Minimum Charge	Job		157		157	
3' x 7', 18 Gauge						Cost includes material and labor to
Demolish	Ea.		6.20		6.20	install 18 gauge 3' x 7' steel door
Install	Ea.	228	22.50		250.50	only, 1-3/4" thick, hinges.
Demolish and Install	Ea.	228	28.70		256.70	
Reinstall	Ea.		22.40		22.40	
Clean	Ea.	1.20	7.20		8.40	
Paint	Ea.	3.80	23		26.80	
Minimum Charge	Job		157		157	
3' 6" x 7', 18 Gauge						Cost includes material and labor to
Demolish	Ea.		6.20		6.20	install 18 gauge 3' 6" x 7' steel door
Install	Ea.	261	24		285	only, 1-3/4" thick, hinges.
Demolish and Install	Ea.	261	30.20		291.20	
Reinstall	Ea.		24.12		24.12	
Clean	Ea.	1.20	7.20		8.40	
Paint	Ea.	3.80	23		26.80	
Minimum Charge	Job		157		157	

Exterior Doors and Windows

Metal Door		Unit	Material	Labor	Equip.	Total	Specification
Casing Trim							
Single Width							
	Demolish	Ea.		5.55		5.55	Cost includes material and labor to
	Install	Opng.	15.90	53		68.90	install 11/16" x 2-1/2" pine ranch
	Demolish and Install	Opng.	15.90	58.55		74.45	style casing for one side of a standard
	Reinstall	Opng.		42.52		42.52	door opening.
	Clean	Opng.	.41	2.30		2.71	
	Paint	Ea.	3.08	5.50		8.58	
	Minimum Charge	Job		157		157	
Double Width							
	Demolish	Ea.		6.55		6.55	Cost includes material and labor to
	Install	Opng.	18.70	62.50		81.20	install 11/16" x 2-1/2" pine ranch
	Demolish and Install	Opng.	18.70	69.05		87.75	style door casing for one side of a
	Reinstall	Opng.		50.18		50.18	double door opening.
	Clean	Opng.	.48	4.59		5.07	
	Paint	Ea.	3.08	6.25		9.33	
	Minimum Charge	Job		157		157	

Solid Core Door		Unit	Material	Labor	Equip.	Total	Specification
Exterior Type							
Flush (1-3/4")							
	Demolish	Ea.		29		29	Cost includes material and labor to
	Install	Ea.	305	39		344	install 3' x 6' 8" pre-hung 1-3/4"
	Demolish and Install	Ea.	305	68		373	door, including casing and stop,
	Reinstall	Ea.		39.20		39.20	hinges, jamb, aluminum sill,
	Clean	Ea.	2.88	9.55		12.43	weatherstripped.
	Paint	Ea.	8.95	39.50		48.45	
	Minimum Charge	Job		157		157	
Raised Panel							
	Demolish	Ea.		29		29	Cost includes material and labor to
	Install	Ea.	370	17.40		387.40	install 3' x 6' 8" pre-hung 1-3/4"
	Demolish and Install	Ea.	370	46.40		416.40	door, jamb, including casing and
	Reinstall	Ea.		17.42		17.42	stop, hinges, aluminum sill,
	Clean	Ea.	2.88	9.55		12.43	weatherstripping.
	Paint	Ea.	8.95	92		100.95	
	Minimum Charge	Job		157		157	
Entry w / Lights							
	Demolish	Ea.		29		29	Cost includes material and labor to
	Install	Ea.	340	18.45		358.45	install 3' x 6' 8" pre-hung 1-3/4"
	Demolish and Install	Ea.	340	47.45		387.45	door, solid wood lower panel, with up
	Reinstall	Ea.		18.45		18.45	to 9-light glass panels, casing, sill,
	Clean	Ea.	2.88	9.55		12.43	weatherstripped.
	Paint	Ea.	2.10	92		94.10	
	Minimum Charge	Job		157		157	
Custom Entry							
	Demolish	Ea.		29		29	Includes material and labor to install
	Install	Ea.	1000	18.45		1018.45	prehung, fir, single glazed with lead
	Demolish and Install	Ea.	1000	47.45		1047.45	caming, 1-3/4" x 6' 8" x 3'.
	Reinstall	Ea.		18.45		18.45	
	Clean	Ea.	2.88	9.55		12.43	
	Paint	Ea.	3.80	23		26.80	
	Minimum Charge	Job		157		157	

Exterior Doors and Windows

Solid Core Door		Unit	Material	Labor	Equip.	Total	Specification
Dutch Style							
	Demolish	Ea.		29		29	Includes material and labor to install
	Install	Ea.	655	52.50		707.50	dutch door, pine, 1-3/4" x 6' 8" x 2'
	Demolish and Install	Ea.	655	81.50		736.50	8" wide.
	Reinstall	Ea.		52.27		52.27	
	Clean	Ea.	2.88	9.55		12.43	
	Paint	Ea.	2.10	92		94.10	
	Minimum Charge	Job		157		157	
Miami Style							
	Demolish	Ea.		29		29	Cost includes material and labor to
	Install	Ea.	200	15.70		215.70	install 3' x 6' 8" pre-hung 1-3/4"
	Demolish and Install	Ea.	200	44.70		244.70	lauan door of solid wood.
	Reinstall	Ea.		15.68		15.68	
	Clean	Ea.	2.88	9.55		12.43	
	Paint	Ea.	7.60	39.50		47.10	
	Minimum Charge	Job		157		157	
French Style							
	Demolish	Ea.		32		32	Cost includes material and labor to
	Install	Ea.	615	45		660	install 3' x 6' 8" pre-hung 1-3/4"
	Demolish and Install	Ea.	615	77		692	french style door with 15 lights.
	Reinstall	Ea.		44.80		44.80	
	Clean	Ea.	.20	9.55		9.75	
	Paint	Ea.	5.65	138		143.65	
	Minimum Charge	Job		157		157	
Double							
	Demolish	Ea.		21.50		21.50	Includes material and labor to install a
	Install	Ea.	1050	78.50		1128.50	pre-hung double entry door including
	Demolish and Install	Ea.	1050	100		1150	frame and exterior trim.
	Clean	Ea.	.11	33		33.11	
	Paint	Ea.	18	138		156	
	Minimum Charge	Job		78.50		78.50	
Commercial Grade							
3' x 6' 8"							
	Demolish	Ea.		29		29	Includes material and labor to install
	Install	Ea.	270	18.45		288.45	flush solid core 1-3/4" hardwood
	Demolish and Install	Ea.	270	47.45		317.45	veneer 3' x 6' 8' door only.
	Reinstall	Ea.		18.45		18.45	
	Clean	Ea.	2.88	9.55		12.43	
	Paint	Ea.	6.35	23		29.35	
	Minimum Charge	Job		157		157	
3' 6" x 6' 8"							
	Demolish	Ea.		29		29	Includes material and labor to install
	Install	Ea.	365	18.45		383.45	flush solid core 1-3/4" hardwood
	Demolish and Install	Ea.	365	47.45		412.45	veneer 3' 6" x 6' 8' door only.
	Reinstall	Ea.		18.45		18.45	
	Clean	Ea.	2.88	9.55		12.43	
	Paint	Ea.	6.35	23		29.35	
	Minimum Charge	Job		157		157	
3' x 7'							
	Demolish	Ea.		29		29	Includes material and labor to install
	Install	Ea.	294	45		339	flush solid core 1-3/4" hardwood
	Demolish and Install	Ea.	294	74		368	veneer 3' x 7' door only.
	Reinstall	Ea.		44.80		44.80	
	Clean	Ea.	2.88	9.55		12.43	
	Paint	Ea.	6.35	23		29.35	
	Minimum Charge	Job		157		157	

Solid Core Door		Unit	Material	Labor	Equip.	Total	Specification
3' 6" x 7'							
	Demolish	Ea.		29		29	Includes material and labor to install
	Install	Ea.	390	22.50		412.50	flush solid core 1-3/4" hardwood
	Demolish and Install	Ea.	390	51.50		441.50	veneer 3' 6" x 7' door only.
	Reinstall	Ea.		22.40		22.40	
	Clean	Ea.	2.88	9.55		12.43	
	Paint	Ea.	6.35	23		29.35	
	Minimum Charge	Job		157		157	
3' x 8'							
	Demolish	Ea.		29		29	Includes material and labor to install
	Install	Ea.	340	39		379	flush solid core 1-3/4" hardwood
	Demolish and Install	Ea.	340	68		408	veneer 3' x 8' door only.
	Reinstall	Ea.		39.20		39.20	
	Clean	Ea.	3.47	14.35		17.82	
	Paint	Ea.	6.35	23		29.35	
	Minimum Charge	Job		157		157	
3' 6" x 8'							
	Demolish	Ea.		29		29	Includes material and labor to install
	Install	Ea.	420	39		459	flush solid core 1-3/4" hardwood
	Demolish and Install	Ea.	420	68		488	veneer 3' 6" x 8' door only.
	Reinstall	Ea.		39.20		39.20	
	Clean	Ea.	3.47	14.35		17.82	
	Paint	Ea.	6.35	23		29.35	
	Minimum Charge	Job		157		157	
3' x 9'							
	Demolish	Ea.		29		29	Includes material and labor to install
	Install	Ea.	435	45		480	flush solid core 1-3/4" hardwood
	Demolish and Install	Ea.	435	74		509	veneer 3' x 9' door only.
	Reinstall	Ea.		44.80		44.80	
	Clean	Ea.	3.47	14.35		17.82	
	Paint	Ea.	6.35	23		29.35	
	Minimum Charge	Job		157		157	
Stain							
	Install	Ea.	5.80	30.50		36.30	Includes labor and material to stain
	Minimum Charge	Job		138		138	exterior type door and trim on both
							sides.
Door Only (SC)							
2' 4" x 6' 8"							
	Demolish	Ea.		6.20		6.20	Cost includes material and labor to
	Install	Ea.	180	17.40		197.40	install 2' 4" x 6' 8" lauan solid core
	Demolish and Install	Ea.	180	23.60		203.60	door slab.
	Reinstall	Ea.		17.42		17.42	
	Clean	Ea.	1.20	7.20		8.40	
	Paint	Ea.	2.37	34.50		36.87	
	Minimum Charge	Job		157		157	
2' 8" x 6' 8"							
	Demolish	Ea.		6.20		6.20	Cost includes material and labor to
	Install	Ea.	185	17.40		202.40	install 2' 8" x 6' 8" lauan solid core
	Demolish and Install	Ea.	185	23.60		208.60	door slab.
	Reinstall	Ea.		17.42		17.42	
	Clean	Ea.	1.20	7.20		8.40	
	Paint	Ea.	2.37	34.50		36.87	
	Minimum Charge	Job		157		157	

Exterior Doors and Windows

Solid Core Door		Unit	Material	Labor	Equip.	Total	Specification
3' x 6' 8"							
	Demolish	Ea.		6.20		6.20	Cost includes material and labor to
	Install	Ea.	191	18.45		209.45	install 3' x 6' 8" lauan solid core door
	Demolish and Install	Ea.	191	24.65		215.65	slab.
	Reinstall	Ea.		18.45		18.45	
	Clean	Ea.	1.20	7.20		8.40	
	Paint	Ea.	2.37	34.50		36.87	
	Minimum Charge	Job		157		157	
3' 6" x 6' 8"							
	Demolish	Ea.		6.20		6.20	Cost includes material and labor to
	Install	Ea.	218	19.60		237.60	install 3' 6" x 6' 8" lauan solid core
	Demolish and Install	Ea.	218	25.80		243.80	door slab.
	Reinstall	Ea.		19.60		19.60	
	Clean	Ea.	1.20	7.20		8.40	
	Paint	Ea.	2.37	34.50		36.87	
	Minimum Charge	Job		157		157	
Panel							
	Demolish	Ea.		6.20		6.20	Cost includes material and labor to
	Install	Ea.	163	37		200	install 3' x 6' 8" six (6) panel interior
	Demolish and Install	Ea.	163	43.20		206.20	door.
	Reinstall	Ea.		36.89		36.89	
	Clean	Ea.	1.44	7.20		8.64	
	Paint	Ea.	2.37	34.50		36.87	
	Minimum Charge	Job		157		157	
Casing Trim							
Single Width							
	Demolish	Ea.		5.55		5.55	Cost includes material and labor to
	Install	Opng.	15.90	53		68.90	install 11/16" x 2-1/2" pine ranch
	Demolish and Install	Opng.	15.90	58.55		74.45	style casing for one side of a standard
	Reinstall	Opng.		42.52		42.52	door opening.
	Clean	Opng.	.41	2.30		2.71	
	Paint	Ea.	3.08	5.50		8.58	
	Minimum Charge	Job		157		157	
Double Width							
	Demolish	Ea.		6.55		6.55	Cost includes material and labor to
	Install	Opng.	18.70	62.50		81.20	install 11/16" x 2-1/2" pine ranch
	Demolish and Install	Opng.	18.70	69.05		87.75	style door casing for one side of a
	Reinstall	Opng.		50.18		50.18	double door opening.
	Clean	Opng.	.48	4.59		5.07	
	Paint	Ea.	3.08	6.25		9.33	
	Minimum Charge	Job		157		157	

Components		Unit	Material	Labor	Equip.	Total	Specification
Hardware							
Doorknob							
	Demolish	Ea.		11.60		11.60	Includes material and labor to install
	Install	Ea.	17.65	19.60		37.25	residential passage lockset, keyless
	Demolish and Install	Ea.	17.65	31.20		48.85	bored type.
	Reinstall	Ea.		15.68		15.68	
	Clean	Ea.	.29	7.20		7.49	
	Minimum Charge	Job		157		157	

Exterior Doors and Windows

Components		Unit	Material	Labor	Equip.	Total	Specification
Doorknob w / Lock							Includes material and labor to install privacy lockset.
	Demolish	Ea.		11.60		11.60	
	Install	Ea.	20	19.60		39.60	
	Demolish and Install	Ea.	20	31.20		51.20	
	Reinstall	Ea.		15.68		15.68	
	Clean	Ea.	.29	7.20		7.49	
	Minimum Charge	Job		157		157	
Lever Handle							Includes material and labor to install residential passage lockset, keyless bored type with lever handle.
	Demolish	Ea.		11.60		11.60	
	Install	Ea.	27	31.50		58.50	
	Demolish and Install	Ea.	27	43.10		70.10	
	Reinstall	Ea.		25.09		25.09	
	Clean	Ea.	.29	7.20		7.49	
	Minimum Charge	Job		157		157	
Deadbolt							Includes material and labor to install a deadbolt.
	Demolish	Ea.		10.90		10.90	
	Install	Ea.	37.50	22.50		60	
	Demolish and Install	Ea.	37.50	33.40		70.90	
	Reinstall	Ea.		17.92		17.92	
	Clean	Ea.	.29	7.20		7.49	
	Minimum Charge	Job		157		157	
Grip Handle Entry							Includes material and labor to install residential keyed entry lock set with grip type handle.
	Demolish	Ea.		9.95		9.95	
	Install	Ea.	119	35		154	
	Demolish and Install	Ea.	119	44.95		163.95	
	Reinstall	Ea.		27.88		27.88	
	Clean	Ea.	.29	7.20		7.49	
	Minimum Charge	Job		157		157	
Panic Device							Includes material and labor to install touch bar, low profile, exit only.
	Demolish	Ea.		25.50		25.50	
	Install	Ea.	325	52.50		377.50	
	Demolish and Install	Ea.	325	78		403	
	Reinstall	Ea.		41.81		41.81	
	Clean	Ea.	.14	12.75		12.89	
	Minimum Charge	Job		157		157	
Closer							Includes material and labor to install pneumatic heavy duty closer for exterior type doors.
	Demolish	Ea.		6.55		6.55	
	Install	Ea.	153	52.50		205.50	
	Demolish and Install	Ea.	153	59.05		212.05	
	Reinstall	Ea.		41.81		41.81	
	Clean	Ea.	.14	9.55		9.69	
	Minimum Charge	Job		157		157	
Peep-hole							Includes material and labor to install peep-hole for door.
	Demolish	Ea.		5.65		5.65	
	Install	Ea.	13.95	9.80		23.75	
	Demolish and Install	Ea.	13.95	15.45		29.40	
	Reinstall	Ea.		7.84		7.84	
	Clean	Ea.	.02	2.54		2.56	
	Minimum Charge	Job		157		157	
Exit Sign							Includes material and labor to install a wall mounted interior electric exit sign.
	Demolish	Ea.		17.35		17.35	
	Install	Ea.	44	43		87	
	Demolish and Install	Ea.	44	60.35		104.35	
	Reinstall	Ea.		34.36		34.36	
	Clean	Ea.	.02	5.15		5.17	
	Minimum Charge	Job		157		157	

93

Exterior Doors and Windows

Components		Unit	Material	Labor	Equip.	Total	Specification
Kickplate							
	Demolish	Ea.		9.70		9.70	Includes material and labor to install
	Install	Ea.	19.80	21		40.80	aluminum kickplate, 10" x 28".
	Demolish and Install	Ea.	19.80	30.70		50.50	
	Reinstall	Ea.		16.73		16.73	
	Clean	Ea.	.29	7.20		7.49	
	Minimum Charge	Job		157		157	
Threshold							
	Install	L.F.	2.56	3.14		5.70	Includes labor and material to install
	Minimum Charge	Job		78.50		78.50	oak threshold.
Weatherstripping							
	Install	L.F.	.28	.78		1.06	Includes labor and material to install
	Minimum Charge	Job		78.50		78.50	rubber weatherstripping for door.
Metal Jamb							
6' 8" High							
	Demolish	Ea.		17.40		17.40	Cost includes material and labor to
	Install	Ea.	78	19.60		97.60	install 18 gauge hollow metal door
	Demolish and Install	Ea.	78	37		115	frame to fit a 4-1/2" jamb 6' 8" high
	Reinstall	Ea.		19.60		19.60	and up to 3' 6" opening.
	Clean	Ea.	.61	4.78		5.39	
	Paint	Ea.	1.05	5.50		6.55	
	Minimum Charge	Job		157		157	
7' High							
	Demolish	Ea.		17.40		17.40	Cost includes material and labor to
	Install	Ea.	81	21		102	install 18 gauge hollow metal door
	Demolish and Install	Ea.	81	38.40		119.40	frame to fit a 4-1/2" jamb 7' high and
	Reinstall	Ea.		20.91		20.91	up to 3' 6" opening.
	Clean	Ea.	.61	4.78		5.39	
	Paint	Ea.	1.05	5.50		6.55	
	Minimum Charge	Job		157		157	
8' High							
	Demolish	Ea.		17.40		17.40	Cost includes material and labor to
	Install	Ea.	97.50	26		123.50	install 18 gauge hollow metal door
	Demolish and Install	Ea.	97.50	43.40		140.90	frame to fit a 4-1/2" jamb 8' high and
	Reinstall	Ea.		26.13		26.13	up to 3' 6" opening.
	Clean	Ea.	.85	5.75		6.60	
	Paint	Ea.	1.18	6.15		7.33	
	Minimum Charge	Job		157		157	
9' High							
	Demolish	Ea.		17.40		17.40	Cost includes material and labor to
	Install	Ea.	111	31.50		142.50	install 18 gauge hollow metal door
	Demolish and Install	Ea.	111	48.90		159.90	frame to fit a 4-1/2" jamb 9' high and
	Reinstall	Ea.		31.36		31.36	up to 3' 6" opening.
	Clean	Ea.	.85	5.75		6.60	
	Paint	Ea.	1.18	6.15		7.33	
	Minimum Charge	Job		157		157	
Wood Jamb							
	Demolish	Ea.		22		22	Includes material and labor to install
	Install	Ea.	80	8.50		88.50	flat pine jamb with square cut heads
	Demolish and Install	Ea.	80	30.50		110.50	and rabbeted sides for 6' 8" high and
	Clean	Ea.	.09	5.75		5.84	3-9/16" door including trim sets for
	Paint	Ea.	5.60	18.40		24	both sides.
	Minimum Charge	Job		78.50		78.50	
Stain							
	Install	Ea.	3.98	23		26.98	Includes labor and material to stain
	Minimum Charge	Job		138		138	single door and trim on both sides.

Exterior Doors and Windows

Components	Unit	Material	Labor	Equip.	Total	Specification
Shave & Refit						
Install	Ea.		26		26	Includes labor to shave and rework
Minimum Charge	Job		78.50		78.50	door to fit opening at the job site.

Sliding Patio Door	Unit	Material	Labor	Equip.	Total	Specification
Aluminum						
6' Single Pane						
Demolish	Ea.		43.50		43.50	Cost includes material and labor to
Install	Ea.	400	157		557	install 6' x 6' 8" sliding anodized
Demolish and Install	Ea.	400	200.50		600.50	aluminum patio door with single pane
Reinstall	Ea.		125.44		125.44	tempered safety glass, anodized
Clean	Ea.	1.80	14.35		16.15	frame, screen, weatherstripped.
Minimum Charge	Job		157		157	
8' Single Pane						
Demolish	Ea.		43.50		43.50	Cost includes material and labor to
Install	Ea.	455	209		664	install 8' x 6' 8" sliding aluminum
Demolish and Install	Ea.	455	252.50		707.50	patio door with single pane tempered
Reinstall	Ea.		167.25		167.25	safety glass, anodized frame, screen,
Clean	Ea.	1.80	14.35		16.15	weatherstripped.
Minimum Charge	Job		157		157	
10' Single Pane						
Demolish	Ea.		43.50		43.50	Cost includes material and labor to
Install	Ea.	595	315		910	install 10' x 6' 8" sliding aluminum
Demolish and Install	Ea.	595	358.50		953.50	patio door with single pane tempered
Reinstall	Ea.		250.88		250.88	safety glass, anodized frame, screen,
Clean	Ea.	1.80	14.35		16.15	weatherstripped.
Minimum Charge	Job		157		157	
6' Insulated						
Demolish	Ea.		43.50		43.50	Cost includes material and labor to
Install	Ea.	525	157		682	install 6' x 6' 8" sliding aluminum
Demolish and Install	Ea.	525	200.50		725.50	patio door with double insulated
Reinstall	Ea.		125.44		125.44	tempered glass, baked-on enamel
Clean	Ea.	1.80	14.35		16.15	finish, screen, weatherstripped.
Minimum Charge	Job		157		157	
8' Insulated						
Demolish	Ea.		43.50		43.50	Cost includes material and labor to
Install	Ea.	620	209		829	install 8' x 6' 8" sliding aluminum
Demolish and Install	Ea.	620	252.50		872.50	patio door with insulated tempered
Reinstall	Ea.		167.25		167.25	glass, baked-on enamel finish, screen,
Clean	Ea.	1.80	14.35		16.15	weatherstripped.
Minimum Charge	Job		157		157	
10' Insulated						
Demolish	Ea.		43.50		43.50	Cost includes material and labor to
Install	Ea.	695	315		1010	install 10' x 6' 8" sliding aluminum
Demolish and Install	Ea.	695	358.50		1053.50	patio door with double insulated
Reinstall	Ea.		250.88		250.88	tempered glass, baked-on enamel
Clean	Ea.	1.80	14.35		16.15	finish, screen, weatherstripped.
Minimum Charge	Job		157		157	

Sliding Patio Door		Unit	Material	Labor	Equip.	Total	Specification
Vinyl							
6' Single Pane							
	Demolish	Ea.		43.50		43.50	Cost includes material and labor to
	Install	Ea.	380	157		537	install 6' x 6' 8" sliding vinyl patio
	Demolish and Install	Ea.	380	200.50		580.50	door with single pane tempered safety
	Reinstall	Ea.		125.44		125.44	glass, anodized frame, screen,
	Clean	Ea.	1.80	14.35		16.15	weatherstripped.
	Minimum Charge	Job		157		157	
8' Single Pane							
	Demolish	Ea.		43.50		43.50	Cost includes material and labor to
	Install	Ea.	435	209		644	install 8' x 6' 8" sliding vinyl patio
	Demolish and Install	Ea.	435	252.50		687.50	door with single pane tempered safety
	Reinstall	Ea.		167.25		167.25	glass, anodized frame, screen,
	Clean	Ea.	1.80	14.35		16.15	weatherstripped.
	Minimum Charge	Job		157		157	
10' Single Pane							
	Demolish	Ea.		43.50		43.50	Cost includes material and labor to
	Install	Ea.	565	315		880	install 10' x 6' 8" sliding vinyl patio
	Demolish and Install	Ea.	565	358.50		923.50	door with single pane tempered safety
	Reinstall	Ea.		250.88		250.88	glass, anodized frame, screen,
	Clean	Ea.	1.80	14.35		16.15	weatherstripped.
	Minimum Charge	Job		157		157	
6' Insulated							
	Demolish	Ea.		43.50		43.50	Cost includes material and labor to
	Install	Ea.	595	157		752	install 6' x 6' 8" sliding vinyl patio
	Demolish and Install	Ea.	595	200.50		795.50	door with double insulated tempered
	Reinstall	Ea.		125.44		125.44	glass, anodized frame, screen,
	Clean	Ea.	1.80	14.35		16.15	weatherstripped.
	Minimum Charge	Job		157		157	
8' Insulated							
	Demolish	Ea.		43.50		43.50	Cost includes material and labor to
	Install	Ea.	775	209		984	install 8' x 6' 8" sliding vinyl patio
	Demolish and Install	Ea.	775	252.50		1027.50	door with double insulated tempered
	Reinstall	Ea.		167.25		167.25	glass, anodized frame, screen,
	Clean	Ea.	1.80	14.35		16.15	weatherstripped.
	Minimum Charge	Job		157		157	
10' Insulated							
	Demolish	Ea.		43.50		43.50	Cost includes material and labor to
	Install	Ea.	855	315		1170	install 10' x 6' 8" sliding vinyl patio
	Demolish and Install	Ea.	855	358.50		1213.50	door with double insulated tempered
	Reinstall	Ea.		250.88		250.88	glass, anodized frame, screen,
	Clean	Ea.	1.80	14.35		16.15	weatherstripped.
	Minimum Charge	Job		157		157	
Wood Framed							
6' Premium Insulated							
	Demolish	Ea.		43.50		43.50	Cost includes material and labor to
	Install	Ea.	1075	157		1232	install 6' x 6' 8" sliding wood patio
	Demolish and Install	Ea.	1075	200.50		1275.50	door with pine frame, double insulated
	Reinstall	Ea.		125.44		125.44	glass, screen, lock and dead bolt,
	Clean	Ea.	1.80	14.35		16.15	weatherstripped. Add for casing and
	Minimum Charge	Job		157		157	finishing.

Exterior Doors and Windows

Sliding Patio Door		Unit	Material	Labor	Equip.	Total	Specification
8' Premium Insulated							Cost includes material and labor to
	Demolish	Ea.		43.50		43.50	install 8' x 6' 8" sliding wood patio
	Install	Ea.	1275	209		1484	door with pine frame, double insulated
	Demolish and Install	Ea.	1275	252.50		1527.50	glass, screen, lock and dead bolt,
	Reinstall	Ea.		167.25		167.25	weatherstripped. Add for casing and
	Clean	Ea.	1.80	14.35		16.15	finishing.
	Minimum Charge	Job		157		157	
9' Premium Insulated							Cost includes material and labor to
	Demolish	Ea.		43.50		43.50	install 9' x 6' 8" sliding wood patio
	Install	Ea.	1275	209		1484	door with pine frame, double insulated
	Demolish and Install	Ea.	1275	252.50		1527.50	glass, screen, lock and dead bolt,
	Reinstall	Ea.		167.25		167.25	weatherstripped. Add for casing and
	Clean	Ea.	1.80	14.35		16.15	finishing.
	Minimum Charge	Job		157		157	
Stain							Includes labor and material to stain
	Install	Ea.	5.80	30.50		36.30	exterior type door and trim on both
	Minimum Charge	Job		138		138	sides.
Casing Trim							Cost includes material and labor to
	Demolish	Ea.		6.55		6.55	install 11/16" x 2-1/2" pine ranch
	Install	Opng.	18.70	62.50		81.20	style door casing for one side of a
	Demolish and Install	Opng.	18.70	69.05		87.75	double door opening.
	Reinstall	Opng.		50.18		50.18	
	Clean	Opng.	.48	4.59		5.07	
	Paint	Ea.	3.08	6.25		9.33	
	Minimum Charge	Job		157		157	
Sliding Screen							Includes material and labor to install
	Demolish	Ea.		6.20		6.20	sliding screen door, 6' 8" x 3'.
	Install	Ea.	100	15.70		115.70	
	Demolish and Install	Ea.	100	21.90		121.90	
	Reinstall	Ea.		12.54		12.54	
	Clean	Ea.	.61	4.78		5.39	
	Minimum Charge	Job		157		157	

Screen Door		Unit	Material	Labor	Equip.	Total	Specification
Aluminum							Includes material and labor to install
	Demolish	Ea.		29		29	aluminum frame screen door with
	Install	Ea.	103	45		148	fiberglass screen cloth, closer, hinges,
	Demolish and Install	Ea.	103	74		177	latch.
	Reinstall	Ea.		35.84		35.84	
	Clean	Ea.	2.88	9.55		12.43	
	Paint	Ea.	2.96	11.05		14.01	
	Minimum Charge	Job		157		157	
Wood							Includes material and labor to install
	Demolish	Ea.		29		29	wood screen door with aluminum cloth
	Install	Ea.	132	52.50		184.50	screen. Frame and hardware not
	Demolish and Install	Ea.	132	81.50		213.50	included.
	Reinstall	Ea.		41.81		41.81	
	Clean	Ea.	2.88	9.55		12.43	
	Paint	Ea.	2.96	11.05		14.01	
	Minimum Charge	Job		157		157	
Shave & Refit							Includes labor to shave and rework
	Install	Ea.		26		26	door to fit opening at the job site.
	Minimum Charge	Job		157		157	

Exterior Doors and Windows

Storm Door		Unit	Material	Labor	Equip.	Total	Specification
Good Grade							
	Demolish	Ea.		29		29	Includes material and labor to install
	Install	Ea.	156	45		201	aluminum frame storm door with
	Demolish and Install	Ea.	156	74		230	fiberglass screen cloth, upper glass
	Reinstall	Ea.		35.84		35.84	panel, closer, hinges, latch.
	Clean	Ea.	2.88	9.55		12.43	
	Paint	Ea.	2.96	11.05		14.01	
	Minimum Charge	Job		157		157	
Better Grade							
	Demolish	Ea.		29		29	Includes material and labor to install
	Install	Ea.	205	45		250	aluminum frame storm door with
	Demolish and Install	Ea.	205	74		279	fiberglass screen cloth, upper and
	Reinstall	Ea.		35.84		35.84	lower glass panel, closer, latch.
	Clean	Ea.	2.88	9.55		12.43	
	Paint	Ea.	2.96	11.05		14.01	
	Minimum Charge	Job		157		157	
Premium Grade							
	Demolish	Ea.		29		29	Includes material and labor to install
	Install	Ea.	254	45		299	aluminum frame storm door with
	Demolish and Install	Ea.	254	74		328	fiberglass screen cloth, upper and
	Reinstall	Ea.		35.84		35.84	lower glass panel, closer, latch.
	Clean	Ea.	2.88	9.55		12.43	
	Paint	Ea.	2.96	11.05		14.01	
	Minimum Charge	Job		157		157	
Insulated							
	Demolish	Ea.		29		29	Includes material and labor to install
	Install	Ea.	234	22.50		256.50	aluminum frame storm door with
	Demolish and Install	Ea.	234	51.50		285.50	fiberglass screen cloth, upper and
	Reinstall	Ea.		17.92		17.92	lower glass panel, closer, latch.
	Clean	Ea.	2.88	9.55		12.43	
	Paint	Ea.	2.96	11.05		14.01	
	Minimum Charge	Job		157		157	

Garage Door		Unit	Material	Labor	Equip.	Total	Specification
9' One Piece Metal							
	Demolish	Ea.		73.50		73.50	Includes material and labor to install
	Install	Ea.	350	78.50		428.50	one piece steel garage door 9' x 7'
	Demolish and Install	Ea.	350	152		502	primed including hardware.
	Reinstall	Ea.		62.72		62.72	
	Clean	Ea.	3.61	28.50		32.11	
	Paint	Ea.	6.10	69		75.10	
	Minimum Charge	Job		157		157	
16' One Piece Metal							
	Demolish	Ea.		87		87	Includes material and labor to install
	Install	Ea.	555	105		660	one piece steel garage door 16' x 7'
	Demolish and Install	Ea.	555	192		747	primed including hardware.
	Reinstall	Ea.		83.63		83.63	
	Clean	Ea.	7.20	38.50		45.70	
	Paint	Ea.	12.15	79		91.15	
	Minimum Charge	Job		157		157	

Garage Door		Unit	Material	Labor	Equip.	Total	Specification
9' Metal Sectional							
	Demolish	Ea.		73.50		73.50	Includes material and labor to install
	Install	Ea.	480	119		599	sectional metal garage door 9' x 7'
	Demolish and Install	Ea.	480	192.50		672.50	primed including hardware.
	Reinstall	Ea.		95.03		95.03	
	Clean	Ea.	3.70	33		36.70	
	Paint	Ea.	12.15	79		91.15	
	Minimum Charge	Job		157		157	
16' Metal Sectional							
	Demolish	Ea.		87		87	Includes material and labor to install
	Install	Ea.	605	209		814	sectional metal garage door 16' x 7'
	Demolish and Install	Ea.	605	296		901	primed including hardware.
	Reinstall	Ea.		167.25		167.25	
	Clean	Ea.	7.20	38.50		45.70	
	Paint	Ea.	12.15	79		91.15	
	Minimum Charge	Job		157		157	
Metal / Steel Roll Up							
8' x 8' Steel							
	Demolish	Ea.		99.50		99.50	Cost includes material and labor to
	Install	Ea.	800	480		1280	install 8' x 8' steel overhead roll-up,
	Demolish and Install	Ea.	800	579.50		1379.50	chain hoist operated service door.
	Reinstall	Ea.		384.80		384.80	
	Clean	Ea.	3.70	33		36.70	
	Paint	Ea.	13	92		105	
	Minimum Charge	Job		770		770	
10' x 10' Steel							
	Demolish	Ea.		116		116	Cost includes material and labor to
	Install	Ea.	1050	550		1600	install 10' x 10' steel overhead roll-up,
	Demolish and Install	Ea.	1050	666		1716	chain hoist operated service door.
	Reinstall	Ea.		439.77		439.77	
	Clean	Ea.	5.80	46		51.80	
	Paint	Ea.	20.50	92		112.50	
	Minimum Charge	Job		770		770	
12' x 12' Steel							
	Demolish	Ea.		139		139	Cost includes material and labor to
	Install	Ea.	1375	640		2015	install 12' x 12' steel overhead roll-up,
	Demolish and Install	Ea.	1375	779		2154	chain hoist operated service door.
	Reinstall	Ea.		513.07		513.07	
	Clean	Ea.	8.30	51		59.30	
	Paint	Ea.	29	110		139	
	Minimum Charge	Job		770		770	
14' x 14' Steel							
	Demolish	Ea.		174		174	Cost includes material and labor to
	Install	Ea.	1825	960		2785	install 14' x 14' steel overhead roll-up,
	Demolish and Install	Ea.	1825	1134		2959	chain hoist operated service door.
	Reinstall	Ea.		769.60		769.60	
	Clean	Ea.	11.35	65.50		76.85	
	Paint	Ea.	34	138		172	
	Minimum Charge	Job		770		770	
18' x 18' Steel							
	Demolish	Ea.		194		194	Cost includes material and labor to
	Install	Opng.	2400	1275		3675	install 18' x 18' steel overhead roll-up,
	Demolish and Install	Opng.	2400	1469		3869	chain hoist operated service door.
	Reinstall	Opng.		1026.13		1026.13	
	Clean	Ea.	11.35	38.50		49.85	
	Paint	Ea.	65.50	276		341.50	
	Minimum Charge	Job		770		770	
Automatic Motor							
	Install	Ea.	390	31.50		421.50	Includes labor and material to install
	Minimum Charge	Job		157		157	electric opener motor, 1/3 HP.

Exterior Doors and Windows

Garage Door		Unit	Material	Labor	Equip.	Total	Specification
9' One Piece Wood							
	Demolish	Ea.		73.50		73.50	Includes material and labor to install
	Install	Ea.	385	78.50		463.50	one piece wood garage door 9' x 7'
	Demolish and Install	Ea.	385	152		537	primed including hardware.
	Reinstall	Ea.		62.72		62.72	
	Clean	Ea.	3.61	28.50		32.11	
	Paint	Ea.	6.10	69		75.10	
	Minimum Charge	Job		157		157	
9' Wood Sectional							
	Demolish	Ea.		73.50		73.50	Includes material and labor to install
	Install	Ea.	485	78.50		563.50	sectional wood garage door 9' x 7'
	Demolish and Install	Ea.	485	152		637	unfinished including hardware.
	Reinstall	Ea.		62.72		62.72	
	Clean	Ea.	3.61	28.50		32.11	
	Paint	Ea.	6.10	69		75.10	
	Minimum Charge	Job		157		157	
16' Wood Sectional							
	Demolish	Ea.		87		87	Includes material and labor to install
	Install	Ea.	1050	105		1155	sectional fiberglass garage door 16' x
	Demolish and Install	Ea.	1050	192		1242	7' unfinished including hardware.
	Reinstall	Ea.		83.63		83.63	
	Clean	Ea.	7.20	38.50		45.70	
	Paint	Ea.	12.15	79		91.15	
	Minimum Charge	Job		157		157	
Electric Opener							
	Demolish	Ea.		51		51	Includes material and labor to install
	Install	Ea.	264	59		323	electric opener, 1/3 HP economy.
	Demolish and Install	Ea.	264	110		374	
	Reinstall	Ea.		47.07		47.07	
	Clean	Ea.	.14	12.75		12.89	
	Minimum Charge	Job		157		157	
Springs							
	Install	Pr.	35	78.50		113.50	Includes labor and material to replace
	Minimum Charge	Job		157		157	springs for garage door.

Commercial Metal Door		Unit	Material	Labor	Equip.	Total	Specification
Pre-hung Steel Door							
2' 8" x 6' 8"							
	Demolish	Ea.		43.50		43.50	Cost includes material and labor to
	Install	Ea.	335	19.60		354.60	install 2' 8" x 6' 8" pre-hung 1-3/4"
	Demolish and Install	Ea.	335	63.10		398.10	18 gauge steel door with steel frame,
	Reinstall	Ea.		19.60		19.60	hinges, aluminum sill,
	Clean	Ea.	2.88	9.55		12.43	weather-stripped.
	Paint	Ea.	3.80	23		26.80	
	Minimum Charge	Job		157		157	
3' x 6' 8"							
	Demolish	Ea.		43.50		43.50	Cost includes material and labor to
	Install	Ea.	330	21		351	install 3' x 6' 8" pre-hung 1-3/4" 18
	Demolish and Install	Ea.	330	64.50		394.50	gauge steel door with steel frame,
	Reinstall	Ea.		20.91		20.91	hinges, aluminum sill,
	Clean	Ea.	2.88	9.55		12.43	weather-stripped.
	Paint	Ea.	3.80	23		26.80	
	Minimum Charge	Job		157		157	

Commercial Metal Door		Unit	Material	Labor	Equip.	Total	Specification
3' 6" x 6' 8"							
	Demolish	Ea.		43.50		43.50	Cost includes material and labor to
	Install	Ea.	380	24		404	install 3' 6" x 6' 8" pre-hung 1-3/4"
	Demolish and Install	Ea.	380	67.50		447.50	18 gauge steel door with steel frame,
	Reinstall	Ea.		24.12		24.12	hinges, aluminum sill,
	Clean	Ea.	2.88	9.55		12.43	weather-stripped.
	Paint	Ea.	3.80	23		26.80	
	Minimum Charge	Job		157		157	
4' x 6' 8"							
	Demolish	Ea.		43.50		43.50	Cost includes material and labor to
	Install	Ea.	415	31.50		446.50	install 4' x 6' 8" pre-hung 1-3/4" 18
	Demolish and Install	Ea.	415	75		490	gauge steel door with steel frame,
	Reinstall	Ea.		31.36		31.36	hinges, aluminum sill,
	Clean	Ea.	2.88	9.55		12.43	weather-stripped.
	Paint	Ea.	3.80	23		26.80	
	Minimum Charge	Job		157		157	
Metal Jamb							
6' 8" High							
	Demolish	Ea.		17.40		17.40	Cost includes material and labor to
	Install	Ea.	78	19.60		97.60	install 18 gauge hollow metal door
	Demolish and Install	Ea.	78	37		115	frame to fit a 4-1/2" jamb 6' 8" high
	Reinstall	Ea.		19.60		19.60	and up to 3' 6" opening.
	Clean	Ea.	.61	4.78		5.39	
	Paint	Ea.	1.05	5.50		6.55	
	Minimum Charge	Job		157		157	
7' High							
	Demolish	Ea.		17.40		17.40	Cost includes material and labor to
	Install	Ea.	81	21		102	install 18 gauge hollow metal door
	Demolish and Install	Ea.	81	38.40		119.40	frame to fit a 4-1/2" jamb 7' high and
	Reinstall	Ea.		20.91		20.91	up to 3' 6" opening.
	Clean	Ea.	.61	4.78		5.39	
	Paint	Ea.	1.05	5.50		6.55	
	Minimum Charge	Job		157		157	
8' High							
	Demolish	Ea.		17.40		17.40	Cost includes material and labor to
	Install	Ea.	97.50	26		123.50	install 18 gauge hollow metal door
	Demolish and Install	Ea.	97.50	43.40		140.90	frame to fit a 4-1/2" jamb 8' high and
	Reinstall	Ea.		26.13		26.13	up to 3' 6" opening.
	Clean	Ea.	.85	5.75		6.60	
	Paint	Ea.	1.18	6.15		7.33	
	Minimum Charge	Job		157		157	
9' High							
	Demolish	Ea.		17.40		17.40	Cost includes material and labor to
	Install	Ea.	111	31.50		142.50	install 18 gauge hollow metal door
	Demolish and Install	Ea.	111	48.90		159.90	frame to fit a 4-1/2" jamb 9' high and
	Reinstall	Ea.		31.36		31.36	up to 3' 6" opening.
	Clean	Ea.	.85	5.75		6.60	
	Paint	Ea.	1.18	6.15		7.33	
	Minimum Charge	Job		157		157	
Hardware							
Doorknob w / Lock							
	Demolish	Ea.		11.60		11.60	Includes material and labor to install
	Install	Ea.	100	26		126	keyed entry lock set, bored type
	Demolish and Install	Ea.	100	37.60		137.60	including knob trim, strike and strike
	Reinstall	Ea.		20.91		20.91	box for standard commercial
	Clean	Ea.	.29	7.20		7.49	application.
	Minimum Charge	Job		157		157	

Exterior Doors and Windows

Commercial Metal Door		Unit	Material	Labor	Equip.	Total	Specification
Lever Handle							
	Demolish	Ea.		11.60		11.60	Includes material and labor to install
	Install	Ea.	55	31.50		86.50	passage lockset, keyless bored type
	Demolish and Install	Ea.	55	43.10		98.10	with lever handle, non-locking
	Reinstall	Ea.		25.09		25.09	standard commercial latchset.
	Clean	Ea.	.29	7.20		7.49	
	Minimum Charge	Job		157		157	
Deadbolt							
	Demolish	Ea.		10.90		10.90	Includes material and labor to install a
	Install	Ea.	37.50	22.50		60	deadbolt.
	Demolish and Install	Ea.	37.50	33.40		70.90	
	Reinstall	Ea.		17.92		17.92	
	Clean	Ea.	.29	7.20		7.49	
	Minimum Charge	Job		157		157	
Panic Device							
	Demolish	Ea.		25.50		25.50	Includes material and labor to install
	Install	Ea.	325	52.50		377.50	touch bar, low profile, exit only.
	Demolish and Install	Ea.	325	78		403	
	Reinstall	Ea.		41.81		41.81	
	Clean	Ea.	.14	12.75		12.89	
	Minimum Charge	Job		157		157	
Closer							
	Demolish	Ea.		6.55		6.55	Includes material and labor to install
	Install	Ea.	163	52.50		215.50	pneumatic heavy duty model for
	Demolish and Install	Ea.	163	59.05		222.05	exterior type doors.
	Reinstall	Ea.		41.81		41.81	
	Clean	Ea.	.14	9.55		9.69	
	Minimum Charge	Job		157		157	
Peep-Hole							
	Demolish	Ea.		5.65		5.65	Includes material and labor to install
	Install	Ea.	13.95	9.80		23.75	peep-hole for door.
	Demolish and Install	Ea.	13.95	15.45		29.40	
	Reinstall	Ea.		7.84		7.84	
	Clean	Ea.	.02	2.54		2.56	
	Minimum Charge	Job		157		157	
Exit Sign							
	Demolish	Ea.		17.35		17.35	Includes material and labor to install a
	Install	Ea.	44	43		87	wall mounted interior electric exit sign.
	Demolish and Install	Ea.	44	60.35		104.35	
	Reinstall	Ea.		34.36		34.36	
	Clean	Ea.	.02	5.15		5.17	
	Minimum Charge	Job		157		157	
Kickplate							
	Demolish	Ea.		9.70		9.70	Includes material and labor to install
	Install	Ea.	44.50	21		65.50	stainless steel kickplate, 10" x 36".
	Demolish and Install	Ea.	44.50	30.70		75.20	
	Reinstall	Ea.		16.73		16.73	
	Clean	Ea.	.29	7.20		7.49	
	Minimum Charge	Job		157		157	

Aluminum Window		Unit	Material	Labor	Equip.	Total	Specification
Single Hung Two Light							
2' x 2'							
	Demolish	Ea.		15.95		15.95	Cost includes material and labor to
	Install	Ea.	58	70		128	install 2' x 2' single hung, two light,
	Demolish and Install	Ea.	58	85.95		143.95	double glazed aluminum window
	Reinstall	Ea.		55.97		55.97	including screen.
	Clean	Ea.	.03	5.90		5.93	
	Minimum Charge	Job		192		192	

Exterior Doors and Windows

Aluminum Window		Unit	Material	Labor	Equip.	Total	Specification
2' x 2' 6"							Cost includes material and labor to
	Demolish	Ea.		15.95		15.95	install 2' x 2' 6" single hung, two
	Install	Ea.	62.50	70		132.50	light, double glazed aluminum
	Demolish and Install	Ea.	62.50	85.95		148.45	window including screen.
	Reinstall	Ea.		55.97		55.97	
	Clean	Ea.	.03	5.90		5.93	
	Minimum Charge	Job		192		192	
3' x 1' 6"							Cost includes material and labor to
	Demolish	Ea.		15.95		15.95	install 3' x 1' 6" single hung, two
	Install	Ea.	59	77		136	light, double glazed aluminum
	Demolish and Install	Ea.	59	92.95		151.95	window including screen.
	Reinstall	Ea.		61.57		61.57	
	Clean	Ea.	.02	5.15		5.17	
	Minimum Charge	Job		192		192	
3' x 2'							Cost includes material and labor to
	Demolish	Ea.		15.95		15.95	install 3' x 2' single hung, two light,
	Install	Ea.	71	77		148	double glazed aluminum window
	Demolish and Install	Ea.	71	92.95		163.95	including screen.
	Reinstall	Ea.		61.57		61.57	
	Clean	Ea.	.02	5.15		5.17	
	Minimum Charge	Job		192		192	
3' x 2' 6"							Cost includes material and labor to
	Demolish	Ea.		15.95		15.95	install 3' x 2' 6" single hung, two
	Install	Ea.	74.50	77		151.50	light, double glazed aluminum
	Demolish and Install	Ea.	74.50	92.95		167.45	window including screen.
	Reinstall	Ea.		61.57		61.57	
	Clean	Ea.	.03	5.90		5.93	
	Minimum Charge	Job		192		192	
3' x 3'							Cost includes material and labor to
	Demolish	Ea.		15.95		15.95	install 3' x 3' single hung, two light,
	Install	Ea.	83	77		160	double glazed aluminum window
	Demolish and Install	Ea.	83	92.95		175.95	including screen.
	Reinstall	Ea.		61.57		61.57	
	Clean	Ea.	.03	5.90		5.93	
	Minimum Charge	Job		192		192	
3' x 3' 6"							Cost includes material and labor to
	Demolish	Ea.		15.95		15.95	install 3' x 3' 6" single hung, two
	Install	Ea.	87.50	77		164.50	light, double glazed aluminum
	Demolish and Install	Ea.	87.50	92.95		180.45	window including screen.
	Reinstall	Ea.		61.57		61.57	
	Clean	Ea.	.04	6.85		6.89	
	Minimum Charge	Job		192		192	
3' x 4'							Cost includes material and labor to
	Demolish	Ea.		15.95		15.95	install 3' x 4' single hung, two light,
	Install	Ea.	95.50	77		172.50	double glazed aluminum window
	Demolish and Install	Ea.	95.50	92.95		188.45	including screen.
	Reinstall	Ea.		61.57		61.57	
	Clean	Ea.	.04	6.85		6.89	
	Minimum Charge	Job		192		192	
3' x 5'							Cost includes material and labor to
	Demolish	Ea.		23		23	install 3' x 5' single hung, two light,
	Install	Ea.	106	77		183	double glazed aluminum window
	Demolish and Install	Ea.	106	100		206	including screen.
	Reinstall	Ea.		61.57		61.57	
	Clean	Ea.	.06	8.25		8.31	
	Minimum Charge	Job		192		192	

Exterior Doors and Windows

Aluminum Window	Unit	Material	Labor	Equip.	Total	Specification
3′ x 6′						
Demolish	Ea.		23		23	Cost includes material and labor to
Install	Ea.	114	85.50		199.50	install 3′ x 6′ single hung, two light,
Demolish and Install	Ea.	114	108.50		222.50	double glazed aluminum window
Reinstall	Ea.		68.41		68.41	including screen.
Clean	Ea.	.07	9.70		9.77	
Minimum Charge	Job		192		192	
Double Hung Two Light						
2′ x 2′						
Demolish	Ea.		15.95		15.95	Cost includes material and labor to
Install	Ea.	99.50	70		169.50	install 2′ x 2′ double hung, two light,
Demolish and Install	Ea.	99.50	85.95		185.45	double glazed aluminum window
Reinstall	Ea.		55.97		55.97	including screen.
Clean	Ea.	.02	5.15		5.17	
Minimum Charge	Job		192		192	
2′ x 2′ 6″						
Demolish	Ea.		15.95		15.95	Cost includes material and labor to
Install	Ea.	108	70		178	install 2′ x 2′ 6″ double hung, two
Demolish and Install	Ea.	108	85.95		193.95	light, double glazed aluminum
Reinstall	Ea.		55.97		55.97	window including screen.
Clean	Ea.	.02	5.15		5.17	
Minimum Charge	Job		192		192	
3′ x 1′ 6″						
Demolish	Ea.		15.95		15.95	Cost includes material and labor to
Install	Ea.	256	77		333	install 3′ x 1′ 6″ double hung, two
Demolish and Install	Ea.	256	92.95		348.95	light, double glazed aluminum
Reinstall	Ea.		61.57		61.57	window including screen.
Clean	Ea.	.02	5.15		5.17	
Minimum Charge	Job		192		192	
3′ x 2′						
Demolish	Ea.		15.95		15.95	Cost includes material and labor to
Install	Ea.	263	77		340	install 3′ x 2′ double hung, two light,
Demolish and Install	Ea.	263	92.95		355.95	double glazed aluminum window
Reinstall	Ea.		61.57		61.57	including screen.
Clean	Ea.	.02	5.15		5.17	
Minimum Charge	Job		192		192	
3′ x 2′ 6″						
Demolish	Ea.		15.95		15.95	Cost includes material and labor to
Install	Ea.	261	77		338	install 3′ x 2′ 6″ double hung, two
Demolish and Install	Ea.	261	92.95		353.95	light, double glazed aluminum
Reinstall	Ea.		61.57		61.57	window including screen.
Clean	Ea.	.03	5.90		5.93	
Minimum Charge	Job		192		192	
3′ x 3′						
Demolish	Ea.		15.95		15.95	Cost includes material and labor to
Install	Ea.	282	77		359	install 3′ x 3′ double hung, two light,
Demolish and Install	Ea.	282	92.95		374.95	double glazed aluminum window
Reinstall	Ea.		61.57		61.57	including screen.
Clean	Ea.	.03	5.90		5.93	
Minimum Charge	Job		192		192	
3′ x 3′ 6″						
Demolish	Ea.		15.95		15.95	Cost includes material and labor to
Install	Ea.	310	77		387	install 3′ x 3′ 6″ double hung, two
Demolish and Install	Ea.	310	92.95		402.95	light, double glazed aluminum
Reinstall	Ea.		61.57		61.57	window including screen.
Clean	Ea.	.04	6.85		6.89	
Minimum Charge	Job		192		192	

Aluminum Window		Unit	Material	Labor	Equip.	Total	Specification
3' x 4'							
	Demolish	Ea.		15.95		15.95	Cost includes material and labor to
	Install	Ea.	320	77		397	install 3' x 4' double hung, two light,
	Demolish and Install	Ea.	320	92.95		412.95	double glazed aluminum window
	Reinstall	Ea.		61.57		61.57	including screen.
	Clean	Ea.	.04	6.85		6.89	
	Minimum Charge	Job		192		192	
3' x 5'							
	Demolish	Ea.		23		23	Cost includes material and labor to
	Install	Ea.	360	77		437	install 3' x 5' double hung, two light,
	Demolish and Install	Ea.	360	100		460	double glazed aluminum window
	Reinstall	Ea.		61.57		61.57	including screen.
	Clean	Ea.	.06	8.25		8.31	
	Minimum Charge	Job		192		192	
3' x 6'							
	Demolish	Ea.		23		23	Cost includes material and labor to
	Install	Ea.	470	85.50		555.50	install 3' x 6' double hung, two light,
	Demolish and Install	Ea.	470	108.50		578.50	double glazed aluminum window
	Reinstall	Ea.		68.41		68.41	including screen.
	Clean	Ea.	.07	9.70		9.77	
	Minimum Charge	Job		192		192	
Sliding Sash							
1' 6" x 3'							
	Demolish	Ea.		15.95		15.95	Cost includes material and labor to
	Install	Ea.	55.50	70		125.50	install 1' 6" x 3' sliding sash, two
	Demolish and Install	Ea.	55.50	85.95		141.45	light, double glazed aluminum
	Reinstall	Ea.		55.97		55.97	window including screen.
	Clean	Ea.	.02	5.15		5.17	
	Minimum Charge	Job		192		192	
2' x 2'							
	Demolish	Ea.		15.95		15.95	Cost includes material and labor to
	Install	Ea.	50.50	77		127.50	install 2' x 2' sliding sash, two light,
	Demolish and Install	Ea.	50.50	92.95		143.45	double glazed aluminum window
	Reinstall	Ea.		61.57		61.57	including screen.
	Clean	Ea.	.02	5.15		5.17	
	Minimum Charge	Job		192		192	
3' x 2'							
	Demolish	Ea.		15.95		15.95	Cost includes material and labor to
	Install	Ea.	64.50	77		141.50	install 3' x 2' sliding sash, two light,
	Demolish and Install	Ea.	64.50	92.95		157.45	double glazed aluminum window
	Reinstall	Ea.		61.57		61.57	including screen.
	Clean	Ea.	.02	5.15		5.17	
	Minimum Charge	Job		192		192	
3' x 2' 6"							
	Demolish	Ea.		15.95		15.95	Cost includes material and labor to
	Install	Ea.	68	77		145	install 3' x 2' 6" sliding sash, two
	Demolish and Install	Ea.	68	92.95		160.95	light, double glazed aluminum
	Reinstall	Ea.		61.57		61.57	window including screen.
	Clean	Ea.	.03	5.90		5.93	
	Minimum Charge	Job		192		192	
3' x 3'							
	Demolish	Ea.		15.95		15.95	Cost includes material and labor to
	Install	Ea.	69	77		146	install 3' x 3' sliding sash, two light,
	Demolish and Install	Ea.	69	92.95		161.95	double glazed aluminum window
	Reinstall	Ea.		61.57		61.57	including screen.
	Clean	Ea.	.03	5.90		5.93	
	Minimum Charge	Job		192		192	

Exterior Doors and Windows

Aluminum Window	Unit	Material	Labor	Equip.	Total	Specification
3' x 3' 6"						
Demolish	Ea.		15.95		15.95	Cost includes material and labor to
Install	Ea.	81.50	77		158.50	install 3' x 3' 6" sliding sash, two
Demolish and Install	Ea.	81.50	92.95		174.45	light, double glazed aluminum
Reinstall	Ea.		61.57		61.57	window including screen.
Clean	Ea.	.04	6.85		6.89	
Minimum Charge	Job		192		192	
3' x 4'						
Demolish	Ea.		15.95		15.95	Cost includes material and labor to
Install	Ea.	89	85.50		174.50	install 3' x 4' sliding sash, two light,
Demolish and Install	Ea.	89	101.45		190.45	double glazed aluminum window
Reinstall	Ea.		68.41		68.41	including screen.
Clean	Ea.	.04	6.85		6.89	
Minimum Charge	Job		192		192	
3' x 5'						
Demolish	Ea.		23		23	Cost includes material and labor to
Install	Ea.	113	85.50		198.50	install 3' x 5' sliding sash, two light,
Demolish and Install	Ea.	113	108.50		221.50	double glazed aluminum window
Reinstall	Ea.		68.41		68.41	including screen.
Clean	Ea.	.06	8.25		8.31	
Minimum Charge	Job		192		192	
4' x 5'						
Demolish	Ea.		23		23	Cost includes material and labor to
Install	Ea.	114	110		224	install 4' x 5' sliding sash, two light,
Demolish and Install	Ea.	114	133		247	double glazed aluminum window
Reinstall	Ea.		87.95		87.95	including screen.
Clean	Ea.	.09	11.75		11.84	
Minimum Charge	Job		192		192	
5' x 5'						
Demolish	Ea.		23		23	Cost includes material and labor to
Install	Ea.	156	110		266	install 5' x 5' sliding sash, two light,
Demolish and Install	Ea.	156	133		289	double glazed aluminum window
Reinstall	Ea.		87.95		87.95	including screen.
Clean	Ea.	.09	11.75		11.84	
Minimum Charge	Job		192		192	
Awning Sash						
26" x 26", 2 Light						
Demolish	Ea.		15.95		15.95	Cost includes material and labor to
Install	Ea.	241	85.50		326.50	install 26" x 26" two light awning,
Demolish and Install	Ea.	241	101.45		342.45	double glazed aluminum window
Reinstall	Ea.		68.41		68.41	including screen.
Clean	Ea.	.03	5.90		5.93	
Minimum Charge	Job		192		192	
26" x 37", 2 Light						
Demolish	Ea.		15.95		15.95	Cost includes material and labor to
Install	Ea.	274	85.50		359.50	install 26" x 37" two light awning,
Demolish and Install	Ea.	274	101.45		375.45	double glazed aluminum window
Reinstall	Ea.		68.41		68.41	including screen.
Clean	Ea.	.03	5.90		5.93	
Minimum Charge	Job		192		192	
38" x 26", 3 Light						
Demolish	Ea.		15.95		15.95	Cost includes material and labor to
Install	Ea.	305	85.50		390.50	install 38" x 26" three light awning,
Demolish and Install	Ea.	305	101.45		406.45	double glazed aluminum window
Reinstall	Ea.		68.41		68.41	including screen.
Clean	Ea.	.04	6.85		6.89	
Minimum Charge	Job		192		192	

Aluminum Window		Unit	Material	Labor	Equip.	Total	Specification
38" x 37", 3 Light							Cost includes material and labor to
	Demolish	Ea.		15.95		15.95	install 38" x 37" three light awning,
	Install	Ea.	315	96		411	double glazed aluminum window
	Demolish and Install	Ea.	315	111.95		426.95	including screen.
	Reinstall	Ea.		76.96		76.96	
	Clean	Ea.	.04	6.85		6.89	
	Minimum Charge	Job		192		192	
38" x 53", 3 Light							Cost includes material and labor to
	Demolish	Ea.		23		23	install 38" x 53" three light awning,
	Install	Ea.	505	96		601	double glazed aluminum window
	Demolish and Install	Ea.	505	119		624	including screen.
	Reinstall	Ea.		76.96		76.96	
	Clean	Ea.	.06	8.25		8.31	
	Minimum Charge	Job		192		192	
50" x 37", 4 Light							Cost includes material and labor to
	Demolish	Ea.		23		23	install 50" x 37" four light awning,
	Install	Ea.	510	96		606	double glazed aluminum window
	Demolish and Install	Ea.	510	119		629	including screen.
	Reinstall	Ea.		76.96		76.96	
	Clean	Ea.	.06	8.25		8.31	
	Minimum Charge	Job		192		192	
50" x 53", 4 Light							Cost includes material and labor to
	Demolish	Ea.		23		23	install 50" x 53" four light awning,
	Install	Ea.	835	110		945	double glazed aluminum window
	Demolish and Install	Ea.	835	133		968	including screen.
	Reinstall	Ea.		87.95		87.95	
	Clean	Ea.	.07	9.70		9.77	
	Minimum Charge	Job		192		192	
63" x 37", 5 Light							Cost includes material and labor to
	Demolish	Ea.		23		23	install 63" x 37" five light awning,
	Install	Ea.	665	110		775	double glazed aluminum window
	Demolish and Install	Ea.	665	133		798	including screen.
	Reinstall	Ea.		87.95		87.95	
	Clean	Ea.	.07	9.70		9.77	
	Minimum Charge	Job		192		192	
63" x 53", 5 Light							Cost includes material and labor to
	Demolish	Ea.		23		23	install 63" x 53" five light awning,
	Install	Ea.	900	110		1010	double glazed aluminum window
	Demolish and Install	Ea.	900	133		1033	including screen.
	Reinstall	Ea.		87.95		87.95	
	Clean	Ea.	.09	11.75		11.84	
	Minimum Charge	Job		192		192	
Mullion Bars							Includes labor and material to install
	Install	Ea.	9.15	31		40.15	mullions for sliding aluminum window.
	Minimum Charge	Job		192		192	
Glass Louvered							
2' x 2'							Cost includes material and labor to
	Demolish	Ea.		15.95		15.95	install 2' x 2' aluminum window
	Install	Ea.	45.50	70		115.50	including 4" glass louvers, hardware,
	Demolish and Install	Ea.	45.50	85.95		131.45	screen.
	Reinstall	Ea.		55.97		55.97	
	Clean	Ea.	.02	5.15		5.17	
	Minimum Charge	Job		192		192	

Exterior Doors and Windows

Aluminum Window		Unit	Material	Labor	Equip.	Total	Specification
2' x 2' 6"							
	Demolish	Ea.		15.95		15.95	Cost includes material and labor to
	Install	Ea.	55.50	70		125.50	install 2' x 2' 6" aluminum window
	Demolish and Install	Ea.	55.50	85.95		141.45	including 4" glass louvers, hardware,
	Reinstall	Ea.		55.97		55.97	screen.
	Clean	Ea.	.02	5.15		5.17	
	Minimum Charge	Job		192		192	
3' x 1' 6"							
	Demolish	Ea.		15.95		15.95	Cost includes material and labor to
	Install	Ea.	58.50	77		135.50	install 3' x 1' 6" aluminum window
	Demolish and Install	Ea.	58.50	92.95		151.45	including 4" glass louvers, hardware,
	Reinstall	Ea.		61.57		61.57	screen.
	Clean	Ea.	.02	5.15		5.17	
	Minimum Charge	Job		192		192	
3' x 2'							
	Demolish	Ea.		15.95		15.95	Cost includes material and labor to
	Install	Ea.	67	77		144	install 3' x 2' aluminum window
	Demolish and Install	Ea.	67	92.95		159.95	including 4" glass louvers, hardware,
	Reinstall	Ea.		61.57		61.57	screen.
	Clean	Ea.	.02	5.15		5.17	
	Minimum Charge	Job		192		192	
3' x 2' 6"							
	Demolish	Ea.		15.95		15.95	Cost includes material and labor to
	Install	Ea.	75	77		152	install 3' x 2' 6" aluminum window
	Demolish and Install	Ea.	75	92.95		167.95	including 4" glass louvers, hardware,
	Reinstall	Ea.		61.57		61.57	screen.
	Clean	Ea.	.03	5.90		5.93	
	Minimum Charge	Job		192		192	
3' x 3'							
	Demolish	Ea.		15.95		15.95	Cost includes material and labor to
	Install	Ea.	85.50	85.50		171	install 3' x 3' aluminum window
	Demolish and Install	Ea.	85.50	101.45		186.95	including 4" glass louvers, hardware,
	Reinstall	Ea.		68.41		68.41	screen.
	Clean	Ea.	.03	5.90		5.93	
	Minimum Charge	Job		192		192	
3' x 3' 6"							
	Demolish	Ea.		15.95		15.95	Cost includes material and labor to
	Install	Ea.	97	85.50		182.50	install 3' x 3' 6" aluminum window
	Demolish and Install	Ea.	97	101.45		198.45	including 4" glass louvers, hardware,
	Reinstall	Ea.		68.41		68.41	screen.
	Clean	Ea.	.04	6.85		6.89	
	Minimum Charge	Job		192		192	
3' x 4'							
	Demolish	Ea.		15.95		15.95	Cost includes material and labor to
	Install	Ea.	108	85.50		193.50	install 3' x 4' aluminum window
	Demolish and Install	Ea.	108	101.45		209.45	including 4" glass louvers, hardware,
	Reinstall	Ea.		68.41		68.41	screen.
	Clean	Ea.	.04	6.85		6.89	
	Minimum Charge	Job		192		192	
3' x 5'							
	Demolish	Ea.		23		23	Cost includes material and labor to
	Install	Ea.	138	96		234	install 3' x 5' aluminum window with
	Demolish and Install	Ea.	138	119		257	4" glass louvers, hardware, screen.
	Reinstall	Ea.		76.96		76.96	
	Clean	Ea.	.06	8.25		8.31	
	Minimum Charge	Job		192		192	

Aluminum Window		Unit	Material	Labor	Equip.	Total	Specification
3' x 6'							Cost includes material and labor to
	Demolish	Ea.		23		23	install 3' x 6' aluminum window with
	Install	Ea.	160	96		256	4" glass louvers, hardware, screen.
	Demolish and Install	Ea.	160	119		279	
	Reinstall	Ea.		76.96		76.96	
	Clean	Ea.	.07	9.70		9.77	
	Minimum Charge	Job		192		192	
Bay							
4' 8" x 6' 4"							Cost includes material and labor to
	Demolish	Ea.		17		17	install 4' 8" x 6' 4" angle bay double
	Install	Ea.	1050	78.50		1128.50	hung unit, pine frame with aluminum
	Demolish and Install	Ea.	1050	95.50		1145.50	cladding, picture center sash,
	Reinstall	Ea.		62.72		62.72	hardware, screens.
	Clean	Ea.	.12	13.75		13.87	
	Minimum Charge	Job		192		192	
4' 8" x 7' 2"							Cost includes material and labor to
	Demolish	Ea.		17		17	install 4' 8" x 7' 2" angle bay double
	Install	Ea.	1075	78.50		1153.50	hung unit, pine frame with aluminum
	Demolish and Install	Ea.	1075	95.50		1170.50	cladding, picture center sash,
	Reinstall	Ea.		62.72		62.72	hardware, screens.
	Clean	Ea.	.12	13.75		13.87	
	Minimum Charge	Job		192		192	
4' 8" x 7' 10"							Cost includes material and labor to
	Demolish	Ea.		17		17	install 4' 8" x 7' 10" angle bay
	Install	Ea.	1100	78.50		1178.50	double hung unit, pine frame with
	Demolish and Install	Ea.	1100	95.50		1195.50	aluminum cladding, picture center
	Reinstall	Ea.		62.72		62.72	sash, hardware, screens.
	Clean	Ea.	.12	13.75		13.87	
	Minimum Charge	Job		192		192	
4' 8" x 8' 6"							Cost includes material and labor to
	Demolish	Ea.		19.60		19.60	install 4' 8" x 8' 6" angle bay double
	Install	Ea.	1250	89.50		1339.50	hung unit, pine frame with aluminum
	Demolish and Install	Ea.	1250	109.10		1359.10	cladding, picture center sash,
	Reinstall	Ea.		71.68		71.68	hardware, screens.
	Clean	Ea.	.02	18.30		18.32	
	Minimum Charge	Job		192		192	
4' 8" x 8' 10"							Cost includes material and labor to
	Demolish	Ea.		19.60		19.60	install 4' 8" x 8' 10" angle bay
	Install	Ea.	1275	89.50		1364.50	double hung unit, pine frame with
	Demolish and Install	Ea.	1275	109.10		1384.10	aluminum cladding, picture center
	Reinstall	Ea.		71.68		71.68	sash, hardware, screens.
	Clean	Ea.	.02	18.30		18.32	
	Minimum Charge	Job		192		192	
4' 8" x 9' 2"							Cost includes material and labor to
	Demolish	Ea.		19.60		19.60	install 4' 8" x 9' 2" angle bay double
	Install	Ea.	1325	157		1482	hung unit, pine frame with aluminum
	Demolish and Install	Ea.	1325	176.60		1501.60	cladding, picture center sash,
	Reinstall	Ea.		125.44		125.44	hardware, screens.
	Clean	Ea.	.02	18.30		18.32	
	Minimum Charge	Job		192		192	
4' 8" x 9' 10"							Cost includes material and labor to
	Demolish	Ea.		19.60		19.60	install 4' 8" x 9' 10" angle bay
	Install	Ea.	1400	157		1557	double hung unit, pine frame with
	Demolish and Install	Ea.	1400	176.60		1576.60	aluminum cladding, picture center
	Reinstall	Ea.		125.44		125.44	sash, hardware, screens.
	Clean	Ea.	.02	18.30		18.32	
	Minimum Charge	Job		192		192	

Exterior Doors and Windows

Aluminum Window		Unit	Material	Labor	Equip.	Total	Specification
Casing							
	Demolish	Ea.		7.25		7.25	Cost includes material and labor to
	Install	Opng.	22	24		46	install 11/16" x 2-1/2" pine ranch
	Demolish and Install	Opng.	22	31.25		53.25	style window casing.
	Reinstall	Opng.		19.30		19.30	
	Clean	Opng.	.41	2.30		2.71	
	Paint	Ea.	3.08	5.50		8.58	
	Minimum Charge	Job		192		192	
Storm Window							
2' x 3' 6"							
	Demolish	Ea.		9.45		9.45	Includes material and labor to install
	Install	Ea.	73	21		94	residential aluminum storm window
	Demolish and Install	Ea.	73	30.45		103.45	with mill finish, 2' x 3' 5" high.
	Reinstall	Ea.		16.73		16.73	
	Clean	Ea.	.02	5.15		5.17	
	Minimum Charge	Job		192		192	
2' 6" x 5'							
	Demolish	Ea.		9.45		9.45	Includes material and labor to install
	Install	Ea.	81.50	22.50		104	residential aluminum storm window
	Demolish and Install	Ea.	81.50	31.95		113.45	with mill finish, 2' 6" x 5' high.
	Reinstall	Ea.		17.92		17.92	
	Clean	Ea.	.04	6.85		6.89	
	Minimum Charge	Job		192		192	
4' x 6'							
	Demolish	Ea.		18.20		18.20	Includes material and labor to install
	Install	Ea.	97	25		122	residential aluminum storm window, 4'
	Demolish and Install	Ea.	97	43.20		140.20	x 6' high.
	Reinstall	Ea.		20.07		20.07	
	Clean	Ea.	.09	10.30		10.39	
	Minimum Charge	Job		192		192	
Hardware							
Weatherstripping							
	Install	L.F.	.26	2.61		2.87	Includes labor and material to install
	Minimum Charge	Job		105		105	vinyl V strip weatherstripping for
							double-hung window.
Trim Set							
	Demolish	Ea.		7.25		7.25	Cost includes material and labor to
	Install	Opng.	22	24		46	install 11/16" x 2-1/2" pine ranch
	Demolish and Install	Opng.	22	31.25		53.25	style window casing.
	Reinstall	Opng.		19.30		19.30	
	Clean	Opng.	.41	2.30		2.71	
	Paint	Ea.	3.08	5.50		8.58	
	Minimum Charge	Job		157		157	

Vinyl Window		Unit	Material	Labor	Equip.	Total	Specification
Double Hung							
2' x 2' 6"							
	Demolish	Ea.		11.60		11.60	Cost includes material and labor to
	Install	Ea.	163	42		205	install 2' x 2' 6" double hung, two
	Demolish and Install	Ea.	163	53.60		216.60	light, vinyl window with screen.
	Reinstall	Ea.		33.45		33.45	
	Clean	Ea.	.02	5.15		5.17	
	Minimum Charge	Job		315		315	

Exterior Doors and Windows

Vinyl Window		Unit	Material	Labor	Equip.	Total	Specification
2' x 3' 6"							
	Demolish	Ea.		11.60		11.60	Cost includes material and labor to
	Install	Ea.	167	45		212	install 2' x 3' 6" double hung, two
	Demolish and Install	Ea.	167	56.60		223.60	light, vinyl window with screen.
	Reinstall	Ea.		35.84		35.84	
	Clean	Ea.	.03	5.90		5.93	
	Minimum Charge	Job		315		315	
2' 6" x 4' 6"							
	Demolish	Ea.		11.60		11.60	Cost includes material and labor to
	Install	Ea.	204	48.50		252.50	install 2' 6" x 4' 6" double hung, two
	Demolish and Install	Ea.	204	60.10		264.10	light, vinyl window with screen.
	Reinstall	Ea.		38.60		38.60	
	Clean	Ea.	.04	6.85		6.89	
	Minimum Charge	Job		315		315	
3' x 4'							
	Demolish	Ea.		11.60		11.60	Includes material and labor to install
	Install	Ea.	215	62.50		277.50	medium size double hung, 3' x 4',
	Demolish and Install	Ea.	215	74.10		289.10	two light, vinyl window with screen.
	Reinstall	Ea.		50.18		50.18	
	Clean	Ea.	.04	6.85		6.89	
	Minimum Charge	Job		315		315	
3' x 4' 6"							
	Demolish	Ea.		14.15		14.15	Cost includes material and labor to
	Install	Ea.	232	69.50		301.50	install 3' x 4' 6" double hung, two
	Demolish and Install	Ea.	232	83.65		315.65	light, vinyl window with screen.
	Reinstall	Ea.		55.75		55.75	
	Clean	Ea.	.06	8.25		8.31	
	Minimum Charge	Job		315		315	
4' x 4' 6"							
	Demolish	Ea.		14.15		14.15	Cost includes material and labor to
	Install	Ea.	251	78.50		329.50	install 4' x 4' 6" double hung, two
	Demolish and Install	Ea.	251	92.65		343.65	light, vinyl window with screen.
	Reinstall	Ea.		62.72		62.72	
	Clean	Ea.	.07	9.70		9.77	
	Minimum Charge	Job		315		315	
4' x 6'							
	Demolish	Ea.		14.15		14.15	Includes material and labor to install
	Install	Ea.	290	89.50		379.50	large double hung, 4' x 6', two light,
	Demolish and Install	Ea.	290	103.65		393.65	vinyl window with screen.
	Reinstall	Ea.		71.68		71.68	
	Clean	Ea.	.09	11.75		11.84	
	Minimum Charge	Job		315		315	
Trim Set							
	Demolish	Ea.		7.25		7.25	Cost includes material and labor to
	Install	Opng.	22	24		46	install 11/16" x 2-1/2" pine ranch
	Demolish and Install	Opng.	22	31.25		53.25	style window casing.
	Reinstall	Opng.		19.30		19.30	
	Clean	Opng.	.41	2.30		2.71	
	Paint	Ea.	3.08	5.50		8.58	
	Minimum Charge	Job		192		192	

Wood Window		Unit	Material	Labor	Equip.	Total	Specification
2' 2" x 3' 4"							
	Demolish	Ea.		11.60		11.60	Cost includes material and labor to
	Install	Ea.	178	42		220	install 2' 2" x 3' 4" double hung, two
	Demolish and Install	Ea.	178	53.60		231.60	light, double glazed wood window
	Reinstall	Ea.		33.45		33.45	with screen.
	Clean	Ea.	.03	5.90		5.93	
	Paint	Ea.	1.93	11.50		13.43	
	Minimum Charge	Job		157		157	

Wood Window		Unit	Material	Labor	Equip.	Total	Specification
2' 2" x 4' 4"							
	Demolish	Ea.		11.60		11.60	Cost includes material and labor to
	Install	Ea.	200	45		245	install 2' 2" x 4' 4" double hung, two
	Demolish and Install	Ea.	200	56.60		256.60	light, double glazed wood window
	Reinstall	Ea.		35.84		35.84	with screen.
	Clean	Ea.	.03	5.90		5.93	
	Paint	Ea.	1.93	11.50		13.43	
	Minimum Charge	Job		157		157	
2' 6" x 3' 4"							
	Demolish	Ea.		11.60		11.60	Cost includes material and labor to
	Install	Ea.	188	48.50		236.50	install 2' 6" x 3' 4" double hung, two
	Demolish and Install	Ea.	188	60.10		248.10	light, double glazed wood window
	Reinstall	Ea.		38.60		38.60	with screen.
	Clean	Ea.	.04	6.85		6.89	
	Paint	Ea.	2.46	13.80		16.26	
	Minimum Charge	Job		157		157	
2' 6" x 4'							
	Demolish	Ea.		11.60		11.60	Cost includes material and labor to
	Install	Ea.	202	52.50		254.50	install 2' 6" x 4' double hung, two
	Demolish and Install	Ea.	202	64.10		266.10	light, double glazed wood window
	Reinstall	Ea.		41.81		41.81	with screen.
	Clean	Ea.	.04	6.85		6.89	
	Paint	Ea.	2.46	13.80		16.26	
	Minimum Charge	Job		157		157	
2' 6" x 4' 8"							
	Demolish	Ea.		11.60		11.60	Cost includes material and labor to
	Install	Ea.	223	52.50		275.50	install 2' 6" x 4' 8" double hung, two
	Demolish and Install	Ea.	223	64.10		287.10	light, double glazed wood window
	Reinstall	Ea.		41.81		41.81	with screen.
	Clean	Ea.	.04	6.85		6.89	
	Paint	Ea.	2.46	13.80		16.26	
	Minimum Charge	Job		157		157	
2' 10" x 3' 4"							
	Demolish	Ea.		11.60		11.60	Cost includes material and labor to
	Install	Ea.	198	62.50		260.50	install 2' 10" x 3' 4" double hung, two
	Demolish and Install	Ea.	198	74.10		272.10	light, double glazed wood window
	Reinstall	Ea.		50.18		50.18	with screen.
	Clean	Ea.	.04	6.85		6.89	
	Paint	Ea.	2.46	13.80		16.26	
	Minimum Charge	Job		157		157	
2' 10" x 4'							
	Demolish	Ea.		11.60		11.60	Cost includes material and labor to
	Install	Ea.	217	62.50		279.50	install 2' 10" x 4' double hung, two
	Demolish and Install	Ea.	217	74.10		291.10	light, double glazed wood window
	Reinstall	Ea.		50.18		50.18	with screen.
	Clean	Ea.	.04	6.85		6.89	
	Paint	Ea.	2.46	13.80		16.26	
	Minimum Charge	Job		157		157	
3' 7" x 3' 4"							
	Demolish	Ea.		11.60		11.60	Cost includes material and labor to
	Install	Ea.	223	69.50		292.50	install 3' 7" x 3' 4" double hung, two
	Demolish and Install	Ea.	223	81.10		304.10	light, double glazed wood window
	Reinstall	Ea.		55.75		55.75	with screen.
	Clean	Ea.	.04	6.85		6.89	
	Paint	Ea.	2.46	13.80		16.26	
	Minimum Charge	Job		157		157	

Exterior Doors and Windows

Wood Window		Unit	Material	Labor	Equip.	Total	Specification
3' 7" x 5' 4"							Cost includes material and labor to
	Demolish	Ea.		14.15		14.15	install 3' 7" x 5' 4" double hung, two
	Install	Ea.	286	69.50		355.50	light, double glazed wood window
	Demolish and Install	Ea.	286	83.65		369.65	with screen.
	Reinstall	Ea.		55.75		55.75	
	Clean	Ea.	.09	11.75		11.84	
	Paint	Ea.	4.11	23		27.11	
	Minimum Charge	Job		157		157	
4' 3" x 5' 4"							Cost includes material and labor to
	Demolish	Ea.		14.15		14.15	install 4' 3" x 5' 4" double hung, two
	Install	Ea.	450	78.50		528.50	light, double glazed wood window
	Demolish and Install	Ea.	450	92.65		542.65	with screen.
	Reinstall	Ea.		62.72		62.72	
	Clean	Ea.	.09	11.75		11.84	
	Paint	Ea.	2.65	27.50		30.15	
	Minimum Charge	Job		157		157	
Casement							
2' x 3' 4"							Cost includes material and labor to
	Demolish	Ea.		11.60		11.60	install 2' x 3' 4 " casement window
	Install	Ea.	187	31.50		218.50	with 3/4" insulated glass, screens,
	Demolish and Install	Ea.	187	43.10		230.10	weatherstripping, hardware.
	Reinstall	Ea.		25.09		25.09	
	Clean	Ea.	.03	5.90		5.93	
	Paint	Ea.	1.93	11.50		13.43	
	Minimum Charge	Job		315		315	
2' x 4'							Cost includes material and labor to
	Demolish	Ea.		11.60		11.60	install 2' x 4' casement window with
	Install	Ea.	210	35		245	3/4" insulated glass, screens,
	Demolish and Install	Ea.	210	46.60		256.60	weatherstripping, hardware.
	Reinstall	Ea.		27.88		27.88	
	Clean	Ea.	.03	5.90		5.93	
	Paint	Ea.	1.93	11.50		13.43	
	Minimum Charge	Job		315		315	
2' x 5'							Cost includes material and labor to
	Demolish	Ea.		11.60		11.60	install 2' x 5' casement window with
	Install	Ea.	235	37		272	3/4" insulated glass, screens,
	Demolish and Install	Ea.	235	48.60		283.60	weatherstripping, hardware.
	Reinstall	Ea.		29.52		29.52	
	Clean	Ea.	.04	6.85		6.89	
	Paint	Ea.	2.46	13.80		16.26	
	Minimum Charge	Job		315		315	
2' x 6'							Cost includes material and labor to
	Demolish	Ea.		11.60		11.60	install 2' x 6' casement window with
	Install	Ea.	265	39		304	3/4" insulated glass, screens,
	Demolish and Install	Ea.	265	50.60		315.60	weatherstripping, hardware.
	Reinstall	Ea.		31.36		31.36	
	Clean	Ea.	.04	6.85		6.89	
	Paint	Ea.	2.46	13.80		16.26	
	Minimum Charge	Job		315		315	
4' x 3' 4"							Cost includes material and labor to
	Demolish	Ea.		14.15		14.15	install 4' x 3' 4 " casement windows
	Install	Ea.	540	42		582	with 3/4" insulated glass, screens,
	Demolish and Install	Ea.	540	56.15		596.15	weatherstripping, hardware.
	Reinstall	Ea.		33.45		33.45	
	Clean	Ea.	.04	6.85		6.89	
	Paint	Ea.	1.33	13.80		15.13	
	Minimum Charge	Job		315		315	

Wood Window		Unit	Material	Labor	Equip.	Total	Specification
4' x 4'							
	Demolish	Ea.		14.15		14.15	Cost includes material and labor to
	Install	Ea.	595	42		637	install 4' x 4' casement windows with
	Demolish and Install	Ea.	595	56.15		651.15	3/4" insulated glass, screens,
	Reinstall	Ea.		33.45		33.45	weatherstripping, hardware.
	Clean	Ea.	.07	9.70		9.77	
	Paint	Ea.	1.56	17.25		18.81	
	Minimum Charge	Job		315		315	
4' x 5'							
	Demolish	Ea.		14.15		14.15	Cost includes material and labor to
	Install	Ea.	670	45		715	install 4' x 5' casement windows with
	Demolish and Install	Ea.	670	59.15		729.15	3/4" insulated glass, screens,
	Reinstall	Ea.		35.84		35.84	weatherstripping, hardware.
	Clean	Ea.	.09	11.75		11.84	
	Paint	Ea.	2.05	23		25.05	
	Minimum Charge	Job		315		315	
4' x 6'							
	Demolish	Ea.		14.15		14.15	Cost includes material and labor to
	Install	Ea.	760	52.50		812.50	install 4' x 6' casement windows with
	Demolish and Install	Ea.	760	66.65		826.65	3/4" insulated glass, screens,
	Reinstall	Ea.		41.81		41.81	weatherstripping, hardware.
	Clean	Ea.	.09	11.75		11.84	
	Paint	Ea.	2.41	27.50		29.91	
	Minimum Charge	Job		315		315	
Casement Bow							
4' 8" x 8' 1"							
	Demolish	Ea.		19.60		19.60	Includes material and labor to install
	Install	Ea.	1425	118		1543	casement bow window unit with 4
	Demolish and Install	Ea.	1425	137.60		1562.60	lights, pine frame, 1/2" insulated
	Reinstall	Ea.		94.08		94.08	glass, weatherstripping, hardware,
	Clean	Ea.	.12	13.75		13.87	screens.
	Paint	Ea.	4.47	55		59.47	
	Minimum Charge	Job		315		315	
4' 8" x 10'							
	Demolish	Ea.		19.60		19.60	Includes material and labor to install
	Install	Ea.	1525	157		1682	casement bow window unit with 5
	Demolish and Install	Ea.	1525	176.60		1701.60	lights, pine frame, 1/2" insulated
	Reinstall	Ea.		125.44		125.44	glass, weatherstripping, hardware,
	Clean	Ea.	.02	18.30		18.32	screens.
	Paint	Ea.	5.55	69		74.55	
	Minimum Charge	Job		315		315	
5' 4" x 8' 1"							
	Demolish	Ea.		19.60		19.60	Includes material and labor to install
	Install	Ea.	1575	118		1693	casement bow window unit with 4
	Demolish and Install	Ea.	1575	137.60		1712.60	lights, pine frame, 1/2" insulated
	Reinstall	Ea.		94.08		94.08	glass, weatherstripping, hardware,
	Clean	Ea.	.02	18.30		18.32	screens.
	Paint	Ea.	5.05	55		60.05	
	Minimum Charge	Job		315		315	
5' 4" x 10'							
	Demolish	Ea.		19.60		19.60	Includes material and labor to install
	Install	Ea.	1725	157		1882	casement bow window unit with 5
	Demolish and Install	Ea.	1725	176.60		1901.60	lights, pine frame, 1/2" insulated
	Reinstall	Ea.		125.44		125.44	glass, weatherstripping, hardware,
	Clean	Ea.	.02	18.30		18.32	screens.
	Paint	Ea.	6.40	69		75.40	
	Minimum Charge	Job		315		315	

Wood Window		Unit	Material	Labor	Equip.	Total	Specification
6' x 8' 1"							
	Demolish	Ea.		19.60		19.60	Includes material and labor to install
	Install	Ea.	1525	157		1682	casement bow window unit with 5
	Demolish and Install	Ea.	1525	176.60		1701.60	lights, pine frame, 1/2" insulated
	Reinstall	Ea.		125.44		125.44	glass, weatherstripping, hardware,
	Clean	Ea.	.02	18.30		18.32	screens.
	Paint	Ea.	5.80	69		74.80	
	Minimum Charge	Job		315		315	
6' x 10'							
	Demolish	Ea.		19.60		19.60	Includes material and labor to install
	Install	Ea.	1825	157		1982	casement bow window unit with 5
	Demolish and Install	Ea.	1825	176.60		2001.60	lights, pine frame, 1/2" insulated
	Reinstall	Ea.		125.44		125.44	glass, weatherstripping, hardware,
	Clean	Ea.	.02	18.30		18.32	screens.
	Paint	Ea.	7.25	92		99.25	
	Minimum Charge	Job		315		315	
Casing							
	Demolish	Ea.		7.25		7.25	Cost includes material and labor to
	Install	Opng.	22	24		46	install 11/16" x 2-1/2" pine ranch
	Demolish and Install	Opng.	22	31.25		53.25	style window casing.
	Reinstall	Opng.		19.30		19.30	
	Clean	Opng.	.41	2.30		2.71	
	Paint	Ea.	3.08	5.50		8.58	
	Minimum Charge	Job		192		192	
Trim Set							
	Demolish	Ea.		5.55		5.55	Cost includes material and labor to
	Install	Opng.	22	24		46	install 11/16" x 2-1/2" pine ranch
	Demolish and Install	Opng.	22	29.55		51.55	style window casing.
	Reinstall	Opng.		19.30		19.30	
	Clean	Opng.	.41	2.30		2.71	
	Paint	Ea.	3.08	5.50		8.58	
	Minimum Charge	Job		157		157	
Wood Picture Window							
3' 6" x 4'							
	Demolish	Ea.		14.15		14.15	Includes material and labor to install a
	Install	Ea.	310	52.50		362.50	wood framed picture window
	Demolish and Install	Ea.	310	66.65		376.65	including exterior trim.
	Clean	Ea.	.09	11.75		11.84	
	Paint	Ea.	1.69	21		22.69	
	Minimum Charge	Job		315		315	
4' x 4' 6"							
	Demolish	Ea.		14.15		14.15	Includes material and labor to install a
	Install	Ea.	325	57		382	wood framed picture window
	Demolish and Install	Ea.	325	71.15		396.15	including exterior trim.
	Clean	Ea.	.09	11.75		11.84	
	Paint	Ea.	1.69	21		22.69	
	Minimum Charge	Job		315		315	
5' x 4'							
	Demolish	Ea.		14.15		14.15	Includes material and labor to install a
	Install	Ea.	405	57		462	wood framed picture window
	Demolish and Install	Ea.	405	71.15		476.15	including exterior trim.
	Clean	Ea.	.09	11.75		11.84	
	Paint	Ea.	3.61	34.50		38.11	
	Minimum Charge	Job		315		315	

Exterior Doors and Windows

Wood Window

	Unit	Material	Labor	Equip.	Total	Specification
6' x 4' 6"						
Demolish	Ea.		14.15		14.15	Includes material and labor to install a
Install	Ea.	525	62.50		587.50	wood framed picture window
Demolish and Install	Ea.	525	76.65		601.65	including exterior trim.
Clean	Ea.	.09	11.75		11.84	
Paint	Ea.	3.61	34.50		38.11	
Minimum Charge	Job		315		315	

Window Screens

	Unit	Material	Labor	Equip.	Total	Specification
Small						
Demolish	Ea.		2.66		2.66	Includes material and labor to install
Install	Ea.	15.35	9.50		24.85	small window screen, 4 S.F.
Demolish and Install	Ea.	15.35	12.16		27.51	
Reinstall	Ea.		7.60		7.60	
Clean	Ea.	.29	4.78		5.07	
Minimum Charge	Job		78.50		78.50	
Standard Size						
Demolish	Ea.		2.66		2.66	Includes material and labor to install
Install	Ea.	37.50	9.50		47	medium window screen, 12 S.F.
Demolish and Install	Ea.	37.50	12.16		49.66	
Reinstall	Ea.		7.60		7.60	
Clean	Ea.	.29	4.78		5.07	
Minimum Charge	Job		78.50		78.50	
Large						
Demolish	Ea.		2.66		2.66	Includes material and labor to install
Install	Ea.	49.50	9.50		59	large window screen, 16 S.F.
Demolish and Install	Ea.	49.50	12.16		61.66	
Reinstall	Ea.		7.60		7.60	
Clean	Ea.	.29	4.78		5.07	
Minimum Charge	Job		78.50		78.50	

Window Glass

	Unit	Material	Labor	Equip.	Total	Specification
Plate, Clear, 1/4"						
Demolish	S.F.		1.28		1.28	Includes material and labor to install
Install	S.F.	6.55	5		11.55	clear glass.
Demolish and Install	S.F.	6.55	6.28		12.83	
Reinstall	S.F.		4.01		4.01	
Clean	S.F.	.02	.13		.15	
Minimum Charge	Job		150		150	
Insulated						
Demolish	S.F.		1.28		1.28	Includes material and labor to install
Install	S.F.	10.75	8.60		19.35	insulated glass.
Demolish and Install	S.F.	10.75	9.88		20.63	
Reinstall	S.F.		6.87		6.87	
Clean	S.F.	.02	.13		.15	
Minimum Charge	Job		150		150	
Tinted						
Demolish	S.F.		1.28		1.28	Includes material and labor to install
Install	S.F.	6.60	4.62		11.22	tinted glass.
Demolish and Install	S.F.	6.60	5.90		12.50	
Reinstall	S.F.		3.70		3.70	
Clean	S.F.	.02	.13		.15	
Minimum Charge	Job		150		150	

Exterior Doors and Windows

Window Glass		Unit	Material	Labor	Equip.	Total	Specification
Obscure / Florentine							
	Demolish	S.F.		1.28		1.28	Includes material and labor to install
	Install	S.F.	7.35	4.29		11.64	obscure glass.
	Demolish and Install	S.F.	7.35	5.57		12.92	
	Reinstall	S.F.		3.43		3.43	
	Clean	S.F.	.02	.13		.15	
	Minimum Charge	Job		150		150	
Laminated Safety							
	Demolish	S.F.		1.28		1.28	Includes material and labor to install
	Install	S.F.	8.95	6.70		15.65	clear laminated glass.
	Demolish and Install	S.F.	8.95	7.98		16.93	
	Reinstall	S.F.		5.34		5.34	
	Clean	S.F.	.02	.13		.15	
	Minimum Charge	Job		150		150	
Wired							
	Demolish	S.F.		1.28		1.28	Includes material and labor to install
	Install	S.F.	12	4.45		16.45	rough obscure wire glass.
	Demolish and Install	S.F.	12	5.73		17.73	
	Reinstall	S.F.		3.56		3.56	
	Clean	S.F.	.02	.13		.15	
	Minimum Charge	Job		150		150	
Tempered							
	Demolish	S.F.		1.28		1.28	Includes material and labor to install
	Install	S.F.	6.55	5		11.55	clear glass.
	Demolish and Install	S.F.	6.55	6.28		12.83	
	Reinstall	S.F.		4.01		4.01	
	Clean	S.F.	.02	.13		.15	
	Minimum Charge	Job		150		150	
Stained							
	Demolish	S.F.		1.28		1.28	Includes material and labor to install
	Install	S.F.	48	5		53	stained glass.
	Demolish and Install	S.F.	48	6.28		54.28	
	Reinstall	S.F.		4.01		4.01	
	Clean	S.F.	.02	.13		.15	
	Minimum Charge	Job		150		150	
Etched							
	Demolish	S.F.		1.28		1.28	Includes material and labor to install
	Install	S.F.	39	5		44	etched glass.
	Demolish and Install	S.F.	39	6.28		45.28	
	Reinstall	S.F.		4.01		4.01	
	Clean	S.F.	.02	.13		.15	
	Minimum Charge	Job		150		150	
Wall Mirror							
	Demolish	S.F.		1.28		1.28	Includes material and labor to install
	Install	S.F.	7.05	4.81		11.86	distortion-free float glass mirror, 1/4"
	Demolish and Install	S.F.	7.05	6.09		13.14	thick, cut and polished edges.
	Reinstall	S.F.		3.85		3.85	
	Clean	S.F.	.02	.13		.15	
	Minimum Charge	Job		150		150	
Bullet Resistant							
	Demolish	S.F.		1.28		1.28	Includes material and labor to install
	Install	S.F.	86	60		146	bullet resistant clear laminated glass.
	Demolish and Install	S.F.	86	61.28		147.28	
	Reinstall	S.F.		48.06		48.06	
	Clean	S.F.	.02	.13		.15	
	Minimum Charge	Job		150		150	

Exterior Doors and Windows

Storefront System	Unit	Material	Labor	Equip.	Total	Specification
Clear Glass						
Stub Wall to 8'						
Demolish	S.F.		1.99		1.99	Includes material and labor to install
Install	S.F.	16.15	8.30		24.45	commercial grade 4-1/2" section,
Demolish and Install	S.F.	16.15	10.29		26.44	anodized aluminum with clear glass in
Reinstall	S.F.		6.64		6.64	upper section and safety glass in
Clean	S.F.		.27		.27	lower section.
Minimum Charge	Job		192		192	
Floor to 10'						
Demolish	S.F.		1.99		1.99	Includes material and labor to install
Install	S.F.	17.70	8.30		26	commercial grade 4-1/2" section,
Demolish and Install	S.F.	17.70	10.29		27.99	anodized aluminum with clear glass in
Reinstall	S.F.		6.64		6.64	upper section and safety glass in
Clean	S.F.		.27		.27	lower section.
Minimum Charge	Job		192		192	
Polished Plate Glass						
Stub Wall to 9'						
Demolish	S.F.		1.99		1.99	Cost includes material and labor to
Install	S.F.	12.25	8.30		20.55	install 4' x 9' polished plate glass
Demolish and Install	S.F.	12.25	10.29		22.54	window built on 8" stub wall.
Reinstall	S.F.		6.64		6.64	
Clean	S.F.		.27		.27	
Minimum Charge	Job		192		192	
Stub Wall to 13'						
Demolish	S.F.		1.99		1.99	Cost includes material and labor to
Install	S.F.	12.80	8.30		21.10	install 4' x 13' polished plate glass
Demolish and Install	S.F.	12.80	10.29		23.09	window built on 8" stub wall.
Reinstall	S.F.		6.64		6.64	
Clean	S.F.		.27		.27	
Minimum Charge	Job		192		192	
Floor to 9'						
Demolish	S.F.		1.99		1.99	Cost includes material and labor to
Install	S.F.	13.30	8.30		21.60	install 4' x 9' polished plate glass
Demolish and Install	S.F.	13.30	10.29		23.59	window built on floor.
Reinstall	S.F.		6.64		6.64	
Clean	S.F.		.27		.27	
Minimum Charge	Job		192		192	
Floor to 13'						
Demolish	S.F.		1.99		1.99	Cost includes material and labor to
Install	S.F.	14.30	8.30		22.60	install 4' x 13' polished plate glass
Demolish and Install	S.F.	14.30	10.29		24.59	window built on floor.
Reinstall	S.F.		6.64		6.64	
Clean	S.F.		.27		.27	
Minimum Charge	Job		192		192	
Clear Plate Glass						
Floor to 9'						
Demolish	S.F.		1.99		1.99	Cost includes material and labor to
Install	S.F.	19.20	8.30		27.50	install 4' x 9' clear plate glass window
Demolish and Install	S.F.	19.20	10.29		29.49	built on floor.
Reinstall	S.F.		6.64		6.64	
Clean	S.F.		.27		.27	
Minimum Charge	Job		192		192	

Storefront System		Unit	Material	Labor	Equip.	Total	Specification
Floor to 11'							
	Demolish	S.F.		1.99		1.99	Cost includes material and labor to
	Install	S.F.	18.65	8.30		26.95	install 4' x 11' clear plate glass
	Demolish and Install	S.F.	18.65	10.29		28.94	window built on floor.
	Reinstall	S.F.		6.64		6.64	
	Clean	S.F.		.27		.27	
	Minimum Charge	Job		192		192	
Floor to 13'							
	Demolish	S.F.		1.99		1.99	Cost includes material and labor to
	Install	S.F.	18.80	8.30		27.10	install 4' x 13' clear plate glass
	Demolish and Install	S.F.	18.80	10.29		29.09	window built on floor.
	Reinstall	S.F.		6.64		6.64	
	Clean	S.F.		.27		.27	
	Minimum Charge	Job		192		192	
Decorative Type							
	Demolish	S.F.		1.99		1.99	Includes material and labor to install
	Install	S.F.	14.65	6		20.65	decorative storefront system.
	Demolish and Install	S.F.	14.65	7.99		22.64	
	Reinstall	S.F.		4.81		4.81	
	Clean	S.F.		.27		.27	
	Minimum Charge	Job		150		150	
Decorative Curved							
	Demolish	S.F.		1.99		1.99	Includes material and labor to install
	Install	S.F.	27.50	12		39.50	decorative curved storefront system.
	Demolish and Install	S.F.	27.50	13.99		41.49	
	Reinstall	S.F.		9.61		9.61	
	Clean	S.F.		.27		.27	
	Minimum Charge	Job		150		150	
Entrance							
Narrow Stile 3' x 7"							
	Demolish	Ea.		17.40		17.40	Cost includes material and labor to
	Install	Ea.	695	209		904	install 3' x 7' narrow stile door, center
	Demolish and Install	Ea.	695	226.40		921.40	pivot concealed closer.
	Reinstall	Ea.		167.25		167.25	
	Clean	Ea.	3.61	28.50		32.11	
	Minimum Charge	Job		192		192	
Tempered Glass							
	Demolish	Ea.		17.40		17.40	Cost includes material and labor to
	Install	Ea.	2550	415		2965	install 3' x 7' tempered glass door,
	Demolish and Install	Ea.	2550	432.40		2982.40	center pivot concealed closer,.
	Reinstall	Ea.		332.80		332.80	
	Clean	Ea.	3.61	28.50		32.11	
	Minimum Charge	Job		192		192	
Heavy Section							
	Demolish	Ea.		17.40		17.40	Includes material and labor to install
	Install	Ea.	790	385		1175	wide stile door with 3' 6" x 7'
	Demolish and Install	Ea.	790	402.40		1192.40	opening.
	Reinstall	Ea.		307.84		307.84	
	Clean	Ea.	.73	9.55		10.28	
	Minimum Charge	Job		600		600	

Exterior Doors and Windows

Curtain Wall System

Curtain Wall System		Unit	Material	Labor	Equip.	Total	Specification
Regular Weight							
Bronze Anodized							
	Demolish	S.F.		1.99		1.99	Includes material and labor to install
	Install	S.F.	30.50	8.30		38.80	structural curtain wall system with
	Demolish and Install	S.F.	30.50	10.29		40.79	bronze anodized frame of heavy
	Clean	S.F.		.27		.27	weight, float glass.
	Minimum Charge	Job		157		157	
Black Anodized							
	Demolish	S.F.		1.99		1.99	Includes material and labor to install
	Install	S.F.	32.50	8.30		40.80	structural curtain wall system with
	Demolish and Install	S.F.	32.50	10.29		42.79	black anodized frame of regular
	Clean	S.F.		.27		.27	weight, float glass.
	Minimum Charge	Job		157		157	
Heavy Weight							
Bronze Anodized							
	Demolish	S.F.		1.99		1.99	Includes material and labor to install
	Install	S.F.	30.50	8.30		38.80	structural curtain wall system with
	Demolish and Install	S.F.	30.50	10.29		40.79	bronze anodized frame of heavy
	Clean	S.F.		.27		.27	weight, float glass.
	Minimum Charge	Job		157		157	
Black Anodized							
	Demolish	S.F.		1.99		1.99	Includes material and labor to install
	Install	S.F.	31	8.30		39.30	structural curtain wall system with
	Demolish and Install	S.F.	31	10.29		41.29	black anodized frame of heavy
	Clean	S.F.		.27		.27	weight, float glass.
	Minimum Charge	Job		157		157	
Glass							
1" Insulated							
	Install	S.F.	12.85	8		20.85	Cost includes labor and material to
	Minimum Charge	Job		150		150	install 1" insulated glass.
Heat Absorbing							
	Install	S.F.	7.25	8		15.25	Includes labor an material to install
	Minimum Charge	Job		150		150	heat absorbing glass.
Low Transmission							
	Install	S.F.	4.77	8		12.77	Includes labor an material to install
	Minimum Charge	Job		150		150	low transmission glass.

Skylight

Skylight		Unit	Material	Labor	Equip.	Total	Specification
Dome							
22" x 22"							
	Demolish	Ea.		60.50		60.50	Cost includes material and labor to
	Install	Ea.	48	95.50		143.50	install 22" x 22" fixed dome skylight.
	Demolish and Install	Ea.	48	156		204	
	Reinstall	Ea.		76.33		76.33	
	Clean	Ea.	.33	8.20		8.53	
	Minimum Charge	Job		570		570	
22" x 46"							
	Demolish	Ea.		60.50		60.50	Cost includes material and labor to
	Install	Ea.	107	115		222	install 22" x 46" fixed dome skylight.
	Demolish and Install	Ea.	107	175.50		282.50	
	Reinstall	Ea.		91.60		91.60	
	Clean	Ea.		12.75		12.75	
	Minimum Charge	Job		570		570	

Exterior Doors and Windows

Skylight		Unit	Material	Labor	Equip.	Total	Specification
30″ x 30″							
	Demolish	Ea.		60.50		60.50	Cost includes material and labor to
	Install	Ea.	89	95.50		184.50	install 30″ x 30″ fixed dome skylight.
	Demolish and Install	Ea.	89	156		245	
	Reinstall	Ea.		76.33		76.33	
	Clean	Ea.		12.75		12.75	
	Minimum Charge	Job		570		570	
30″ x 46″							
	Demolish	Ea.		60.50		60.50	Cost includes material and labor to
	Install	Ea.	155	115		270	install 30″ x 46″ fixed dome skylight.
	Demolish and Install	Ea.	155	175.50		330.50	
	Reinstall	Ea.		91.60		91.60	
	Clean	Ea.		12.75		12.75	
	Minimum Charge	Job		570		570	
Fixed							
22″ x 27″							
	Demolish	Ea.		60.50		60.50	Includes material and labor to install
	Install	Ea.	228	95.50		323.50	double glazed skylight, 22″ x 27″,
	Demolish and Install	Ea.	228	156		384	fixed.
	Reinstall	Ea.		76.33		76.33	
	Clean	Ea.	.33	8.20		8.53	
	Minimum Charge	Job		570		570	
22″ x 46″							
	Demolish	Ea.		60.50		60.50	Includes material and labor to install
	Install	Ea.	286	115		401	double glazed skylight, 22″ x 46″,
	Demolish and Install	Ea.	286	175.50		461.50	fixed.
	Reinstall	Ea.		91.60		91.60	
	Clean	Ea.		12.75		12.75	
	Minimum Charge	Job		570		570	
44″ x 46″							
	Demolish	Ea.		60.50		60.50	Includes material and labor to install
	Install	Ea.	405	115		520	double glazed skylight, 44″ x 46″,
	Demolish and Install	Ea.	405	175.50		580.50	fixed.
	Reinstall	Ea.		91.60		91.60	
	Clean	Ea.		14.35		14.35	
	Minimum Charge	Job		570		570	
Operable							
22″ x 27″							
	Demolish	Ea.		60.50		60.50	Includes material and labor to install
	Install	Ea.	340	95.50		435.50	double glazed skylight, 22″ x 27″,
	Demolish and Install	Ea.	340	156		496	operable.
	Reinstall	Ea.		76.33		76.33	
	Clean	Ea.	.33	8.20		8.53	
	Minimum Charge	Job		570		570	
22″ x 46″							
	Demolish	Ea.		60.50		60.50	Includes material and labor to install
	Install	Ea.	395	115		510	double glazed skylight, 22″ x 46″,
	Demolish and Install	Ea.	395	175.50		570.50	operable.
	Reinstall	Ea.		91.60		91.60	
	Clean	Ea.		12.75		12.75	
	Minimum Charge	Job		570		570	
44″ x 46″							
	Demolish	Ea.		60.50		60.50	Includes material and labor to install
	Install	Ea.	495	115		610	double glazed skylight, 44″ x 46″,
	Demolish and Install	Ea.	495	175.50		670.50	operable.
	Reinstall	Ea.		91.60		91.60	
	Clean	Ea.		14.35		14.35	
	Minimum Charge	Job		570		570	

Exterior Doors and Windows

Industrial Steel Window	Unit	Material	Labor	Equip.	Total	Specification
100% Fixed						
Demolish	S.F.		1.64		1.64	Includes material and labor to install
Install	S.F.	26.50	3.85		30.35	commercial grade steel windows,
Demolish and Install	S.F.	26.50	5.49		31.99	including glazing.
Reinstall	S.F.		3.08		3.08	
Clean	Ea.	.13	.86		.99	
Paint	S.F.	.11	.61		.72	
Minimum Charge	Job		192		192	
50% Vented						
Demolish	S.F.		1.64		1.64	Includes material and labor to install
Install	S.F.	38	3.85		41.85	commercial grade steel windows,
Demolish and Install	S.F.	38	5.49		43.49	including glazing.
Reinstall	S.F.		3.08		3.08	
Clean	Ea.	.13	.86		.99	
Paint	S.F.	.11	.61		.72	
Minimum Charge	Job		192		192	

Masonry

Brick Veneer

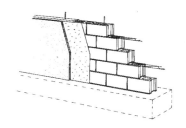

Concrete Masonry Units

Masonry Wall	Unit	Material	Labor	Equip.	Total	Specification
Unreinforced CMU						
4″ x 8″ x 16″						
Demolish	S.F.		1.42		1.42	Includes material and labor to install
Install	S.F.	1.07	3.12		4.19	normal weight 4″ x 8″ x 16″ block.
Demolish and Install	S.F.	1.07	4.54		5.61	
Clean	S.F.	.03	.35		.38	
Paint	S.F.	.13	.63		.76	
Minimum Charge	Job		160		160	
6″ x 8″ x 16″						
Demolish	S.F.		1.50		1.50	Includes material and labor to install
Install	S.F.	1.56	3.26		4.82	normal weight 6″ x 8″ x 16″ block.
Demolish and Install	S.F.	1.56	4.76		6.32	
Clean	S.F.	.03	.35		.38	
Paint	S.F.	.13	.63		.76	
Minimum Charge	Job		160		160	
8″ x 8″ x 16″						
Demolish	S.F.		1.70		1.70	Includes material and labor to install
Install	S.F.	1.69	3.58		5.27	normal weight 8″ x 8″ x 16″ block.
Demolish and Install	S.F.	1.69	5.28		6.97	
Clean	S.F.	.03	.35		.38	
Paint	S.F.	.13	.63		.76	
Minimum Charge	Job		160		160	
12″ x 8″ x 16″						
Demolish	S.F.		1.70		1.70	Includes material and labor to install
Install	S.F.	2.44	5.40		7.84	normal weight 12″ x 8″ x 16″ block.
Demolish and Install	S.F.	2.44	7.10		9.54	
Clean	S.F.	.03	.35		.38	
Paint	S.F.	.13	.63		.76	
Minimum Charge	Job		160		160	
Reinforced CMU						
4″ x 8″ x 16″						
Demolish	S.F.		1.42	.63	2.05	Includes material and labor to install
Install	S.F.	1.16	3.19		4.35	normal weight 4″ x 8″ x 16″ block,
Demolish and Install	S.F.	1.16	4.61	.63	6.40	mortar and 1/2″ re-bar installed
Clean	S.F.	.03	.35		.38	vertically every 24″ O.C.
Paint	S.F.	.13	.63		.76	
Minimum Charge	Job		279		279	
6″ x 8″ x 16″						
Demolish	S.F.		1.49	.31	1.80	Includes material and labor to install
Install	S.F.	1.65	3.33		4.98	normal weight 6″ x 8″ x 16″ block,
Demolish and Install	S.F.	1.65	4.82	.31	6.78	mortar and 1/2″ re-bar installed
Clean	S.F.	.03	.35		.38	vertically every 24″ O.C.
Paint	S.F.	.13	.63		.76	
Minimum Charge	Job		279		279	

Masonry

Masonry Wall		Unit	Material	Labor	Equip.	Total	Specification
8" x 8" x 16"							
	Demolish	S.F.		1.57	.32	1.89	Includes material and labor to install
	Install	S.F.	1.78	3.63		5.41	normal weight 8" x 8" x 16" block,
	Demolish and Install	S.F.	1.78	5.20	.32	7.30	mortar and 1/2" re-bar installed
	Clean	S.F.	.03	.35		.38	vertically every 24" O.C.
	Paint	S.F.	.13	.63		.76	
	Minimum Charge	Job		279		279	
12" x 8" x 16"							
	Demolish	S.F.		1.62	.33	1.95	Includes material and labor to install
	Install	S.F.	2.54	5.55		8.09	normal weight 12" x 8" x 16" block,
	Demolish and Install	S.F.	2.54	7.17	.33	10.04	mortar and 1/2" re-bar installed
	Clean	S.F.	.03	.35		.38	vertically every 24" O.C.
	Paint	S.F.	.13	.63		.76	
	Minimum Charge	Job		279		279	
Structural Brick							
4" Single Row							
	Demolish	S.F.		2.04		2.04	Cost includes material and labor to
	Install	S.F.	2.99	6.65		9.64	install 4" red brick wall, running bond
	Demolish and Install	S.F.	2.99	8.69		11.68	with 3/8" concave joints, mortar and
	Clean	S.F.	.03	.30		.33	ties.
	Paint	S.F.	.13	.63		.76	
	Minimum Charge	Job		279		279	
8" Double Row							
	Demolish	S.F.		2.04		2.04	Cost includes material and labor to
	Install	S.F.	4.77	10.60		15.37	install 8" red brick wall, running bond,
	Demolish and Install	S.F.	4.77	12.64		17.41	with 3/8" concave joints, mortar and
	Clean	S.F.	.03	.30		.33	wall ties.
	Paint	S.F.	.13	.63		.76	
	Minimum Charge	Job		279		279	
12" Triple Row							
	Demolish	S.F.		2.83		2.83	Cost includes material and labor to
	Install	S.F.	7.20	15.10		22.30	install 12" red brick wall, running
	Demolish and Install	S.F.	7.20	17.93		25.13	bond, with 3/8" concave joints,
	Clean	S.F.	.03	.30		.33	mortar and wall ties.
	Paint	S.F.	.13	.63		.76	
	Minimum Charge	Job		279		279	
Brick Veneer							
4" Red Common							
	Demolish	S.F.		2.04		2.04	Cost includes material and labor to
	Install	S.F.	2.27	6.25		8.52	install 4" common red face brick,
	Demolish and Install	S.F.	2.27	8.29		10.56	running bond with 3/8" concave
	Clean	S.F.	.03	.30		.33	joints, mortar and wall ties.
	Paint	S.F.	.13	.63		.76	
	Minimum Charge	Job		160		160	
4" Used							
	Demolish	S.F.		2.04		2.04	Cost includes material and labor to
	Install	S.F.	2.99	6.65		9.64	install 4" red brick wall, running bond
	Demolish and Install	S.F.	2.99	8.69		11.68	with 3/8" concave joints, mortar and
	Clean	S.F.	.03	.30		.33	ties.
	Paint	S.F.	.13	.63		.76	
	Minimum Charge	Job		160		160	

Masonry Wall		Unit	Material	Labor	Equip.	Total	Specification
6" Jumbo							
	Demolish	S.F.		2.04		2.04	Cost includes material and labor to
	Install	S.F.	3.97	3.30		7.27	install 6" red jumbo brick veneer,
	Demolish and Install	S.F.	3.97	5.34		9.31	running bond with 3/8" concave
	Clean	S.F.	.03	.30		.33	joints, mortar and ties.
	Paint	S.F.	.13	.63		.76	
	Minimum Charge	Job		160		160	
Norman (11-1/2")							
	Demolish	S.F.		2.04		2.04	Cost includes material and labor to
	Install	S.F.	4.21	4.48		8.69	install 4" norman red brick wall,
	Demolish and Install	S.F.	4.21	6.52		10.73	running bond with 3/8" concave
	Clean	S.F.	.03	.30		.33	joints, mortar and ties.
	Paint	S.F.	.13	.63		.76	
	Minimum Charge	Job		160		160	
Glazed							
	Demolish	S.F.		2.04		2.04	Includes material, labor and
	Install	S.F.	7.70	6.85		14.55	equipment to install glazed brick.
	Demolish and Install	S.F.	7.70	8.89		16.59	
	Clean	S.F.	.03	.30		.33	
	Paint	S.F.	.13	.63		.76	
	Minimum Charge	Job		160		160	
Slumpstone							
6" x 4"							
	Demolish	S.F.		1.42		1.42	Cost includes material and labor to
	Install	S.F.	4.41	3.48		7.89	install 6" x 4" x 16" concrete slump
	Demolish and Install	S.F.	4.41	4.90		9.31	block.
	Clean	S.F.	.03	.35		.38	
	Paint	S.F.	.13	.63		.76	
	Minimum Charge	Job		160		160	
6" x 6"							
	Demolish	S.F.		1.50		1.50	Cost includes material and labor to
	Install	S.F.	4.08	3.60		7.68	install 6" x 6" x 16" concrete slump
	Demolish and Install	S.F.	4.08	5.10		9.18	block.
	Clean	S.F.	.03	.35		.38	
	Paint	S.F.	.13	.63		.76	
	Minimum Charge	Job		160		160	
8" x 6"							
	Demolish	S.F.		1.50		1.50	Cost includes material and labor to
	Install	S.F.	5.35	3.72		9.07	install 8" x 6" x 16" concrete slump
	Demolish and Install	S.F.	5.35	5.22		10.57	block.
	Clean	S.F.	.03	.35		.38	
	Paint	S.F.	.13	.63		.76	
	Minimum Charge	Job		160		160	
12" x 4"							
	Demolish	S.F.		1.42		1.42	Cost includes material and labor to
	Install	S.F.	9.40	4.29		13.69	install 12" x 4" x 16" concrete slump
	Demolish and Install	S.F.	9.40	5.71		15.11	block.
	Clean	S.F.	.03	.35		.38	
	Paint	S.F.	.13	.63		.76	
	Minimum Charge	Job		160		160	
12" x 6"							
	Demolish	S.F.		1.50		1.50	Cost includes material and labor to
	Install	S.F.	8.55	4.64		13.19	install 12" x 6" x 16" concrete slump
	Demolish and Install	S.F.	8.55	6.14		14.69	block.
	Clean	S.F.	.03	.35		.38	
	Paint	S.F.	.13	.63		.76	
	Minimum Charge	Job		160		160	

Masonry Wall		Unit	Material	Labor	Equip.	Total	Specification
Stucco							
Three Coat							
	Demolish	S.Y.		1.76		1.76	Includes material and labor to install
	Install	S.Y.	3.58	6.55	.50	10.63	stucco on exterior masonry walls.
	Demolish and Install	S.Y.	3.58	8.31	.50	12.39	
	Clean	S.Y.		3.19		3.19	
	Paint	S.F.	.12	.25		.37	
	Minimum Charge	Job		139		139	
Three Coat w / Wire Mesh							
	Demolish	S.Y.		3.19		3.19	Includes material and labor to install
	Install	S.Y.	4.35	25	1.57	30.92	stucco on wire mesh over wood
	Demolish and Install	S.Y.	4.35	28.19	1.57	34.11	framing.
	Clean	S.Y.		3.19		3.19	
	Paint	S.F.	.12	.25		.37	
	Minimum Charge	Job		139		139	
Fill Cracks							
	Install	S.F.	.40	2.22		2.62	Includes labor and material to fill-in
	Minimum Charge	Job		139		139	hairline cracks up to 1/8" with filler, including sand and fill per S.F.
Furring							
1" x 2"							
	Demolish	L.F.		.10		.10	Cost includes material and labor to
	Install	L.F.	.21	.63		.84	install 1" x 2" furring on masonry wall.
	Demolish and Install	L.F.	.21	.73		.94	
	Minimum Charge	Job		157		157	
2" x 2"							
	Demolish	L.F.		.11		.11	Cost includes material and labor to
	Install	L.F.	.20	.70		.90	install 2" x 2" furring on masonry wall.
	Demolish and Install	L.F.	.20	.81		1.01	
	Minimum Charge	Job		157		157	
2" x 4"							
	Demolish	L.F.		.12		.12	Cost includes material and labor to
	Install	L.F.	.34	.79		1.13	install 2" x 4" furring on masonry wall.
	Demolish and Install	L.F.	.34	.91		1.25	
	Reinstall	L.F.		.64		.64	
	Minimum Charge	Job		157		157	
Maintenance / Cleaning							
Repoint							
	Install	S.F.	.30	3.99		4.29	Includes labor and material to cut and
	Minimum Charge	Hr.		40		40	repoint brick, hard mortar, running bond.
Acid Etch							
	Install	S.F.	.02	.57		.59	Includes labor and material to acid
	Minimum Charge	Job		590	88	678	wash smooth brick.
Pressure Wash							
	Clean	S.F.		.78	.12	.90	Includes labor and material to high
	Minimum Charge	Job		160		160	pressure water wash masonry, average.
Waterproof							
	Install	S.F.	.18	.58		.76	Includes labor and material to install
	Minimum Charge	Job		144		144	masory waterproofing.

Chimney		Unit	Material	Labor	Equip.	Total	Specification
Demo & Haul Exterior							
	Demolish	V.L.F.		23		23	Includes minimum labor and
	Minimum Charge	Job		128		128	equipment to remove masonry and haul debris to truck or dumpster.
Demo & Haul Interior							
	Demolish	V.L.F.		28.50		28.50	Includes minimum labor and
	Minimum Charge	Job		128		128	equipment to remove masonry and haul debris to truck or dumpster.
Install New (w / flue)							
Brick							
	Install	V.L.F.	26	35		61	Includes material and labor to install
	Minimum Charge	Job		160		160	brick chimney with flue from ground level up. Foundation not included.
Natural Stone							
	Install	L.F.	64	55.50		119.50	Includes labor and material to install
	Minimum Charge	Job		160		160	natural stone chimney.
Install From Roof Up							
Brick							
	Install	V.L.F.	26	35		61	Includes material and labor to install
	Minimum Charge	Job		160		160	brick chimney with flue from ground level up. Foundation not included.
Natural Stone							
	Install	L.F.	64	46.50		110.50	Includes labor and material to install
	Minimum Charge	Job		160		160	natural stone chimney.
Install Base Pad							
	Install	Ea.	98.50	172	1.21	271.71	Includes labor and material to replace
	Minimum Charge	Job		730	28.50	758.50	spread footing, 5' square x 16" deep.
Metal Cap							
	Install	Ea.	39.50	19.95		59.45	Includes labor and materials to install
	Minimum Charge	Job		157		157	bird and squirrel screens in a chimney flue.

Fireplace		Unit	Material	Labor	Equip.	Total	Specification
Common Brick							
30"							
	Demolish	Ea.		910		910	Includes material and labor to install a
	Install	Ea.	1350	2775		4125	complete unit which includes common
	Demolish and Install	Ea.	1350	3685		5035	brick, foundation, damper, flue lining
	Minimum Charge	Job		160		160	stack to 15' high and a 30" fire box.
36"							
	Demolish	Ea.		910		910	Includes material and labor to install a
	Install	Ea.	1425	3100		4525	complete unit which includes common
	Demolish and Install	Ea.	1425	4010		5435	brick, foundation, damper, flue lining
	Minimum Charge	Job		160		160	stack to 15' high and a 36" fire box.
42"							
	Demolish	Ea.		910		910	Includes material and labor to install a
	Install	Ea.	1575	3475		5050	complete unit which includes common
	Demolish and Install	Ea.	1575	4385		5960	brick, foundation, damper, flue lining
	Minimum Charge	Job		160		160	stack to 15' high and a 42" fire box.

Fireplace		Unit	Material	Labor	Equip.	Total	Specification
48"							Includes material and labor to install a complete unit which includes common brick, foundation, damper, flue lining stack to 15' high and a 48" fire box.
	Demolish	Ea.		910		910	
	Install	Ea.	1650	3975		5625	
	Demolish and Install	Ea.	1650	4885		6535	
	Minimum Charge	Job		160		160	
Raised Hearth							
Brick Hearth							Includes material and labor to install a brick fireplace hearth based on 30" x 29" opening.
	Demolish	S.F.		3.99		3.99	
	Install	S.F.	12.80	23		35.80	
	Demolish and Install	S.F.	12.80	26.99		39.79	
	Clean	S.F.	.03	.30		.33	
	Minimum Charge	Job		160		160	
Marble Hearth / Facing							Includes material and labor to install marble fireplace hearth.
	Demolish	Ea.		25.50		25.50	
	Install	Ea.	585	94.50		679.50	
	Demolish and Install	Ea.	585	120		705	
	Reinstall	Ea.		75.60		75.60	
	Clean	Ea.	22	4.59		26.59	
	Minimum Charge	Job		160		160	
Breast (face) Brick							Includes material and labor to install standard red brick, running bond.
	Demolish	S.F.		3.99		3.99	
	Install	S.F.	3.04	6.50		9.54	
	Demolish and Install	S.F.	3.04	10.49		13.53	
	Clean	S.F.		.78	.12	.90	
	Minimum Charge	Hr.		40		40	
Fire Box							Includes material and labor to install fireplace box only (110 brick).
	Demolish	Ea.		204		204	
	Install	Ea.	138	279		417	
	Demolish and Install	Ea.	138	483		621	
	Clean	Ea.	.11	10.30		10.41	
	Minimum Charge	Hr.		40		40	
Natural Stone							Includes material and labor to install high priced stone, split or rock face.
	Demolish	S.F.		3.99		3.99	
	Install	S.F.	13.05	11.95		25	
	Demolish and Install	S.F.	13.05	15.94		28.99	
	Clean	S.F.	.06	.70		.76	
	Minimum Charge	Job		279		279	
Slumpstone Veneer							Cost includes material and labor to install 8" x 6" x 16" concrete slump block.
	Demolish	S.F.		1.50		1.50	
	Install	S.F.	5.35	3.72		9.07	
	Demolish and Install	S.F.	5.35	5.22		10.57	
	Clean	S.F.	.03	.35		.38	
	Paint	S.F.	.13	.63		.76	
	Minimum Charge	Job		160		160	
Prefabricated							
Firebox							Includes material and labor to install up to 43" radiant heat, zero clearance, prefabricated fireplace. Flues, doors or blowers not included.
	Demolish	Ea.		36.50		36.50	
	Install	Ea.	1400	285		1685	
	Demolish and Install	Ea.	1400	321.50		1721.50	
	Reinstall	Ea.		228.07		228.07	
	Clean	Ea.	1.44	14.35		15.79	
	Minimum Charge	Job		157		157	

Masonry

Fireplace		Unit	Material	Labor	Equip.	Total	Specification
Flue							
	Demolish	V.L.F.		4.25		4.25	Includes material and labor to install
	Install	V.L.F.	11.15	11.85		23	prefabricated chimney vent, 8"
	Demolish and Install	V.L.F.	11.15	16.10		27.25	exposed metal flue.
	Reinstall	V.L.F.		9.50		9.50	
	Clean	L.F.	.57	2.87		3.44	
	Minimum Charge	Job		310		310	
Blower							
	Demolish	Ea.		3.40		3.40	Includes material and labor to install
	Install	Ea.	102	21.50		123.50	blower for metal fireplace flue.
	Demolish and Install	Ea.	102	24.90		126.90	
	Reinstall	Ea.		17.14		17.14	
	Clean	Ea.	2.88	28.50		31.38	
	Minimum Charge	Job		310		310	
Accessories							
Wood Mantle							
	Demolish	L.F.		4.25		4.25	Includes material and labor to install
	Install	L.F.	6.60	8.95		15.55	wood fireplace mantel.
	Demolish and Install	L.F.	6.60	13.20		19.80	
	Reinstall	L.F.		8.96		8.96	
	Clean	Ea.	.02	.63		.65	
	Paint	L.F.	.07	7.50		7.57	
	Minimum Charge	Job		157		157	
Colonial Style							
	Demolish	Ea.		39		39	Includes material and labor to install
	Install	Opng.	2350	157		2507	wood fireplace mantel.
	Demolish and Install	Opng.	2350	196		2546	
	Reinstall	Opng.		156.80		156.80	
	Clean	Ea.		.46		.46	
	Paint	Ea.	11.40	103		114.40	
	Minimum Charge	Job		157		157	
Modern Design							
	Demolish	Ea.		39		39	Includes material and labor to install
	Install	Opng.	185	62.50		247.50	wood fireplace mantel.
	Demolish and Install	Opng.	185	101.50		286.50	
	Reinstall	Opng.		62.72		62.72	
	Clean	Ea.		.46		.46	
	Paint	Ea.	11.40	103		114.40	
	Minimum Charge	Job		157		157	
Glass Door Unit							
	Demolish	Ea.		5.30		5.30	Includes material and labor to install a
	Install	Ea.	128	6.65		134.65	fireplace glass door unit.
	Demolish and Install	Ea.	128	11.95		139.95	
	Reinstall	Ea.		6.65		6.65	
	Clean	Ea.	.34	8.85		9.19	
	Minimum Charge	Job		157		157	
Fire Screen							
	Demolish	Ea.		2.66		2.66	Includes material and labor to install a
	Install	Ea.	39	3.32		42.32	fireplace fire screen.
	Demolish and Install	Ea.	39	5.98		44.98	
	Reinstall	Ea.		3.32		3.32	
	Clean	Ea.	.18	7.65		7.83	
	Minimum Charge	Job		157		157	

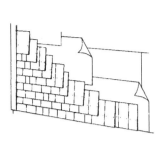

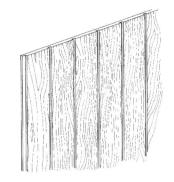

| Shingles | Board & Batten | Texture 1-11 |

Exterior Siding	Unit	Material	Labor	Equip.	Total	Specification
Trim						
1″ x 2″						
Demolish	L.F.		.21		.21	Cost includes material and labor to
Install	L.F.	.35	.95		1.30	install 1″ x 2″ pine trim.
Demolish and Install	L.F.	.35	1.16		1.51	
Clean	L.F.	.01	.15		.16	
Paint	L.F.	.02	.43		.45	
Minimum Charge	Job		157		157	
1″ x 3″						
Demolish	L.F.		.21		.21	Cost includes material and labor to
Install	L.F.	.23	1.05		1.28	install 1″ x 3″ pine trim.
Demolish and Install	L.F.	.23	1.26		1.49	
Clean	L.F.	.01	.15		.16	
Paint	L.F.	.02	.43		.45	
Minimum Charge	Job		157		157	
1″ x 4″						
Demolish	L.F.		.21		.21	Cost includes material and labor to
Install	L.F.	.73	1.25		1.98	install 1″ x 4″ pine trim.
Demolish and Install	L.F.	.73	1.46		2.19	
Clean	L.F.	.01	.15		.16	
Paint	L.F.	.02	.43		.45	
Minimum Charge	Job		157		157	
1″ x 6″						
Demolish	L.F.		.21		.21	Cost includes material and labor to
Install	L.F.	1.27	1.25		2.52	install 1″ x 6″ pine trim.
Demolish and Install	L.F.	1.27	1.46		2.73	
Clean	L.F.	.01	.15		.16	
Paint	L.F.	.07	.43		.50	
Minimum Charge	Job		157		157	
1″ x 8″						
Demolish	L.F.		.23		.23	Cost includes material and labor to
Install	L.F.	1.56	1.57		3.13	install 1″ x 8″ pine trim.
Demolish and Install	L.F.	1.56	1.80		3.36	
Clean	L.F.	.02	.16		.18	
Paint	L.F.	.07	.43		.50	
Minimum Charge	Job		157		157	
Aluminum						
8″ Non-Insulated						
Demolish	S.F.		.57		.57	Includes material and labor to install
Install	S.F.	1.35	1.24		2.59	horizontal, colored clapboard, 8″ to
Demolish and Install	S.F.	1.35	1.81		3.16	10″ wide, plain.
Reinstall	S.F.		1.24		1.24	
Clean	S.F.		.23		.23	
Minimum Charge	Job		157		157	

Siding

Exterior Siding		Unit	Material	Labor	Equip.	Total	Specification
8″ Insulated							
	Demolish	S.F.		.57		.57	Includes material and labor to install
	Install	S.F.	1.53	1.24		2.77	horizontal, colored insulated
	Demolish and Install	S.F.	1.53	1.81		3.34	clapboard, 8″ wide, plain.
	Reinstall	S.F.		1.24		1.24	
	Clean	S.F.		.23		.23	
	Minimum Charge	Job		157		157	
12″ Non-Insulated							
	Demolish	S.F.		.57		.57	Includes material and labor to install
	Install	S.F.	1.42	1.06		2.48	horizontal, colored clapboard, plain.
	Demolish and Install	S.F.	1.42	1.63		3.05	
	Reinstall	S.F.		1.06		1.06	
	Clean	S.F.		.23		.23	
	Minimum Charge	Job		157		157	
12″ Insulated							
	Demolish	S.F.		.57		.57	Includes material and labor to install
	Install	S.F.	1.67	1.06		2.73	horizontal, colored insulated
	Demolish and Install	S.F.	1.67	1.63		3.30	clapboard, plain.
	Reinstall	S.F.		1.06		1.06	
	Clean	S.F.		.23		.23	
	Minimum Charge	Job		157		157	
Clapboard 8″ to 10″							
	Demolish	S.F.		.57		.57	Includes material and labor to install
	Install	S.F.	1.35	1.24		2.59	horizontal, colored clapboard, 8″ to
	Demolish and Install	S.F.	1.35	1.81		3.16	10″ wide, plain.
	Reinstall	S.F.		1.24		1.24	
	Clean	S.F.		.23		.23	
	Minimum Charge	Job		157		157	
Corner Strips							
	Demolish	L.F.		.30		.30	Includes material and labor to install
	Install	V.L.F.	1.87	1.24		3.11	corners for horizontal lapped
	Demolish and Install	V.L.F.	1.87	1.54		3.41	aluminum siding.
	Reinstall	V.L.F.		1.24		1.24	
	Clean	L.F.		.15		.15	
	Minimum Charge	Job		157		157	
Vinyl							
Non-Insulated							
	Demolish	S.F.		.45		.45	Cost includes material and labor to
	Install	S.F.	.76	1.23		1.99	install 10″ non-insulated solid vinyl
	Demolish and Install	S.F.	.76	1.68		2.44	siding.
	Reinstall	S.F.		.98		.98	
	Clean	S.F.		.23		.23	
	Minimum Charge	Job		157		157	
Insulated							
	Demolish	S.F.		.45		.45	Includes material and labor to install
	Install	S.F.	.84	1.23		2.07	insulated solid vinyl siding panels.
	Demolish and Install	S.F.	.84	1.68		2.52	
	Reinstall	S.F.		.98		.98	
	Clean	S.F.		.23		.23	
	Minimum Charge	Job		157		157	
Corner Strips							
	Demolish	L.F.		.28		.28	Includes material and labor to install
	Install	L.F.	1.27	1.53		2.80	corner posts for vinyl siding panels.
	Demolish and Install	L.F.	1.27	1.81		3.08	
	Reinstall	L.F.		1.22		1.22	
	Clean	L.F.		.15		.15	
	Minimum Charge	Job		157		157	

Siding

Exterior Siding	Unit	Material	Labor	Equip.	Total	Specification
Hardboard						
4' x 8' Panel						
Demolish	S.F.		.47		.47	Cost includes material and labor to
Install	S.F.	1.02	.78		1.80	install 4' x 8' masonite siding 7/16"
Demolish and Install	S.F.	1.02	1.25		2.27	thick.
Clean	S.F.		.23		.23	
Paint	S.F.	.24	.52		.76	
Minimum Charge	Job		157		157	
Lap						
Demolish	S.F.		.47		.47	Cost includes material and labor to
Install	S.F.	.93	.81		1.74	install 8" or 12" masonite lap siding
Demolish and Install	S.F.	.93	1.28		2.21	7/16" thick with wood grain finish.
Clean	S.F.		.23		.23	
Paint	S.F.	.24	.52		.76	
Minimum Charge	Job		157		157	
Plywood Panel						
Douglas Fir (T-1-11)						
Demolish	S.F.		.35		.35	Includes material and labor to install
Install	S.F.	1.02	.94		1.96	Douglas fir 4' x 8' plywood exterior
Demolish and Install	S.F.	1.02	1.29		2.31	siding plain or patterns.
Reinstall	S.F.		.94		.94	
Clean	S.F.		.23		.23	
Paint	S.F.	.24	.52		.76	
Minimum Charge	Job		157		157	
Redwood						
Demolish	S.F.		.35		.35	Includes material and labor to install
Install	S.F.	1.98	.94		2.92	redwood 4' x 8' plywood exterior
Demolish and Install	S.F.	1.98	1.29		3.27	siding plain or patterns.
Reinstall	S.F.		.94		.94	
Clean	S.F.		.23		.23	
Paint	S.F.	.24	.52		.76	
Minimum Charge	Job		157		157	
Cedar						
Demolish	S.F.		.35		.35	Cost includes material and labor to
Install	S.F.	2.55	.94		3.49	install 4' x 8' cedar plywood exterior
Demolish and Install	S.F.	2.55	1.29		3.84	siding plain or patterns.
Reinstall	S.F.		.94		.94	
Clean	S.F.		.23		.23	
Paint	S.F.	.24	.52		.76	
Minimum Charge	Job		157		157	
Tongue and Groove						
Fir						
Demolish	S.F.		.64		.64	Includes material and labor to install
Install	S.F.	1.39	.84		2.23	fir tongue and groove exterior siding
Demolish and Install	S.F.	1.39	1.48		2.87	1" x 8".
Reinstall	S.F.		.67		.67	
Clean	S.F.		.23		.23	
Paint	S.F.	.24	.52		.76	
Minimum Charge	Job		157		157	
Pine						
Demolish	S.F.		.64		.64	Includes material and labor to install
Install	S.F.	.67	1.70		2.37	pine tongue and groove exterior
Demolish and Install	S.F.	.67	2.34		3.01	siding 1" x 8".
Reinstall	S.F.		1.70		1.70	
Clean	S.F.		.23		.23	
Paint	S.F.	.24	.52		.76	
Minimum Charge	Job		157		157	

Siding

Exterior Siding		Unit	Material	Labor	Equip.	Total	Specification
Cedar							
	Demolish	S.F.		.64		.64	Includes material and labor to install
	Install	S.F.	3.04	.84		3.88	cedar tongue and groove exterior
	Demolish and Install	S.F.	3.04	1.48		4.52	siding 1" x 8".
	Reinstall	S.F.		.67		.67	
	Clean	S.F.		.23		.23	
	Paint	S.F.	.24	.52		.76	
	Minimum Charge	Job		157		157	
Redwood							
	Demolish	S.F.		.64		.64	Includes material and labor to install
	Install	S.F.	2.93	2.13		5.06	redwood tongue and groove exterior
	Demolish and Install	S.F.	2.93	2.77		5.70	siding.
	Reinstall	S.F.		2.13		2.13	
	Clean	S.F.		.23		.23	
	Paint	S.F.	.24	.52		.76	
	Minimum Charge	Job		157		157	
Board and Batten							
Fir							
	Demolish	S.F.		.64		.64	Cost includes material and labor to
	Install	S.F.	1.49	.78		2.27	install 1" x 8" fir boards with 1" x 3"
	Demolish and Install	S.F.	1.49	1.42		2.91	battens.
	Reinstall	S.F.		.63		.63	
	Clean	S.F.		.23		.23	
	Paint	S.F.	.24	.52		.76	
	Minimum Charge	Job		157		157	
Pine							
	Demolish	S.F.		.64		.64	Cost includes material and labor to
	Install	S.F.	.77	.78		1.55	install 1" x 10" pine boards with 1" x
	Demolish and Install	S.F.	.77	1.42		2.19	3" battens.
	Reinstall	S.F.		.63		.63	
	Clean	S.F.		.23		.23	
	Paint	S.F.	.24	.52		.76	
	Minimum Charge	Job		157		157	
Cedar							
	Demolish	S.F.		.64		.64	Cost includes material and labor to
	Install	S.F.	2.18	.78		2.96	install 1" x 10" cedar boards with 1" x
	Demolish and Install	S.F.	2.18	1.42		3.60	3" battens.
	Reinstall	S.F.		.63		.63	
	Clean	S.F.		.23		.23	
	Paint	S.F.	.24	.52		.76	
	Minimum Charge	Job		157		157	
Redwood							
	Demolish	S.F.		.64		.64	Cost includes material and labor to
	Install	S.F.	3.95	.78		4.73	install 1" x 10" redwood boards with
	Demolish and Install	S.F.	3.95	1.42		5.37	1" x 3" battens.
	Reinstall	S.F.		.63		.63	
	Clean	S.F.		.23		.23	
	Paint	S.F.	.24	.52		.76	
	Minimum Charge	Job		157		157	
Wood Shingle							
	Demolish	Sq.		59.50		59.50	Includes material and labor to install
	Install	Sq.	130	139		269	wood shingles/shakes.
	Demolish and Install	Sq.	130	198.50		328.50	
	Clean	Sq.		24		24	
	Paint	Sq.	15.05	46		61.05	
	Minimum Charge	Job		157		157	

Siding

Exterior Siding	Unit	Material	Labor	Equip.	Total	Specification
Synthetic Stucco System						
Demolish	S.F.		2.13		2.13	Cost includes material and labor to
Install	S.F.	2.85	5.95	.45	9.25	install 2 coats of adhesive mixture,
Demolish and Install	S.F.	2.85	8.08	.45	11.38	glass fiber mesh, 1″ insulation board,
Clean	S.F.		.21		.21	and textured finish coat.
Paint	S.F.	.12	.25		.37	
Minimum Charge	Job		655	49.50	704.50	

Front-end Loader   **Bulldozer**

Grading

		Unit	Material	Labor	Equip.	Total	Specification
Hand (fine)							Includes labor and material to install fine grade for slab on grade, hand grade.
	Install	S.F.		.46		.46	
	Minimum Charge	Job		115		115	
Machine (fine)							Includes labor and material to install fine grade for roadway, base or leveling course.
	Install	S.Y.		.39	.35	.74	
	Minimum Charge	Job		268	238	506	

Asphalt Paving

		Unit	Material	Labor	Equip.	Total	Specification
Seal Coat							Includes labor and material to install sealcoat for a small area.
	Install	S.F.	.06	.08		.14	
	Minimum Charge	Job		115		115	
Wearing Course							Cost includes labor and material to install 1″ thick wearing course.
	Install	S.F.	.07	.24	.18	.49	
	Minimum Charge	Job		115		115	

Masonry Pavers

		Unit	Material	Labor	Equip.	Total	Specification
Natural Concrete							
Sand Base							Includes material and labor to install concrete pavers on sand base.
	Demolish	S.F.		2.55		2.55	
	Install	S.F.	2.05	5.30		7.35	
	Demolish and Install	S.F.	2.05	7.85		9.90	
	Reinstall	S.F.		4.26		4.26	
	Clean	S.F.		.92		.92	
	Minimum Charge	Job		160		160	
Mortar Base							Includes material and labor to install concrete pavers on mortar base.
	Demolish	S.F.		4.18	1.27	5.45	
	Install	S.F.	2.63	4.43		7.06	
	Demolish and Install	S.F.	2.63	8.61	1.27	12.51	
	Reinstall	S.F.		3.55		3.55	
	Clean	S.F.		.83		.83	
	Minimum Charge	Job		160		160	

Masonry Pavers		Unit	Material	Labor	Equip.	Total	Specification
Adobe							
Sand Base							
	Demolish	S.F.		2.55		2.55	Includes material and labor to install
	Install	S.F.	1.62	3.55		5.17	adobe pavers on sand base.
	Demolish and Install	S.F.	1.62	6.10		7.72	
	Reinstall	S.F.		2.84		2.84	
	Clean	S.F.		.92		.92	
	Minimum Charge	Job		160		160	
Mortar Base							
	Demolish	S.F.		4.18	1.27	5.45	Includes material and labor to install
	Install	S.F.	2.23	4.43		6.66	adobe pavers on mortar base.
	Demolish and Install	S.F.	2.23	8.61	1.27	12.11	
	Reinstall	S.F.		3.55		3.55	
	Clean	S.F.		.83		.83	
	Minimum Charge	Job		160		160	
Standard Brick							
Sand Base							
	Demolish	S.F.		2.55		2.55	Includes material and labor to install
	Install	S.F.	2.62	5.55		8.17	standard brick pavers on sand base.
	Demolish and Install	S.F.	2.62	8.10		10.72	
	Reinstall	S.F.		4.46		4.46	
	Clean	S.F.		.92		.92	
	Minimum Charge	Job		160		160	
Mortar Base							
	Demolish	S.F.		4.18	1.27	5.45	Includes material and labor to install
	Install	S.F.	4.77	8		12.77	standard brick pavers on mortar base.
	Demolish and Install	S.F.	4.77	12.18	1.27	18.22	
	Reinstall	S.F.		6.38		6.38	
	Clean	S.F.		.83		.83	
	Minimum Charge	Job		160		160	
Deluxe Brick							
Sand Base							
	Demolish	S.F.		2.55		2.55	Includes material and labor to install
	Install	S.F.	4.76	4.43		9.19	deluxe brick pavers on sand base.
	Demolish and Install	S.F.	4.76	6.98		11.74	
	Reinstall	S.F.		3.55		3.55	
	Clean	S.F.		.92		.92	
	Minimum Charge	Job		160		160	
Mortar Base							
	Demolish	S.F.		4.18	1.27	5.45	Includes material and labor to install
	Install	S.F.	5.15	8		13.15	deluxe brick pavers on mortar base.
	Demolish and Install	S.F.	5.15	12.18	1.27	18.60	
	Reinstall	S.F.		6.38		6.38	
	Clean	S.F.		.83		.83	
	Minimum Charge	Job		160		160	

Fencing		Unit	Material	Labor	Equip.	Total	Specification
Block w / Footing							
4"							
	Demolish	S.F.		1.42		1.42	Cost includes material and labor to
	Install	S.F.	1.02	3.33		4.35	install 4" non-reinforced concrete
	Demolish and Install	S.F.	1.02	4.75		5.77	block.
	Clean	S.F.	.13	.98	.15	1.26	
	Paint	S.F.	.17	.37		.54	
	Minimum Charge	Job		160		160	

Site Work

Fencing		Unit	Material	Labor	Equip.	Total	Specification
4″ Reinforced							Cost includes material and labor to install 4″ reinforced concrete block.
	Demolish	S.F.		1.42	.63	2.05	
	Install	S.F.	1.11	3.37		4.48	
	Demolish and Install	S.F.	1.11	4.79	.63	6.53	
	Clean	S.F.	.13	.98	.15	1.26	
	Paint	S.F.	.17	.37		.54	
	Minimum Charge	Job		160		160	
6″							Cost includes material and labor to install 6″ non-reinforced concrete block.
	Demolish	S.F.		1.50		1.50	
	Install	S.F.	1.52	3.58		5.10	
	Demolish and Install	S.F.	1.52	5.08		6.60	
	Clean	S.F.	.13	.98	.15	1.26	
	Paint	S.F.	.17	.37		.54	
	Minimum Charge	Job		160		160	
6″ Reinforced							Cost includes material and labor to install 6″ reinforced concrete block.
	Demolish	S.F.		1.49	.31	1.80	
	Install	S.F.	1.61	3.63		5.24	
	Demolish and Install	S.F.	1.61	5.12	.31	7.04	
	Clean	S.F.	.13	.98	.15	1.26	
	Paint	S.F.	.17	.37		.54	
	Minimum Charge	Job		160		160	
Cap for Block Wall							Includes labor and equipment to install precast concrete coping.
	Demolish	L.F.		1.59		1.59	
	Install	L.F.	19.30	7.45		26.75	
	Demolish and Install	L.F.	19.30	9.04		28.34	
	Clean	L.F.	.01	2.39		2.40	
	Paint	S.F.	.17	.37		.54	
	Minimum Charge	Job		160		160	
Block Pilaster							Includes material and labor to install block pilaster for fence.
	Demolish	Ea.		109		109	
	Install	Ea.	58.50	111		169.50	
	Demolish and Install	Ea.	58.50	220		278.50	
	Reinstall	Ea.		111.46		111.46	
	Paint	Ea.	3.82	12.25		16.07	
	Minimum Charge	Job		160		160	
Block w / o Footing							
4″							Cost includes material and labor to install 4″ non-reinforced hollow lightweight block.
	Demolish	S.F.		1.42		1.42	
	Install	S.F.	1.22	2.53		3.75	
	Demolish and Install	S.F.	1.22	3.95		5.17	
	Clean	S.F.	.13	.98	.15	1.26	
	Paint	S.F.	.17	.37		.54	
	Minimum Charge	Job		160		160	
4″ Reinforced							Cost includes material and labor to install 4″ reinforced hollow lightweight block.
	Demolish	S.F.		1.42	.63	2.05	
	Install	S.F.	1.36	3.30		4.66	
	Demolish and Install	S.F.	1.36	4.72	.63	6.71	
	Clean	S.F.	.13	.98	.15	1.26	
	Paint	S.F.	.17	.37		.54	
	Minimum Charge	Job		160		160	
6″							Cost includes material and labor to install 6″ non-reinforced hollow lightweight block.
	Demolish	S.F.		1.50		1.50	
	Install	S.F.	1.66	2.79		4.45	
	Demolish and Install	S.F.	1.66	4.29		5.95	
	Clean	S.F.	.13	.98	.15	1.26	
	Paint	S.F.	.17	.37		.54	
	Minimum Charge	Job		160		160	

Fencing		Unit	Material	Labor	Equip.	Total	Specification
6" Reinforced							
	Demolish	S.F.		1.49	.31	1.80	Cost includes material and labor to
	Install	S.F.	1.72	3.54		5.26	install 6" reinforced hollow lightweight
	Demolish and Install	S.F.	1.72	5.03	.31	7.06	block.
	Clean	S.F.	.13	.98	.15	1.26	
	Paint	S.F.	.17	.37		.54	
	Minimum Charge	Job		160		160	
Cap for Block Wall							
	Demolish	L.F.		1.59		1.59	Includes labor and equipment to install
	Install	L.F.	19.30	7.45		26.75	precast concrete coping.
	Demolish and Install	L.F.	19.30	9.04		28.34	
	Clean	L.F.	.01	2.39		2.40	
	Paint	S.F.	.17	.37		.54	
	Minimum Charge	Job		160		160	
Block Pilaster							
	Demolish	Ea.		109		109	Includes material and labor to install
	Install	Ea.	58.50	111		169.50	block pilaster for fence.
	Demolish and Install	Ea.	58.50	220		278.50	
	Reinstall	Ea.		111.46		111.46	
	Paint	Ea.	3.82	12.25		16.07	
	Minimum Charge	Job		160		160	
Chain Link							
3' High w / o Top Rail							
	Demolish	L.F.		1.97	.60	2.57	Includes material and labor to install
	Install	L.F.	2.45	1.36		3.81	residential galvanized fence 3' high
	Demolish and Install	L.F.	2.45	3.33	.60	6.38	with 1-3/8" steel post 10' O.C. set
	Reinstall	L.F.		1.09		1.09	without concrete.
	Clean	L.F.	.09	.83		.92	
	Minimum Charge	Job		360		360	
4' High w / o Top Rail							
	Demolish	L.F.		1.97	.60	2.57	Includes material and labor to install
	Install	L.F.	2.61	1.68		4.29	residential galvanized fence 4' high
	Demolish and Install	L.F.	2.61	3.65	.60	6.86	with 1-3/8" steel post 10' O.C. set
	Reinstall	L.F.		1.35		1.35	without concrete.
	Clean	L.F.	.11	.92		1.03	
	Minimum Charge	Job		360		360	
5' High w / o Top Rail							
	Demolish	L.F.		1.97	.60	2.57	Includes material and labor to install
	Install	L.F.	3.06	2.20		5.26	residential galvanized fence 5' high
	Demolish and Install	L.F.	3.06	4.17	.60	7.83	with 1-3/8" steel post 10' O.C. set
	Reinstall	L.F.		1.76		1.76	without concrete.
	Clean	L.F.	.14	1.02		1.16	
	Minimum Charge	Job		360		360	
3' High w / Top Rail							
	Demolish	L.F.		2.09	.63	2.72	Includes material and labor to install
	Install	L.F.	3.03	1.43		4.46	residential galvanized fence 3' high
	Demolish and Install	L.F.	3.03	3.52	.63	7.18	with 1-3/8" steel post 10' O.C. set
	Reinstall	L.F.		1.15		1.15	without concrete.
	Clean	L.F.	.09	.85		.94	
	Minimum Charge	Job		360		360	
4' High w / Top Rail							
	Demolish	L.F.		2.09	.63	2.72	Includes material and labor to install
	Install	L.F.	3.43	1.79		5.22	residential galvanized fence 4' high
	Demolish and Install	L.F.	3.43	3.88	.63	7.94	with 1-5/8" steel post 10' O.C. set
	Reinstall	L.F.		1.43		1.43	without concrete and 1-5/8" top rail
	Clean	L.F.	.11	.94		1.05	with sleeves.
	Minimum Charge	Job		360		360	

Fencing		Unit	Material	Labor	Equip.	Total	Specification
5' High w / Top Rail							Includes material and labor to install
	Demolish	L.F.		2.09	.63	2.72	residential galvanized fence 5' high
	Install	L.F.	3.89	2.39		6.28	with 1-5/8" steel post 10' O.C. set
	Demolish and Install	L.F.	3.89	4.48	.63	9	without concrete and 1-5/8" top rail
	Reinstall	L.F.		1.91		1.91	with sleeves.
	Clean	L.F.	.14	1.04		1.18	
	Minimum Charge	Job		360		360	
6' High w / Top Rail							Includes material and labor to install
	Demolish	L.F.		2.09	.63	2.72	residential galvanized fence 6' high
	Install	L.F.	4.17	3.58		7.75	with 2" steel post 10' O.C. set without
	Demolish and Install	L.F.	4.17	5.67	.63	10.47	concrete and 1-3/8" top rail with
	Reinstall	L.F.		2.86		2.86	sleeves.
	Clean	L.F.	.17	1.18		1.35	
	Minimum Charge	Job		360		360	
Barbed Wire							Cost includes material and labor to
	Demolish	L.F.		1.19		1.19	install 3 strand galvanized barbed
	Install	L.F.	.26	.46	.12	.84	wire fence.
	Demolish and Install	L.F.	.26	1.65	.12	2.03	
	Reinstall	L.F.		.37	.09	.46	
	Minimum Charge	Job		360	285	645	
6' Redwood							Cost includes material and labor to
	Demolish	L.F.		7.95		7.95	install 6' high redwood fencing.
	Install	L.F.	10.05	5.75		15.80	
	Demolish and Install	L.F.	10.05	13.70		23.75	
	Reinstall	L.F.		4.58		4.58	
	Clean	L.F.	.17	1.07		1.24	
	Paint	L.F.	.62	2.91		3.53	
	Minimum Charge	Job		360		360	
6' Red Cedar							Cost includes material and labor to
	Demolish	L.F.		7.95		7.95	install 6' cedar fence.
	Install	L.F.	15.30	5.75		21.05	
	Demolish and Install	L.F.	15.30	13.70		29	
	Reinstall	L.F.		4.58		4.58	
	Clean	L.F.	.17	1.07		1.24	
	Paint	L.F.	.62	2.91		3.53	
	Minimum Charge	Job		360		360	
6' Shadowbox							Cost includes material and labor to
	Demolish	L.F.		7.95		7.95	install 6' pressure treated pine fence.
	Install	L.F.	11.70	4.77		16.47	
	Demolish and Install	L.F.	11.70	12.72		24.42	
	Reinstall	L.F.		3.82		3.82	
	Clean	L.F.	.17	1.07		1.24	
	Paint	L.F.	.62	2.91		3.53	
	Minimum Charge	Job		360		360	
3' Picket							Includes material, labor and
	Demolish	L.F.		7.95		7.95	equipment to install pressure treated
	Install	L.F.	4.44	5.10		9.54	pine picket fence.
	Demolish and Install	L.F.	4.44	13.05		17.49	
	Reinstall	L.F.		4.09		4.09	
	Clean	L.F.	.09	.92		1.01	
	Paint	L.F.	.31	1.58		1.89	
	Minimum Charge	Job		360		360	

Fencing	Unit	Material	Labor	Equip.	Total	Specification
Wrought Iron						
Demolish	L.F.		17		17	Cost includes material and labor to
Install	L.F.	17.20	9.55		26.75	install 5' high wrought iron fence with
Demolish and Install	L.F.	17.20	26.55		43.75	pickets at 4-1/2" O.C.
Reinstall	L.F.		7.64		7.64	
Clean	L.F.		.92		.92	
Paint	L.F.	.56	3.25		3.81	
Minimum Charge	Job		360		360	
Post						
Dirt Based Wood						
Demolish	Ea.		19.90		19.90	Includes material and labor to install
Install	Ea.	10.85	89.50		100.35	wood post set in earth.
Demolish and Install	Ea.	10.85	109.40		120.25	
Reinstall	Ea.		71.59		71.59	
Clean	Ea.		2.87		2.87	
Paint	Ea.	.68	1.10		1.78	
Minimum Charge	Job		360		360	
Concrete Based Wood						
Demolish	Ea.		28	8.45	36.45	Includes material and labor to install
Install	Ea.	10.85	119		129.85	wood post set in concrete.
Demolish and Install	Ea.	10.85	147	8.45	166.30	
Clean	Ea.		2.87		2.87	
Paint	Ea.	.68	1.10		1.78	
Minimum Charge	Job		360		360	
Block Pilaster						
Demolish	Ea.		109		109	Includes material and labor to install
Install	Ea.	58.50	111		169.50	block pilaster for fence.
Demolish and Install	Ea.	58.50	220		278.50	
Reinstall	Ea.		111.46		111.46	
Paint	Ea.	3.82	12.25		16.07	
Minimum Charge	Job		160		160	
Gate						
Chain Link						
Demolish	Ea.		26.50		26.50	Cost includes material and labor to
Install	Ea.	98.50	40		138.50	install 3' wide chain link pass gate
Demolish and Install	Ea.	98.50	66.50		165	and hardware.
Reinstall	Ea.		31.82		31.82	
Clean	Ea.		3.19		3.19	
Minimum Charge	Job		360		360	
Redwood						
Demolish	Ea.		22.50		22.50	Includes material, labor and
Install	Ea.	120	119		239	equipment to install redwood pass
Demolish and Install	Ea.	120	141.50		261.50	gate and hardware.
Reinstall	Ea.		95.46		95.46	
Clean	Ea.		3.19		3.19	
Paint	Ea.	1.02	5.50		6.52	
Minimum Charge	Job		360		360	
Red Cedar						
Demolish	Ea.		22.50		22.50	Cost includes material and labor to
Install	Ea.	67	40		107	install 3' wide red cedar pass gate
Demolish and Install	Ea.	67	62.50		129.50	and hardware.
Reinstall	Ea.		31.82		31.82	
Clean	Ea.		3.19		3.19	
Paint	Ea.	1.02	5.50		6.52	
Minimum Charge	Job		360		360	

Fencing	Unit	Material	Labor	Equip.	Total	Specification
Wrought Iron						
Demolish	Ea.		17		17	Includes material, labor and
Install	Ea.	91	89.50		180.50	equipment to install wrought iron gate
Demolish and Install	Ea.	91	106.50		197.50	with hardware.
Reinstall	Ea.		71.59		71.59	
Clean	Ea.		3.19		3.19	
Paint	Ea.	1.13	5.50		6.63	
Minimum Charge	Job		360		360	

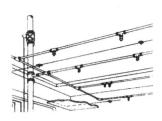

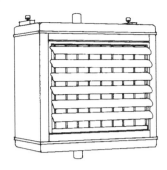

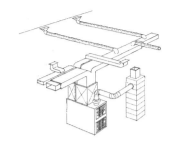

Fire-extinguishing System	Space Heater	Hot-air Furnace

Plumbing		Unit	Material	Labor	Equip.	Total	Specification
Components							
3/4" Supply							
	Demolish	L.F.		1.91		1.91	Includes material and labor to install
	Install	L.F.	1.56	4.52		6.08	type L copper tubing including
	Demolish and Install	L.F.	1.56	6.43		7.99	couplings and hangers at 10' O.C.
	Reinstall	L.F.		4.52		4.52	
	Minimum Charge	Job		172		172	
1/2" Fixture Supply							
	Install	Ea.	19.05	114		133.05	Includes labor and materials to install supply pipe only to a plumbing fixture.
2" Single Vent Stack							
	Demolish	Ea.		32		32	Includes material and labor to install
	Install	Ea.	12.05	172		184.05	PVC vent stack pipe with couplings at
	Demolish and Install	Ea.	12.05	204		216.05	10' O.C.
	Reinstall	Ea.		171.60		171.60	
	Minimum Charge	Job		172		172	
4" Drain							
	Demolish	L.F.		2.54		2.54	Includes material and labor to install
	Install	L.F.	4.96	5.60		10.56	DVW type PVC pipe with couplings at
	Demolish and Install	L.F.	4.96	8.14		13.10	10' O.C. and 3 strap hangers per 10'
	Reinstall	L.F.		5.61		5.61	run.
	Minimum Charge	Job		172		172	
1-1/2" Drain							
	Demolish	L.F.		1.91		1.91	Includes material and labor to install
	Install	L.F.	1.64	3.65		5.29	DVW type PVC pipe with couplings at
	Demolish and Install	L.F.	1.64	5.56		7.20	10' O.C. and 3 strap hangers per 10'
	Reinstall	L.F.		3.65		3.65	run.
	Minimum Charge	Job		172		172	
Sewer / Septic							
	Install	L.F.	2.48	3.20	.97	6.65	Includes labor and material to install sewerage and drainage piping, 6" diameter.
Maintenance							
Flush Out / Unclog							
	Clean	Job		172		172	Minimum labor and equipment charge
	Minimum Charge	Job		172		172	to install water cooler.

Plumbing		Unit	Material	Labor	Equip.	Total	Specification
Vent Cap							
Large							
	Demolish	Ea.		8.65		8.65	Includes labor and material to install a
	Install	Ea.	33	17.15		50.15	PVC plumbing pipe vent cap.
	Demolish and Install	Ea.	33	25.80		58.80	
	Clean	Ea.	.40	5.20		5.60	
	Paint	Ea.	.15	6.25		6.40	
	Minimum Charge	Job		86		86	
Medium							
	Demolish	Ea.		8.65		8.65	Includes material and labor to install a
	Install	Ea.	27.50	15.60		43.10	PVC plumbing pipe vent cap.
	Demolish and Install	Ea.	27.50	24.25		51.75	
	Clean	Ea.	.40	5.20		5.60	
	Paint	Ea.	.15	6.25		6.40	
	Minimum Charge	Job		86		86	
Small							
	Demolish	Ea.		8.65		8.65	Includes material and labor to install a
	Install	Ea.	20.50	14.30		34.80	PVC plumbing pipe vent cap.
	Demolish and Install	Ea.	20.50	22.95		43.45	
	Clean	Ea.	.40	5.20		5.60	
	Paint	Ea.	.15	6.25		6.40	
	Minimum Charge	Job		86		86	
Sump Pump							
1/4 HP							
	Demolish	Ea.		34.50		34.50	Includes material and labor to install a
	Install	Ea.	116	57		173	submersible sump pump.
	Demolish and Install	Ea.	116	91.50		207.50	
	Reinstall	Ea.		45.76		45.76	
	Clean	Ea.	1.01	14.35		15.36	
	Minimum Charge	Job		172		172	
1/3 HP							
	Demolish	Ea.		34.50		34.50	Includes material and labor to install a
	Install	Ea.	140	68.50		208.50	submersible sump pump.
	Demolish and Install	Ea.	140	103		243	
	Reinstall	Ea.		54.91		54.91	
	Clean	Ea.	1.01	14.35		15.36	
	Minimum Charge	Job		172		172	
1/2 HP							
	Demolish	Ea.		34.50		34.50	Includes material and labor to install a
	Install	Ea.	168	68.50		236.50	submersible sump pump.
	Demolish and Install	Ea.	168	103		271	
	Reinstall	Ea.		54.91		54.91	
	Clean	Ea.	1.01	14.35		15.36	
	Minimum Charge	Job		172		172	
Clean and Service							
	Clean	Ea.		21.50		21.50	Includes labor to clean and service a
	Minimum Charge	Job		172		172	sump pump.

Rough Mechanical

Water Heater		Unit	Material	Labor	Equip.	Total	Specification
30 Gallon							
	Demolish	Ea.		107		107	Includes material and labor to install a
	Install	Ea.	310	156		466	30 gallon electric water heater.
	Demolish and Install	Ea.	310	263		573	
	Reinstall	Ea.		124.80		124.80	
	Clean	Ea.	1.01	14.35		15.36	
	Minimum Charge	Job		172		172	
40 Gallon							
	Demolish	Ea.		107		107	Includes material and labor to install a
	Install	Ea.	330	172		502	40 gallon electric water heater.
	Demolish and Install	Ea.	330	279		609	
	Reinstall	Ea.		137.28		137.28	
	Clean	Ea.	1.01	14.35		15.36	
	Minimum Charge	Job		172		172	
52 Gallon							
	Demolish	Ea.		107		107	Includes material and labor to install a
	Install	Ea.	380	172		552	52 gallon electric water heater.
	Demolish and Install	Ea.	380	279		659	
	Reinstall	Ea.		137.28		137.28	
	Clean	Ea.	1.01	14.35		15.36	
	Minimum Charge	Job		172		172	
82 Gallon							
	Demolish	Ea.		107		107	Includes material and labor to install
	Install	Ea.	590	215		805	an 82 gallon electric water heater.
	Demolish and Install	Ea.	590	322		912	
	Reinstall	Ea.		171.60		171.60	
	Clean	Ea.	1.01	14.35		15.36	
	Minimum Charge	Job		172		172	
Gas-fired							
30 Gallon Gas-fired							
	Demolish	Ea.		132		132	Includes material and labor to install a
	Install	Ea.	425	172		597	30 gallon gas water heater.
	Demolish and Install	Ea.	425	304		729	
	Reinstall	Ea.		137.28		137.28	
	Clean	Ea.	1.01	14.35		15.36	
	Minimum Charge	Job		172		172	
30 Gallon Gas-fired, Flue							
	Install	Ea.	35	49		84	Includes labor and materials to install
	Minimum Charge	Job		172		172	a flue for a water heater (up to 50 gallon size).
40 Gallon Gas-fired							
	Demolish	Ea.		132		132	Includes material and labor to install a
	Install	Ea.	445	181		626	40 gallon gas water heater.
	Demolish and Install	Ea.	445	313		758	
	Reinstall	Ea.		144.51		144.51	
	Clean	Ea.	1.01	14.35		15.36	
	Minimum Charge	Job		172		172	
40 Gallon Gas-fired, Flue							
	Install	Ea.	35	49		84	Includes labor and materials to install
	Minimum Charge	Job		172		172	a flue for a water heater (up to 50 gallon size).
50 Gallon Gas-fired							
	Demolish	Ea.		132		132	Includes material and labor to install a
	Install	Ea.	550	191		741	50 gallon gas water heater.
	Demolish and Install	Ea.	550	323		873	
	Reinstall	Ea.		152.53		152.53	
	Clean	Ea.	1.01	14.35		15.36	
	Minimum Charge	Job		172		172	

Rough Mechanical

Water Heater	Unit	Material	Labor	Equip.	Total	Specification
50 Gallon Gas-fired, Flue						
Install	Ea.	35	49		84	Includes labor and materials to install
Minimum Charge	Job		172		172	a flue for a water heater (up to 50 gallon size).
Solar Hot Water System						
Water Heater						
Demolish	Ea.		86		86	Includes material and labor to install a
Install	Ea.	770	229		999	solar hot water heat exchanger
Demolish and Install	Ea.	770	315		1085	storage tank, 100 gallon.
Reinstall	Ea.		183.04		183.04	
Minimum Charge	Job		172		172	
Solar Panel						
Demolish	S.F.		1.25		1.25	Includes material and labor to install
Install	S.F.	18.75	2.26		21.01	solar panel.
Demolish and Install	S.F.	18.75	3.51		22.26	
Reinstall	S.F.		1.81		1.81	
Minimum Charge	Job		172		172	
Repair Plumbing						
Install	Job		172		172	Includes 4 hour minimum labor charge
Minimum Charge	Job		172		172	for a steamfitter or pipefitter.
Insulating Wrap						
Demolish	Ea.		3.97		3.97	Includes material and labor to install
Install	Ea.	71	10.75		81.75	insulating wrap around a 100 gallon
Demolish and Install	Ea.	71	14.72		85.72	solar hot water heat exchanger
Reinstall	Ea.		8.58		8.58	storage tank.
Minimum Charge	Job		172		172	
Water Softener						
Demolish	Ea.		63.50		63.50	Includes material and labor to install a
Install	Ea.	875	154		1029	14 GPM water softener.
Demolish and Install	Ea.	875	217.50		1092.50	
Reinstall	Ea.		123.52		123.52	
Clean	Ea.	1.01	14.35		15.36	
Minimum Charge	Ea.		172		172	

Ductwork	Unit	Material	Labor	Equip.	Total	Specification
Galvanized, Square or Rectangular						
6"						
Demolish	L.F.		1.59		1.59	Includes material and labor to install
Install	L.F.	2.09	14.30		16.39	rectangular ductwork with a greatest
Demolish and Install	L.F.	2.09	15.89		17.98	dimension of 6".
Reinstall	L.F.		11.43		11.43	
Clean	L.F.		1.83		1.83	
Minimum Charge	Job		171		171	
8"						
Demolish	L.F.		1.59		1.59	Includes material and labor to install
Install	L.F.	2.77	14.30		17.07	rectangular ductwork with a greatest
Demolish and Install	L.F.	2.77	15.89		18.66	dimension of 8".
Reinstall	L.F.		11.43		11.43	
Clean	L.F.		2.44		2.44	
Minimum Charge	Job		171		171	

Ductwork		Unit	Material	Labor	Equip.	Total	Specification
10"							
	Demolish	L.F.		1.59		1.59	Includes material and labor to install
	Install	L.F.	3.48	14.30		17.78	rectangular ductwork with a greatest
	Demolish and Install	L.F.	3.48	15.89		19.37	dimension of 10".
	Reinstall	L.F.		11.43		11.43	
	Clean	L.F.		3.05		3.05	
	Minimum Charge	Job		171		171	
Galvanized, Spiral							
4"							
	Demolish	L.F.		1.90		1.90	Cost includes material and labor to
	Install	L.F.	.94	1.71		2.65	install 4" diameter spiral pre-formed
	Demolish and Install	L.F.	.94	3.61		4.55	steel galvanized ductwork.
	Reinstall	L.F.		1.37		1.37	
	Clean	L.F.		.96		.96	
	Minimum Charge	Job		171		171	
6"							
	Demolish	L.F.		1.90		1.90	Cost includes material and labor to
	Install	L.F.	1.35	2.20		3.55	install 6" diameter spiral pre-formed
	Demolish and Install	L.F.	1.35	4.10		5.45	steel galvanized ductwork.
	Reinstall	L.F.		1.76		1.76	
	Clean	L.F.		1.44		1.44	
	Minimum Charge	Job		171		171	
8"							
	Demolish	L.F.		1.90		1.90	Cost includes material and labor to
	Install	L.F.	1.78	3.09		4.87	install 8" diameter spiral pre-formed
	Demolish and Install	L.F.	1.78	4.99		6.77	steel galvanized ductwork.
	Reinstall	L.F.		2.47		2.47	
	Clean	L.F.		1.91		1.91	
	Minimum Charge	Job		171		171	
10"							
	Demolish	L.F.		2.38		2.38	Cost includes material and labor to
	Install	L.F.	2.20	3.86		6.06	install 10" diameter spiral pre-formed
	Demolish and Install	L.F.	2.20	6.24		8.44	steel galvanized ductwork.
	Reinstall	L.F.		3.09		3.09	
	Clean	L.F.		2.39		2.39	
	Minimum Charge	Job		171		171	
12"							
	Demolish	L.F.		3.17		3.17	Cost includes material and labor to
	Install	L.F.	2.71	5.15		7.86	install 12" diameter spiral pre-formed
	Demolish and Install	L.F.	2.71	8.32		11.03	steel galvanized ductwork.
	Reinstall	L.F.		4.12		4.12	
	Clean	L.F.		2.87		2.87	
	Minimum Charge	Job		171		171	
16"							
	Demolish	L.F.		6.35		6.35	Cost includes material and labor to
	Install	L.F.	4.66	10.30		14.96	install 16" diameter spiral pre-formed
	Demolish and Install	L.F.	4.66	16.65		21.31	steel galvanized ductwork.
	Reinstall	L.F.		8.23		8.23	
	Clean	L.F.		3.83		3.83	
	Minimum Charge	Job		171		171	
Flexible Fiberglass							
6" Non-insulated							
	Demolish	L.F.		1.47		1.47	Cost includes material and labor to
	Install	L.F.	1.67	2.20		3.87	install 6" diameter non-insulated
	Demolish and Install	L.F.	1.67	3.67		5.34	flexible fiberglass fabric on corrosion
	Reinstall	L.F.		1.76		1.76	resistant metal ductwork.
	Clean	L.F.		1.44		1.44	
	Minimum Charge	Job		171		171	

Rough Mechanical

Ductwork	Unit	Material	Labor	Equip.	Total	Specification
8″ Non-insulated						
Demolish	L.F.		2.12		2.12	Cost includes material and labor to
Install	L.F.	2.30	3.09		5.39	install 8″ diameter non-insulated
Demolish and Install	L.F.	2.30	5.21		7.51	flexible fiberglass fabric on corrosion
Reinstall	L.F.		2.47		2.47	resistant metal ductwork.
Clean	L.F.		1.91		1.91	
Minimum Charge	Job		171		171	
10″ Non-insulated						
Demolish	L.F.		2.72		2.72	Cost includes material and labor to
Install	L.F.	2.89	3.86		6.75	install 10″ diameter non-insulated
Demolish and Install	L.F.	2.89	6.58		9.47	flexible fiberglass fabric on corrosion
Reinstall	L.F.		3.09		3.09	resistant metal ductwork.
Clean	L.F.		2.39		2.39	
Minimum Charge	Job		171		171	
12″ Non-insulated						
Demolish	L.F.		3.81		3.81	Cost includes material and labor to
Install	L.F.	3.50	5.15		8.65	install 12″ diameter non-insulated
Demolish and Install	L.F.	3.50	8.96		12.46	flexible fiberglass fabric on corrosion
Reinstall	L.F.		4.12		4.12	resistant metal ductwork.
Clean	L.F.		2.87		2.87	
Minimum Charge	Job		171		171	
4″ Insulated						
Demolish	L.F.		1.12		1.12	Cost includes material and labor to
Install	L.F.	1.31	1.82		3.13	install 4″ diameter insulated flexible
Demolish and Install	L.F.	1.31	2.94		4.25	fiberglass fabric on corrosion resistant
Reinstall	L.F.		1.45		1.45	metal ductwork.
Clean	L.F.		.96		.96	
Minimum Charge	Job		171		171	
6″ Insulated						
Demolish	L.F.		1.47		1.47	Cost includes material and labor to
Install	L.F.	1.66	2.37		4.03	install 6″ diameter insulated flexible
Demolish and Install	L.F.	1.66	3.84		5.50	fiberglass fabric on corrosion resistant
Reinstall	L.F.		1.90		1.90	metal ductwork.
Clean	L.F.		1.44		1.44	
Minimum Charge	Job		171		171	
8″ Insulated						
Demolish	L.F.		2.12		2.12	Cost includes material and labor to
Install	L.F.	2.09	3.43		5.52	install 8″ diameter insulated flexible
Demolish and Install	L.F.	2.09	5.55		7.64	fiberglass fabric on corrosion resistant
Reinstall	L.F.		2.74		2.74	metal ductwork.
Clean	L.F.		1.91		1.91	
Minimum Charge	Job		171		171	
10″ Insulated						
Demolish	L.F.		2.72		2.72	Cost includes material and labor to
Install	L.F.	2.56	4.41		6.97	install 10″ diameter insulated flexible
Demolish and Install	L.F.	2.56	7.13		9.69	fiberglass fabric on corrosion resistant
Reinstall	L.F.		3.53		3.53	metal ductwork.
Clean	L.F.		2.39		2.39	
Minimum Charge	Job		171		171	
12″ Insulated						
Demolish	L.F.		3.81		3.81	Cost includes material and labor to
Install	L.F.	3.15	6.15		9.30	install 12″ diameter insulated flexible
Demolish and Install	L.F.	3.15	9.96		13.11	fiberglass fabric on corrosion resistant
Reinstall	L.F.		4.94		4.94	metal ductwork.
Clean	L.F.		2.87		2.87	
Minimum Charge	Job		171		171	

Heating		Unit	Material	Labor	Equip.	Total	Specification
Hot Water Baseboard							
	Demolish	L.F.		5.90		5.90	Includes material and labor to install
	Install	L.F.	5.90	10.35		16.25	baseboard radiation.
	Demolish and Install	L.F.	5.90	16.25		22.15	
	Reinstall	L.F.		8.27		8.27	
	Clean	L.F.	1.01	3.59		4.60	
	Minimum Charge	Job		172		172	
3'							
	Demolish	Ea.		19.05		19.05	Includes material and labor to install
	Install	Ea.	18.70	34.50		53.20	fin tube baseboard radiation in 3'
	Demolish and Install	Ea.	18.70	53.55		72.25	sections.
	Reinstall	Ea.		27.58		27.58	
	Clean	Ea.	.29	10.95		11.24	
	Minimum Charge	Job		172		172	
4'							
	Demolish	Ea.		25.50		25.50	Includes material and labor to install
	Install	Ea.	25	39		64	fin tube baseboard radiation in 4'
	Demolish and Install	Ea.	25	64.50		89.50	sections.
	Reinstall	Ea.		31.02		31.02	
	Clean	Ea.	.57	14.35		14.92	
	Minimum Charge	Job		172		172	
5'							
	Demolish	Ea.		32		32	Includes material and labor to install
	Install	Ea.	31.50	44.50		76	fin tube baseboard radiation in 5'
	Demolish and Install	Ea.	31.50	76.50		108	sections.
	Reinstall	Ea.		35.46		35.46	
	Clean	Ea.	.57	17.65		18.22	
	Minimum Charge	Job		172		172	
6'							
	Demolish	Ea.		38		38	Includes material and labor to install
	Install	Ea.	37.50	51.50		89	fin tube baseboard radiation in 6'
	Demolish and Install	Ea.	37.50	89.50		127	sections.
	Reinstall	Ea.		41.37		41.37	
	Clean	Ea.	.57	21		21.57	
	Minimum Charge	Job		172		172	
7'							
	Demolish	Ea.		42.50		42.50	Includes material and labor to install
	Install	Ea.	43.50	68.50		112	fin tube baseboard radiation in 7'
	Demolish and Install	Ea.	43.50	111		154.50	sections.
	Reinstall	Ea.		54.91		54.91	
	Clean	Ea.	.73	25.50		26.23	
	Minimum Charge	Job		172		172	
9'							
	Demolish	Ea.		63.50		63.50	Includes material and labor to install
	Install	Ea.	56	98		154	fin tube baseboard radiation in 9'
	Demolish and Install	Ea.	56	161.50		217.50	sections.
	Reinstall	Ea.		78.45		78.45	
	Clean	Ea.	1.16	33		34.16	
	Minimum Charge	Job		172		172	
50 MBH							
	Demolish	Ea.		31		31	Includes material and labor to install a
	Install	Ea.	580	88.50		668.50	cabinet space heater with an output of
	Demolish and Install	Ea.	580	119.50		699.50	50 MBH.
	Reinstall	Ea.		70.91		70.91	
	Clean	Ea.	.73	28.50		29.23	
	Minimum Charge	Job		172		172	

Rough Mechanical

Heating

Heating	Unit	Material	Labor	Equip.	Total	Specification
75 MBH						
Demolish	Ea.		31		31	Includes material and labor to install a
Install	Ea.	645	103		748	cabinet space heater with an output of
Demolish and Install	Ea.	645	134		779	75 MBH.
Reinstall	Ea.		82.73		82.73	
Clean	Ea.	.73	28.50		29.23	
Minimum Charge	Job		172		172	
125 MBH						
Demolish	Ea.		37		37	Includes material and labor to install a
Install	Ea.	875	124		999	cabinet space heater with an output of
Demolish and Install	Ea.	875	161		1036	125 MBH.
Reinstall	Ea.		99.28		99.28	
Clean	Ea.	.73	28.50		29.23	
Minimum Charge	Job		172		172	
175 MBH						
Demolish	Ea.		46.50		46.50	Includes material and labor to install a
Install	Ea.	1100	177		1277	cabinet space heater with an output of
Demolish and Install	Ea.	1100	223.50		1323.50	175 MBH.
Reinstall	Ea.		141.82		141.82	
Clean	Ea.	.73	28.50		29.23	
Minimum Charge	Job		172		172	

Furnace Unit

Furnace Unit	Unit	Material	Labor	Equip.	Total	Specification
Wall (gas fired)						
25 MBH w / Thermostat						
Demolish	Ea.		77		77	Includes material and labor to install a
Install	Ea.	395	123		518	gas fired hot air furnace rated at 25
Demolish and Install	Ea.	395	200		595	MBH.
Reinstall	Ea.		98.76		98.76	
Clean	Ea.	1.44	64		65.44	
Minimum Charge	Job		172		172	
35 MBH w / Thermostat						
Demolish	Ea.		77		77	Includes material and labor to install a
Install	Ea.	475	137		612	gas fired hot air furnace rated at 35
Demolish and Install	Ea.	475	214		689	MBH.
Reinstall	Ea.		109.74		109.74	
Clean	Ea.	1.44	64		65.44	
Minimum Charge	Job		172		172	
50 MBH w / Thermostat						
Demolish	Ea.		77		77	Includes material and labor to install a
Install	Ea.	775	154		929	gas fired hot air furnace rated at 50
Demolish and Install	Ea.	775	231		1006	MBH.
Reinstall	Ea.		123.46		123.46	
Clean	Ea.	1.44	64		65.44	
Minimum Charge	Job		172		172	
60 MBH w / Thermostat						
Demolish	Ea.		77		77	Includes material and labor to install a
Install	Ea.	805	171		976	gas fired hot air furnace rated at 60
Demolish and Install	Ea.	805	248		1053	MBH.
Reinstall	Ea.		137.17		137.17	
Clean	Ea.	1.44	64		65.44	
Minimum Charge	Job		172		172	

Furnace Unit		Unit	Material	Labor	Equip.	Total	Specification
Floor (gas fired)							
30 MBH w / Thermostat							
	Demolish	Ea.		77		77	Includes material and labor to install a
	Install	Ea.	845	123		968	gas fired hot air floor furnace rated at
	Demolish and Install	Ea.	845	200		1045	30 MBH.
	Reinstall	Ea.		98.76		98.76	
	Clean	Ea.	1.44	64		65.44	
	Minimum Charge	Job		172		172	
50 MBH w / Thermostat							
	Demolish	Ea.		77		77	Includes material and labor to install a
	Install	Ea.	1075	154		1229	gas fired hot air floor furnace rated at
	Demolish and Install	Ea.	1075	231		1306	50 MBH.
	Reinstall	Ea.		123.46		123.46	
	Clean	Ea.	1.44	64		65.44	
	Minimum Charge	Job		172		172	
Gas Forced Air							
50 MBH							
	Demolish	Ea.		77		77	Includes material and labor to install
	Install	Ea.	755	154		909	an up flow gas fired hot air furnace
	Demolish and Install	Ea.	755	231		986	rated at 50 MBH.
	Reinstall	Ea.		123.46		123.46	
	Clean	Ea.	1.44	64		65.44	
	Minimum Charge	Job		172		172	
80 MBH							
	Demolish	Ea.		77		77	Includes material and labor to install
	Install	Ea.	830	171		1001	an up flow gas fired hot air furnace
	Demolish and Install	Ea.	830	248		1078	rated at 80 MBH.
	Reinstall	Ea.		137.17		137.17	
	Clean	Ea.	1.44	64		65.44	
	Minimum Charge	Job		172		172	
100 MBH							
	Demolish	Ea.		77		77	Includes material and labor to install
	Install	Ea.	950	193		1143	an up flow gas fired hot air furnace
	Demolish and Install	Ea.	950	270		1220	rated at 100 MBH.
	Reinstall	Ea.		154.32		154.32	
	Clean	Ea.	1.44	64		65.44	
	Minimum Charge	Job		172		172	
120 MBH							
	Demolish	Ea.		77		77	Includes material and labor to install
	Install	Ea.	1100	206		1306	an up flow gas fired hot air furnace
	Demolish and Install	Ea.	1100	283		1383	rated at 120 MBH.
	Reinstall	Ea.		164.61		164.61	
	Clean	Ea.	1.44	64		65.44	
	Minimum Charge	Job		172		172	
Electric Forced Air							
31.4 MBH							
	Demolish	Ea.		77		77	Includes material and labor to install
	Install	Ea.	840	175		1015	an electric hot air furnace rated at
	Demolish and Install	Ea.	840	252		1092	31.4 MBH.
	Reinstall	Ea.		140.27		140.27	
	Clean	Ea.	1.44	64		65.44	
	Minimum Charge	Job		172		172	

Rough Mechanical

Furnace Unit		Unit	Material	Labor	Equip.	Total	Specification
47.1 MBH							
	Demolish	Ea.		77		77	Includes material and labor to install
	Install	Ea.	935	183		1118	an electric hot air furnace rated at
	Demolish and Install	Ea.	935	260		1195	47.1 MBH.
	Reinstall	Ea.		146.79		146.79	
	Clean	Ea.	1.44	64		65.44	
	Minimum Charge	Job		172		172	
62.9 MBH							
	Demolish	Ea.		77		77	Includes material and labor to install
	Install	Ea.	1075	197		1272	an electric hot air furnace rated at
	Demolish and Install	Ea.	1075	274		1349	62.9 MBH.
	Reinstall	Ea.		157.80		157.80	
	Clean	Ea.	1.44	64		65.44	
	Minimum Charge	Job		172		172	
78.5 MBH							
	Demolish	Ea.		77		77	Includes material and labor to install
	Install	Ea.	1200	202		1402	an electric hot air furnace rated at
	Demolish and Install	Ea.	1200	279		1479	78.5 MBH.
	Reinstall	Ea.		161.85		161.85	
	Clean	Ea.	1.44	64		65.44	
	Minimum Charge	Job		172		172	
110 MBH							
	Demolish	Ea.		77		77	Includes material and labor to install
	Install	Ea.	1700	213		1913	an electric hot air furnace rated at
	Demolish and Install	Ea.	1700	290		1990	110 MBH.
	Reinstall	Ea.		170.59		170.59	
	Clean	Ea.	1.44	64		65.44	
	Minimum Charge	Job		172		172	
A/C Junction Box							
	Install	Ea.	47	86		133	Includes labor and materials to install
	Minimum Charge	Job		172		172	a heating system emergency shut-off switch. Cost includes up to 40 feet of non-metallic cable.

Boiler		Unit	Material	Labor	Equip.	Total	Specification
Gas Fired							
60 MBH							
	Demolish	Ea.		330		330	Includes material and labor to install a
	Install	Ea.	1600	1125		2725	gas fired cast iron boiler rated at 60
	Demolish and Install	Ea.	1600	1455		3055	MBH.
	Reinstall	Ea.		902.52		902.52	
	Clean	Ea.	2.88	38.50		41.38	
	Minimum Charge	Job		172		172	
Gas Fired							
125 MBH							
	Demolish	Ea.		500		500	Includes material and labor to install a
	Install	Ea.	2525	1375		3900	gas fired cast iron boiler rated at 125
	Demolish and Install	Ea.	2525	1875		4400	MBH.
	Reinstall	Ea.		1103.08		1103.08	
	Clean	Ea.	2.88	38.50		41.38	
	Minimum Charge	Job		172		172	

Rough Mechanical

Boiler		Unit	Material	Labor	Equip.	Total	Specification
Gas Fired							
400 MBH							
	Demolish	Ea.		500		500	Includes material and labor to install a
	Install	Ea.	3225	1950		5175	gas fired cast iron boiler rated at 400
	Demolish and Install	Ea.	3225	2450		5675	MBH.
	Reinstall	Ea.		1939		1939	
	Clean	Ea.	2.88	38.50		41.38	
	Minimum Charge	Job		172		172	

Heat Pump		Unit	Material	Labor	Equip.	Total	Specification
2 Ton							
	Demolish	Ea.		81.50		81.50	Includes material and labor to install a
	Install	Ea.	1075	515		1590	2 ton heat pump.
	Demolish and Install	Ea.	1075	596.50		1671.50	
	Reinstall	Ea.		413.65		413.65	
	Clean	Ea.	2.88	28.50		31.38	
	Minimum Charge	Job		172		172	
3 Ton							
	Demolish	Ea.		154		154	Includes material and labor to install a
	Install	Ea.	1325	775		2100	3 ton heat pump.
	Demolish and Install	Ea.	1325	929		2254	
	Reinstall	Ea.		620.48		620.48	
	Clean	Ea.	2.88	28.50		31.38	
	Minimum Charge	Job		172		172	
4 Ton							
	Demolish	Ea.		280		280	Includes material and labor to install a
	Install	Ea.	1625	1025		2650	4 ton heat pump.
	Demolish and Install	Ea.	1625	1305		2930	
	Reinstall	Ea.		827.31		827.31	
	Clean	Ea.	2.88	28.50		31.38	
	Minimum Charge	Job		172		172	
5 Ton							
	Demolish	Ea.		410		410	Includes material and labor to install a
	Install	Ea.	1875	1250		3125	5 ton heat pump.
	Demolish and Install	Ea.	1875	1660		3535	
	Reinstall	Ea.		992.77		992.77	
	Clean	Ea.	2.88	28.50		31.38	
	Minimum Charge	Job		172		172	
Thermostat							
	Demolish	Ea.		9.60		9.60	Includes material and labor to install
	Install	Ea.	196	43		239	an electric thermostat, 2 set back.
	Demolish and Install	Ea.	196	52.60		248.60	
	Reinstall	Ea.		34.28		34.28	
	Clean	Ea.	.06	7.20		7.26	
	Minimum Charge	Job		172		172	
A/C Junction Box							
	Install	Ea.	360	264		624	Includes labor and materials to install
	Minimum Charge	Job		172		172	a heat pump circuit including up to 40
							feet of non-metallic cable.

Heating

Heating		Unit	Material	Labor	Equip.	Total	Specification
Electric Baseboard							
	Demolish	L.F.		4.77		4.77	Includes material and labor to install
	Install	L.F.	10.75	11.45		22.20	electric baseboard heating elements.
	Demolish and Install	L.F.	10.75	16.22		26.97	
	Reinstall	L.F.		9.16		9.16	
	Clean	L.F.	1.01	3.59		4.60	
	Minimum Charge	Job		172		172	
2' 6"							
	Demolish	Ea.		15.90		15.90	Includes material and labor to install a
	Install	Ea.	32	43		75	2' 6" long section of electric
	Demolish and Install	Ea.	32	58.90		90.90	baseboard heating elements.
	Reinstall	Ea.		34.36		34.36	
	Clean	Ea.	.29	8.85		9.14	
	Minimum Charge	Job		172		172	
3'							
	Demolish	Ea.		17.35		17.35	Includes material and labor to install a
	Install	Ea.	38	43		81	3' long section of electric baseboard
	Demolish and Install	Ea.	38	60.35		98.35	heating elements.
	Reinstall	Ea.		34.36		34.36	
	Clean	Ea.	.29	10.95		11.24	
	Minimum Charge	Job		172		172	
4'							
	Demolish	Ea.		19.10		19.10	Includes material and labor to install a
	Install	Ea.	45	51.50		96.50	4' long section of electric baseboard
	Demolish and Install	Ea.	45	70.60		115.60	heating elements.
	Reinstall	Ea.		41.03		41.03	
	Clean	Ea.	.11	2.39		2.50	
	Minimum Charge	Job		172		172	
5'							
	Demolish	Ea.		21		21	Includes material and labor to install a
	Install	Ea.	53.50	60.50		114	5' long section of electric baseboard
	Demolish and Install	Ea.	53.50	81.50		135	heating elements.
	Reinstall	Ea.		48.22		48.22	
	Clean	Ea.	.57	17.65		18.22	
	Minimum Charge	Job		172		172	
6'							
	Demolish	Ea.		24		24	Includes material and labor to install a
	Install	Ea.	59.50	68.50		128	6' long section of electric baseboard
	Demolish and Install	Ea.	59.50	92.50		152	heating elements.
	Reinstall	Ea.		54.98		54.98	
	Clean	Ea.	.11	2.87		2.98	
	Minimum Charge	Job		172		172	
8'							
	Demolish	Ea.		32		32	Includes material and labor to install
	Install	Ea.	75	86		161	an 8' long section of electric
	Demolish and Install	Ea.	75	118		193	baseboard heating elements.
	Reinstall	Ea.		68.72		68.72	
	Clean	Ea.	.11	3.59		3.70	
	Minimum Charge	Job		172		172	
10'							
	Demolish	Ea.		38		38	Includes material and labor to install a
	Install	Ea.	93	104		197	10' long section of electric baseboard
	Demolish and Install	Ea.	93	142		235	heating elements.
	Reinstall	Ea.		83.30		83.30	
	Clean	Ea.	1.16	36		37.16	
	Minimum Charge	Job		172		172	

Rough Mechanical

Heating	Unit	Material	Labor	Equip.	Total	Specification
Electric Radiant Wall Panel						
Demolish	Ea.		25.50		25.50	Includes material and labor to install
Install	Ea.	390	68.50		458.50	an 18″ x 23″ electric radiant heat wall
Demolish and Install	Ea.	390	94		484	panel.
Reinstall	Ea.		54.98		54.98	
Clean	Ea.	.14	7.20		7.34	
Minimum Charge	Job		172		172	

Heating Control	Unit	Material	Labor	Equip.	Total	Specification
Thermostat						
Demolish	Ea.		9.60		9.60	Includes material and labor to install
Install	Ea.	196	43		239	an electric thermostat, 2 set back.
Demolish and Install	Ea.	196	52.60		248.60	
Reinstall	Ea.		34.28		34.28	
Clean	Ea.	.06	7.20		7.26	
Minimum Charge	Job		172		172	

Air Conditioning	Unit	Material	Labor	Equip.	Total	Specification
2 Ton						
Demolish	Ea.		320		320	Includes material and labor to install a
Install	Ea.	1325	535		1860	rooftop air conditioning unit, 2 ton.
Demolish and Install	Ea.	1325	855		2180	No hoisting is included.
Reinstall	Ea.		535.36		535.36	
Clean	Ea.	2.88	28.50		31.38	
Minimum Charge	Job		172		172	
3 Ton						
Demolish	Ea.		320		320	Includes material and labor to install a
Install	Ea.	1425	705		2130	rooftop air conditioning unit, 3 ton.
Demolish and Install	Ea.	1425	1025		2450	No hoisting is included.
Reinstall	Ea.		705.89		705.89	
Clean	Ea.	2.88	28.50		31.38	
Minimum Charge	Job		172		172	
4 Ton						
Demolish	Ea.		320		320	Includes material and labor to install a
Install	Ea.	1800	820		2620	rooftop air conditioning unit, 4 ton.
Demolish and Install	Ea.	1800	1140		2940	No hoisting is included.
Reinstall	Ea.		818.58		818.58	
Clean	Ea.	2.88	28.50		31.38	
Minimum Charge	Job		172		172	
5 Ton						
Demolish	Ea.		320		320	Includes material and labor to install a
Install	Ea.	2000	885		2885	rooftop air conditioning unit, 5 ton.
Demolish and Install	Ea.	2000	1205		3205	No hoisting is included.
Reinstall	Ea.		883.87		883.87	
Clean	Ea.	2.88	28.50		31.38	
Minimum Charge	Job		172		172	
Compressor						
2 Ton						
Demolish	Ea.		287		287	Includes material and labor to install a
Install	Ea.	535	172		707	refrigeration compressor.
Demolish and Install	Ea.	535	459		994	
Reinstall	Ea.		137.92		137.92	
Clean	Ea.	2.88	28.50		31.38	
Minimum Charge	Job		172		172	

Air Conditioning		Unit	Material	Labor	Equip.	Total	Specification
2.5 Ton							
	Demolish	Ea.		287		287	Includes material and labor to install a
	Install	Ea.	635	187		822	refrigeration compressor.
	Demolish and Install	Ea.	635	474		1109	
	Reinstall	Ea.		149.43		149.43	
	Clean	Ea.	2.88	28.50		31.38	
	Minimum Charge	Job		172		172	
3 Ton							
	Demolish	Ea.		287		287	Includes material and labor to install a
	Install	Ea.	670	195		865	refrigeration compressor.
	Demolish and Install	Ea.	670	482		1152	
	Reinstall	Ea.		155.93		155.93	
	Clean	Ea.	2.88	28.50		31.38	
	Minimum Charge	Job		172		172	
4 Ton							
	Demolish	Ea.		287		287	Includes material and labor to install a
	Install	Ea.	680	157		837	refrigeration compressor.
	Demolish and Install	Ea.	680	444		1124	
	Reinstall	Ea.		125.38		125.38	
	Clean	Ea.	2.88	28.50		31.38	
	Minimum Charge	Job		172		172	
5 Ton							
	Demolish	Ea.		287		287	Includes material and labor to install a
	Install	Ea.	840	214		1054	refrigeration compressor.
	Demolish and Install	Ea.	840	501		1341	
	Reinstall	Ea.		170.80		170.80	
	Clean	Ea.	2.88	28.50		31.38	
	Minimum Charge	Job		172		172	
Condensing Unit							
2 Ton Air-cooled							
	Demolish	Ea.		82		82	Includes material and labor to install a
	Install	Ea.	645	295		940	2 ton condensing unit.
	Demolish and Install	Ea.	645	377		1022	
	Reinstall	Ea.		236.37		236.37	
	Clean	Ea.	2.88	28.50		31.38	
	Minimum Charge	Job		172		172	
2.5 Ton Air-cooled							
	Demolish	Ea.		96		96	Includes material and labor to install a
	Install	Ea.	765	365		1130	2.5 ton condensing unit.
	Demolish and Install	Ea.	765	461		1226	
	Reinstall	Ea.		291.99		291.99	
	Clean	Ea.	2.88	28.50		31.38	
	Minimum Charge	Job		172		172	
3 Ton Air-cooled							
	Demolish	Ea.		115		115	Includes material and labor to install a
	Install	Ea.	910	475		1385	3 ton condensing unit.
	Demolish and Install	Ea.	910	590		1500	
	Reinstall	Ea.		381.83		381.83	
	Clean	Ea.	2.88	28.50		31.38	
	Minimum Charge	Job		172		172	
4 Ton Air-cooled							
	Demolish	Ea.		172		172	Includes material and labor to install a
	Install	Ea.	1175	690		1865	4 ton condensing unit.
	Demolish and Install	Ea.	1175	862		2037	
	Reinstall	Ea.		551.54		551.54	
	Clean	Ea.	2.88	28.50		31.38	
	Minimum Charge	Job		172		172	

Rough Mechanical

Air Conditioning	Unit	Material	Labor	Equip.	Total	Specification
5 Ton Air-cooled						
Demolish	Ea.		259		259	Includes material and labor to install a
Install	Ea.	1425	1025		2450	5 ton condensing unit.
Demolish and Install	Ea.	1425	1284		2709	
Reinstall	Ea.		827.31		827.31	
Clean	Ea.	2.88	28.50		31.38	
Minimum Charge	Job		172		172	
Air Handler						
1200 CFM						
Demolish	Ea.		207		207	Includes material and labor to install a
Install	Ea.	2000	475		2475	1200 CFM capacity air handling unit.
Demolish and Install	Ea.	2000	682		2682	
Reinstall	Ea.		381.83		381.83	
Clean	Ea.	2.88	28.50		31.38	
Minimum Charge	Job		172		172	
2000 CFM						
Demolish	Ea.		207		207	Includes material and labor to install a
Install	Ea.	3350	565		3915	2000 CFM capacity air handling unit.
Demolish and Install	Ea.	3350	772		4122	
Reinstall	Ea.		451.26		451.26	
Clean	Ea.	2.88	28.50		31.38	
Minimum Charge	Job		172		172	
3000 CFM						
Demolish	Ea.		230		230	Includes material and labor to install a
Install	Ea.	4200	620		4820	3000 CFM capacity air handling unit.
Demolish and Install	Ea.	4200	850		5050	
Reinstall	Ea.		496.38		496.38	
Clean	Ea.	2.88	28.50		31.38	
Minimum Charge	Job		172		172	
Evaporative Cooler						
4,300 CFM, 1/3 HP						
Demolish	Ea.		84.50		84.50	Includes material and labor to install
Install	Ea.	655	193		848	an evaporative cooler.
Demolish and Install	Ea.	655	277.50		932.50	
Reinstall	Ea.		154.32		154.32	
Clean	Ea.	1.16	28.50		29.66	
Minimum Charge	Job		172		172	
4,700 CFM, 1/2 HP						
Demolish	Ea.		84.50		84.50	Includes material and labor to install
Install	Ea.	620	199		819	an evaporative cooler.
Demolish and Install	Ea.	620	283.50		903.50	
Reinstall	Ea.		159.30		159.30	
Clean	Ea.	1.16	28.50		29.66	
Minimum Charge	Job		172		172	
5,600 CFM, 3/4 HP						
Demolish	Ea.		84.50		84.50	Includes material and labor to install
Install	Ea.	1075	206		1281	an evaporative cooler.
Demolish and Install	Ea.	1075	290.50		1365.50	
Reinstall	Ea.		164.61		164.61	
Clean	Ea.	1.16	28.50		29.66	
Minimum Charge	Job		172		172	
Master Cool Brand						
Demolish	Ea.		84.50		84.50	Includes material and labor to install
Install	Ea.	655	213		868	an evaporative cooler.
Demolish and Install	Ea.	655	297.50		952.50	
Reinstall	Ea.		170.28		170.28	
Clean	Ea.	1.16	28.50		29.66	
Minimum Charge	Job		172		172	

Rough Mechanical

Air Conditioning	Unit	Material	Labor	Equip.	Total	Specification
Motor						
Demolish	Ea.		47.50		47.50	Includes material and labor to install
Install	Ea.	222	150		372	an evaporative cooler motor.
Demolish and Install	Ea.	222	197.50		419.50	
Reinstall	Ea.		119.97		119.97	
Clean	Ea.	1.16	28.50		29.66	
Minimum Charge	Job		172		172	
Water Pump						
Demolish	Ea.		47.50		47.50	Includes material and labor to install
Install	Ea.	42	85.50		127.50	an evaporative cooler water pump.
Demolish and Install	Ea.	42	133		175	
Reinstall	Ea.		68.56		68.56	
Clean	Ea.	1.16	28.50		29.66	
Minimum Charge	Job		172		172	
Fan Belt						
Demolish	Ea.		9.55		9.55	Includes material and labor to install
Install	Ea.	27.50	17.15		44.65	an evaporative cooler fan belt.
Demolish and Install	Ea.	27.50	26.70		54.20	
Reinstall	Ea.		13.71		13.71	
Minimum Charge	Job		172		172	
Pads (4 ea.)						
Demolish	Ea.		9.55		9.55	Includes material and labor to install
Install	Ea.	29.50	43		72.50	pre-fabricated roof pads for an
Demolish and Install	Ea.	29.50	52.55		82.05	evaporative cooler.
Reinstall	Ea.		34.28		34.28	
Clean	Ea.	1.16	28.50		29.66	
Minimum Charge	Job		172		172	
Service call						
Install	Job		171		171	Includes 4 hour minimum labor charge for an sheetmetal worker.
Window Mounted						
Low Output BTU						
Install	Ea.	390	39		429	Includes material and labor to install a
Reinstall	Ea.		31.36		31.36	window air conditioning unit.
Clean	Ea.	.43	7.20		7.63	
Minimum Charge	Job		172		172	
Mid-range Output BTU						
Demolish	Ea.		27.50		27.50	Includes material and labor to install a
Install	Ea.	450	52.50		502.50	window air conditioning unit.
Demolish and Install	Ea.	450	80		530	
Reinstall	Ea.		41.81		41.81	
Clean	Ea.	.43	7.20		7.63	
Minimum Charge	Job		172		172	
High Output BTU						
Demolish	Ea.		35		35	Includes material and labor to install a
Install	Ea.	535	52.50		587.50	window air conditioning unit.
Demolish and Install	Ea.	535	87.50		622.50	
Reinstall	Ea.		41.81		41.81	
Clean	Ea.	.43	7.20		7.63	
Minimum Charge	Job		172		172	
Hotel-type Hot / Cold						
Demolish	Ea.		58		58	Includes material and labor to install
Install	Ea.	1425	207		1632	packaged air conditioning unit,
Demolish and Install	Ea.	1425	265		1690	15000 BTU cooling, 13900 BTU
Reinstall	Ea.		165.46		165.46	heating.
Clean	Ea.	.57	14.35		14.92	
Minimum Charge	Job		172		172	

Rough Mechanical

Air Conditioning

Air Conditioning	Unit	Material	Labor	Equip.	Total	Specification
Thermostat						
Demolish	Ea.		9.60		9.60	Includes material and labor to install
Install	Ea.	196	43		239	an electric thermostat, 2 set back.
Demolish and Install	Ea.	196	52.60		248.60	
Reinstall	Ea.		34.28		34.28	
Clean	Ea.	.06	7.20		7.26	
Minimum Charge	Job		172		172	
Concrete Pad						
Demolish	S.F.		1.59		1.59	Includes material and labor to install
Install	S.F.	.99	3.57		4.56	air conditioner pad.
Demolish and Install	S.F.	.99	5.16		6.15	
Reinstall	S.F.		2.85		2.85	
Clean	S.F.	.06	.46		.52	
Paint	S.F.	.12	.28		.40	
Minimum Charge	Job		172		172	
A/C Junction Box						
Install	Ea.	156	98		254	Includes labor and materials to install
Minimum Charge	Job		172		172	an air conditioning circuit including up to 40 feet of non-metallic cable.

Sprinkler System

Sprinkler System	Unit	Material	Labor	Equip.	Total	Specification
Exposed						
Wet Type						
Demolish	SF Flr.		.15		.15	Includes material and labor to install
Install	SF Flr.	.86	1.50		2.36	an exposed wet pipe fire sprinkler
Demolish and Install	SF Flr.	.86	1.65		2.51	system.
Reinstall	SF Flr.		1.50		1.50	
Minimum Charge	Job		172		172	
Dry Type						
Demolish	SF Flr.		.15		.15	Includes material and labor to install
Install	SF Flr.	.98	1.50		2.48	an exposed dry pipe fire sprinkler
Demolish and Install	SF Flr.	.98	1.65		2.63	system.
Reinstall	SF Flr.		1.50		1.50	
Minimum Charge	Job		172		172	
Concealed						
Wet Type						
Demolish	SF Flr.		.31		.31	Includes material and labor to install a
Install	SF Flr.	.86	2.05		2.91	concealed wet pipe fire sprinkler
Demolish and Install	SF Flr.	.86	2.36		3.22	system.
Reinstall	SF Flr.		2.05		2.05	
Minimum Charge	Job		172		172	
Dry Type						
Demolish	SF Flr.		.31		.31	Includes material and labor to install a
Install	SF Flr.	.98	2.05		3.03	concealed dry pipe fire sprinkler
Demolish and Install	SF Flr.	.98	2.36		3.34	system.
Reinstall	SF Flr.		2.05		2.05	
Minimum Charge	Job		172		172	

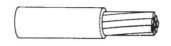

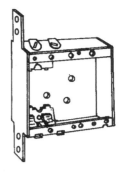

Junction Boxes	Cable	Switch Box

Electrical Per S.F.	Unit	Material	Labor	Equip.	Total	Specification
Residential						
Wiring, Lighting and Fixtures						
Install	S.F.	1.18	1.43		2.61	Includes labor and material to replace
Minimum Charge	Job		172		172	house wiring, lighting fixtures, and
						built-in appliances. Apply S.F. cost to
						floor area including garages but not
						basements.
Wiring Only						
Install	S.F.	.11	.56		.67	Includes labor and material to replace
						house wiring only. Apply S.F. cost to
						floor area including garages but not
						basements.

Main Service	Unit	Material	Labor	Equip.	Total	Specification
100 Amp.						
Demolish	Ea.		141		141	Cost includes material and labor to
Install	Ea.	450	289		739	install 100 Amp. residential service
Demolish and Install	Ea.	450	430		880	panel including 12 breaker interior
Reinstall	Ea.		230.99		230.99	panel, 10 single pole breakers and
Clean	Ea.	.29	57.50		57.79	hook-up. Wiring is not included.
Minimum Charge	Job		172		172	
150 Amp.						
Demolish	Ea.		141		141	Cost includes material and labor to
Install	Ea.	710	335		1045	install 150 Amp. residential service,
Demolish and Install	Ea.	710	476		1186	incl. 20 breaker exterior panel, main
Reinstall	Ea.		266.87		266.87	switch, 1 GFI and 14 single pole
Clean	Ea.	.29	57.50		57.79	breakers, meter socket including
Minimum Charge	Job		172		172	hookup. Wiring is not included.
200 Amp.						
Demolish	Ea.		141		141	Cost includes material and labor to
Install	Ea.	925	380		1305	install 200 Amp. residential service,
Demolish and Install	Ea.	925	521		1446	incl. 20 breaker exterior panel, main
Reinstall	Ea.		305.42		305.42	switch, 1 GFI and 18 single pole
Clean	Ea.	.29	57.50		57.79	breakers, meter socket including
Minimum Charge	Job		172		172	hookup. Wiring is not included.

Rough Electrical

Main Service	Unit	Material	Labor	Equip.	Total	Specification
Masthead						
Demolish	Ea.		24		24	Includes material and labor to install
Install	Ea.	78.50	143		221.50	galvanized steel conduit including
Demolish and Install	Ea.	78.50	167		245.50	masthead.
Reinstall	Ea.		114.53		114.53	
Minimum Charge	Job		172		172	
Allowance to Repair						
Install	Job		172		172	Minimum labor charge for residential
Minimum Charge	Job		172		172	wiring.

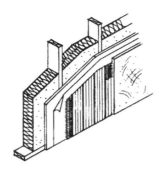

Wall Insulation

Floor Insulation		Unit	Material	Labor	Equip.	Total	Specification
Kraft Faced Batt							
3-1/2″ (R-11)							
	Demolish	S.F.		.11		.11	Cost includes material and labor to
	Install	S.F.	.31	.45		.76	install 3-1/2″ (R-11) kraft paper or foil
	Demolish and Install	S.F.	.31	.56		.87	faced roll and batt fiberglass
	Minimum Charge	Job		157		157	insulation, joists 16″ O.C.
6″ (R-19)							
	Demolish	S.F.		.11		.11	Cost includes material and labor to
	Install	S.F.	.40	.52		.92	install 6″ (R-19) kraft paper or foil
	Demolish and Install	S.F.	.40	.63		1.03	faced roll and batt fiberglass
	Minimum Charge	Job		157		157	insulation, joists 16″ O.C.
10″ (R-30)							
	Demolish	S.F.		.11		.11	Cost includes material and labor to
	Install	S.F.	.75	.63		1.38	install 10″ (R-30) kraft paper-faced roll
	Demolish and Install	S.F.	.75	.74		1.49	and batt fiberglass insulation, joists
	Minimum Charge	Job		157		157	16″ O.C.
12″ (R-38)							
	Demolish	S.F.		.11		.11	Cost includes material and labor to
	Install	S.F.	.80	.66		1.46	install 12″ (R-38) kraft paper-faced roll
	Demolish and Install	S.F.	.80	.77		1.57	and batt fiberglass insulation, joists
	Minimum Charge	Job		157		157	16″ O.C.
Foil Faced Batt							
3-1/2″ (R-11)							
	Demolish	S.F.		.11		.11	Cost includes material and labor to
	Install	S.F.	.31	.45		.76	install 3-1/2″ (R-11) kraft paper or foil
	Demolish and Install	S.F.	.31	.56		.87	faced roll and batt fiberglass
	Minimum Charge	Job		157		157	insulation, joists 16″ O.C.
6″ (R-19)							
	Demolish	S.F.		.11		.11	Cost includes material and labor to
	Install	S.F.	.40	.52		.92	install 6″ (R-19) kraft paper or foil
	Demolish and Install	S.F.	.40	.63		1.03	faced roll and batt fiberglass
	Minimum Charge	Job		157		157	insulation, joists 16″ O.C.
Unfaced Batt							
3-1/2″ (R-11)							
	Demolish	S.F.		.11		.11	Cost includes material and labor to
	Install	S.F.	.23	.52		.75	install 3-1/2″ (R-11) unfaced roll and
	Demolish and Install	S.F.	.23	.63		.86	batt fiberglass insulation, joists 16″
	Minimum Charge	Job		157		157	O.C.

Insulation

Floor Insulation		Unit	Material	Labor	Equip.	Total	Specification
6″ (R-19)							
	Demolish	S.F.		.11		.11	Cost includes material and labor to
	Install	S.F.	.40	.63		1.03	install 6″ (R-19) unfaced roll and batt
	Demolish and Install	S.F.	.40	.74		1.14	fiberglass insulation, joists 16″ O.C.
	Minimum Charge	Job		157		157	
10″ (R-30)							
	Demolish	S.F.		.11		.11	Cost includes material and labor to
	Install	S.F.	.75	.70		1.45	install 10″ (R-30) unfaced roll and batt
	Demolish and Install	S.F.	.75	.81		1.56	fiberglass insulation, joists 16″ O.C.
	Minimum Charge	Job		157		157	
12″ (R-38)							
	Demolish	S.F.		.11		.11	Cost includes material and labor to
	Install	S.F.	.84	.74		1.58	install 12″ (R-38) unfaced roll and batt
	Demolish and Install	S.F.	.84	.85		1.69	fiberglass insulation, joists 16″ O.C.
	Minimum Charge	Job		157		157	
Rigid Board							
1/2″							
	Demolish	S.F.		.11		.11	Includes material and labor to install
	Install	S.F.	.32	.39		.71	rigid foam insulation board foil faced
	Demolish and Install	S.F.	.32	.50		.82	two sides.
	Minimum Charge	Job		157		157	
3/4″							
	Demolish	S.F.		.11		.11	Includes material and labor to install
	Install	S.F.	.26	.39		.65	rigid foam insulation board foil faced
	Demolish and Install	S.F.	.26	.50		.76	two sides.
	Minimum Charge	Job		157		157	
1-1/2″							
	Demolish	S.F.		.11		.11	Includes material and labor to install
	Install	S.F.	.39	.42		.81	rigid foam insulation board foil faced
	Demolish and Install	S.F.	.39	.53		.92	two sides.
	Minimum Charge	Job		157		157	
Support Wire							
	Demolish	S.F.		.04		.04	Includes material and labor to install
	Install	S.F.	.01	.09		.10	friction fit support wire.
	Demolish and Install	S.F.	.01	.13		.14	
	Minimum Charge	Job		157		157	
Vapor Barrier							
	Demolish	S.F.		.06		.06	Cost includes material and labor to
	Install	S.F.	.02	.08		.10	install 4 mil polyethylene vapor barrier
	Demolish and Install	S.F.	.02	.14		.16	10′ wide sheets with 6″ overlaps.
	Minimum Charge	Job		157		157	

Wall Insulation		Unit	Material	Labor	Equip.	Total	Specification
Kraft Faced Batt							
3-1/2″ (R-11)							
	Demolish	S.F.		.10		.10	Cost includes material and labor to
	Install	S.F.	.25	.20		.45	install 3-1/2″ (R-11) kraft paper-faced
	Demolish and Install	S.F.	.25	.30		.55	roll and batt fiberglass insulation, joist
	Minimum Charge	Job		157		157	or studs 16″ O.C.
6″ (R-19)							
	Demolish	S.F.		.10		.10	Cost includes material and labor to
	Install	S.F.	.34	.23		.57	install 6″ (R-19) kraft paper-faced roll
	Demolish and Install	S.F.	.34	.33		.67	and batt fiberglass insulation, joist or
	Minimum Charge	Job		157		157	studs 16″ O.C.

Insulation

Wall Insulation		Unit	Material	Labor	Equip.	Total	Specification
10″ (R-30)							
	Demolish	S.F.		.10		.10	Cost includes material and labor to
	Install	S.F.	.66	.27		.93	install 10″ (R-30) kraft paper-faced roll
	Demolish and Install	S.F.	.66	.37		1.03	and batt fiberglass insulation, joist or
	Minimum Charge	Job		157		157	studs 16″ O.C.
12″ (R-38)							
	Demolish	S.F.		.10		.10	Cost includes material and labor to
	Install	S.F.	.84	.31		1.15	install 12″ (R-38) kraft paper-faced roll
	Demolish and Install	S.F.	.84	.41		1.25	and batt fiberglass insulation, joist or
	Minimum Charge	Job		157		157	studs 16″ O.C.
Foil Faced Batt							
3-1/2″ (R-11)							
	Demolish	S.F.		.10		.10	Cost includes material and labor to
	Install	S.F.	.37	.20		.57	install 3-1/2″ (R-11) foil faced roll and
	Demolish and Install	S.F.	.37	.30		.67	batt fiberglass insulation, joist or studs
	Minimum Charge	Job		157		157	16″ O.C.
6″ (R-19)							
	Demolish	S.F.		.10		.10	Cost includes material and labor to
	Install	S.F.	.45	.23		.68	install 6″ (R-19) foil faced roll and batt
	Demolish and Install	S.F.	.45	.33		.78	fiberglass insulation, joist or studs 16″
	Minimum Charge	Job		157		157	O.C.
Unfaced Batt							
3-1/2″ (R-11)							
	Demolish	S.F.		.10		.10	Cost includes material and labor to
	Install	S.F.	.23	.23		.46	install 3-1/2″ (R-11) unfaced roll and
	Demolish and Install	S.F.	.23	.33		.56	batt fiberglass insulation, joist or studs
	Minimum Charge	Job		157		157	16″ O.C.
6″ (R-19)							
	Demolish	S.F.		.10		.10	Cost includes material and labor to
	Install	S.F.	.40	.27		.67	install 6″ (R-19) unfaced roll and batt
	Demolish and Install	S.F.	.40	.37		.77	fiberglass insulation, joist or studs 16″
	Minimum Charge	Job		157		157	O.C.
10″ (R-30)							
	Demolish	S.F.		.10		.10	Cost includes material and labor to
	Install	S.F.	.66	.31		.97	install 10″ (R-30) unfaced roll and batt
	Demolish and Install	S.F.	.66	.41		1.07	fiberglass insulation, joist or studs 16″
	Minimum Charge	Job		157		157	O.C.
12″ (R-38)							
	Demolish	S.F.		.10		.10	Cost includes material and labor to
	Install	S.F.	.84	.31		1.15	install 12″ (R-38) unfaced roll and batt
	Demolish and Install	S.F.	.84	.41		1.25	fiberglass insulation, joist or studs 16″
	Minimum Charge	Job		157		157	O.C.
Blown							
R-11							
	Demolish	S.F.		.10		.10	Includes material and labor to install
	Install	S.F.	.24	1.04	.33	1.61	fiberglass blown-in insulation in wall
	Demolish and Install	S.F.	.24	1.14	.33	1.71	with wood siding.
	Minimum Charge	Job		360	112	472	
R-19							
	Demolish	S.F.		.10		.10	Includes material and labor to install
	Install	S.F.	.41	1.19	.37	1.97	fiberglass blown-in insulation in wall
	Demolish and Install	S.F.	.41	1.29	.37	2.07	with wood siding.
	Minimum Charge	Job		360	112	472	

Insulation

Wall Insulation

Wall Insulation	Unit	Material	Labor	Equip.	Total	Specification
Rigid Board						
1/2″						
Demolish	S.F.		.13	·	.13	Includes material and labor to install
Install	S.F.	.32	.39		.71	rigid foam insulation board, foil faced,
Demolish and Install	S.F.	.32	.52		.84	two sides.
Minimum Charge	Job		157		157	
3/4″						
Demolish	S.F.		.13		.13	Includes material and labor to install
Install	S.F.	.26	.39		.65	rigid foam insulation board, foil faced,
Demolish and Install	S.F.	.26	.52		.78	two sides.
Minimum Charge	Job		157		157	
1-1/2″						
Demolish	S.F.		.13		.13	Includes material and labor to install
Install	S.F.	.39	.43		.82	rigid foam insulation board, foil faced,
Demolish and Install	S.F.	.39	.56		.95	two sides.
Minimum Charge	Job		157		157	
Vapor Barrier						
Demolish	S.F.		.03		.03	Cost includes material and labor to
Install	S.F.	.02	.08		.10	install 4 mil polyethylene vapor barrier
Demolish and Install	S.F.	.02	.11		.13	10′ wide sheets with 6″ overlaps.
Minimum Charge	Job		157		157	

Ceiling Insulation

Ceiling Insulation	Unit	Material	Labor	Equip.	Total	Specification
Kraft Faced Batt						
3-1/2″ (R-11)						
Demolish	S.F.		.11		.11	Cost includes material and labor to
Install	S.F.	.25	.20		.45	install 3-1/2″ (R-11) kraft paper-faced
Demolish and Install	S.F.	.25	.31		.56	roll and batt fiberglass insulation, joist
Minimum Charge	Job		157		157	or studs 16″ O.C.
6″ (R-19)						
Demolish	S.F.		.11		.11	Cost includes material and labor to
Install	S.F.	.34	.23		.57	install 6″ (R-19) kraft paper-faced roll
Demolish and Install	S.F.	.34	.34		.68	and batt fiberglass insulation, joist or
Minimum Charge	Job		157		157	studs 16″ O.C.
10″ (R-30)						
Demolish	S.F.		.11		.11	Cost includes material and labor to
Install	S.F.	.66	.27		.93	install 10″ (R-30) kraft paper-faced roll
Demolish and Install	S.F.	.66	.38		1.04	and batt fiberglass insulation, joist or
Minimum Charge	Job		157		157	studs 16″ O.C.
12″ (R-38)						
Demolish	S.F.		.11		.11	Cost includes material and labor to
Install	S.F.	.84	.31		1.15	install 12″ (R-38) kraft paper-faced roll
Demolish and Install	S.F.	.84	.42		1.26	and batt fiberglass insulation, joist or
Minimum Charge	Job		157		157	studs 16″ O.C.
Foil Faced Batt						
3-1/2″ (R-11)						
Demolish	S.F.		.11		.11	Cost includes material and labor to
Install	S.F.	.37	.20		.57	install 3-1/2″ (R-11) foil faced roll and
Demolish and Install	S.F.	.37	.31		.68	batt fiberglass insulation, joist or studs
Minimum Charge	Job		157		157	16″ O.C.

Insulation

Ceiling Insulation		Unit	Material	Labor	Equip.	Total	Specification
6" (R-19)							
	Demolish	S.F.		.11		.11	Cost includes material and labor to
	Install	S.F.	.45	.23		.68	install 6" (R-19) foil faced roll and batt
	Demolish and Install	S.F.	.45	.34		.79	fiberglass insulation, joist or studs 16"
	Minimum Charge	Job		157		157	O.C.
Unfaced Batt							
3-1/2" (R-11)							
	Demolish	S.F.		.11		.11	Cost includes material and labor to
	Install	S.F.	.23	.23		.46	install 3-1/2" (R-11) unfaced roll and
	Demolish and Install	S.F.	.23	.34		.57	batt fiberglass insulation, joist or studs
	Minimum Charge	Job		157		157	16" O.C.
6" (R-19)							
	Demolish	S.F.		.11		.11	Cost includes material and labor to
	Install	S.F.	.40	.27		.67	install 6" (R-19) unfaced roll and batt
	Demolish and Install	S.F.	.40	.38		.78	fiberglass insulation, joist or studs 16"
	Minimum Charge	Job		157		157	O.C.
10" (R-30)							
	Demolish	S.F.		.11		.11	Cost includes material and labor to
	Install	S.F.	.66	.31		.97	install 10" (R-30) unfaced roll and batt
	Demolish and Install	S.F.	.66	.42		1.08	fiberglass insulation, joist or studs 16"
	Minimum Charge	Job		157		157	O.C.
12" (R-38)							
	Demolish	S.F.		.11		.11	Cost includes material and labor to
	Install	S.F.	.84	.31		1.15	install 12" (R-38) unfaced roll and batt
	Demolish and Install	S.F.	.84	.42		1.26	fiberglass insulation, joist or studs 16"
	Minimum Charge	Job		157		157	O.C.
Blown							
5"							
	Demolish	S.F.		.32		.32	Includes material and labor to install
	Install	S.F.	.18	.19	.06	.43	fiberglass blown-in insulation.
	Demolish and Install	S.F.	.18	.51	.06	.75	
	Minimum Charge	Job		360	112	472	
6"							
	Demolish	S.F.		.32		.32	Includes material and labor to install
	Install	S.F.	.24	.24	.07	.55	fiberglass blown-in insulation.
	Demolish and Install	S.F.	.24	.56	.07	.87	
	Minimum Charge	Job		360	112	472	
9"							
	Demolish	S.F.		.32		.32	Includes material and labor to install
	Install	S.F.	.34	.33	.10	.77	fiberglass blown-in insulation.
	Demolish and Install	S.F.	.34	.65	.10	1.09	
	Minimum Charge	Job		360	112	472	
12"							
	Demolish	S.F.		.32		.32	Includes material and labor to install
	Install	S.F.	.47	.48	.15	1.10	fiberglass blown-in insulation.
	Demolish and Install	S.F.	.47	.80	.15	1.42	
	Minimum Charge	Job		360	112	472	
Rigid Board							
1/2"							
	Demolish	S.F.		.18		.18	Includes material and labor to install
	Install	S.F.	.32	.39		.71	rigid foam insulation board, foil faced,
	Demolish and Install	S.F.	.32	.57		.89	two sides.
	Minimum Charge	Job		157		157	

Insulation

Ceiling Insulation

	Unit	Material	Labor	Equip.	Total	Specification
3/4"						
Demolish	S.F.		.18		.18	Includes material and labor to install
Install	S.F.	.26	.39		.65	rigid foam insulation board, foil faced,
Demolish and Install	S.F.	.26	.57		.83	two sides.
Minimum Charge	Job		157		157	
1-1/2"						
Demolish	S.F.		.18		.18	Includes material and labor to install
Install	S.F.	.39	.43		.82	rigid foam insulation board, foil faced,
Demolish and Install	S.F.	.39	.61		1	two sides.
Minimum Charge	Job		157		157	

Exterior Insulation

	Unit	Material	Labor	Equip.	Total	Specification
Rigid Board						
1/2"						
Demolish	S.F.		.10		.10	Includes material and labor to install
Install	S.F.	.32	.39		.71	rigid foam insulation board foil faced
Demolish and Install	S.F.	.32	.49		.81	two sides.
Minimum Charge	Job		157		157	
3/4"						
Demolish	S.F.		.10		.10	Includes material and labor to install
Install	S.F.	.26	.39		.65	rigid foam insulation board foil faced
Demolish and Install	S.F.	.26	.49		.75	two sides.
Minimum Charge	Job		157		157	
1-1/2"						
Demolish	S.F.		.10		.10	Includes material and labor to install
Install	S.F.	.39	.42		.81	rigid foam insulation board foil faced
Demolish and Install	S.F.	.39	.52		.91	two sides.
Minimum Charge	Job		157		157	
Vapor Barrier						
Demolish	S.F.		.06		.06	Cost includes material and labor to
Install	S.F.	.02	.08		.10	install 4 mil polyethylene vapor barrier
Demolish and Install	S.F.	.02	.14		.16	10' wide sheets with 6" overlaps.
Minimum Charge	Job		157		157	

Roof Insulation Board

	Unit	Material	Labor	Equip.	Total	Specification
Fiberglass						
3/4" Fiberglass						
Demolish	Sq.		33.50		33.50	Includes material and labor to install
Install	Sq.	50.50	23		73.50	fiberglass board insulation 3/4" thick,
Demolish and Install	Sq.	50.50	56.50		107	R-2.80 and C-0.36 values.
Minimum Charge	Job		157		157	
1" Fiberglass						
Demolish	Sq.		33.50		33.50	Includes material and labor to install
Install	Sq.	72	26		98	fiberglass board insulation 1" thick,
Demolish and Install	Sq.	72	59.50		131.50	R-4.20 and C-0.24 values.
Minimum Charge	Job		157		157	
1-3/8" Fiberglass						
Demolish	Sq.		33.50		33.50	Includes material and labor to install
Install	Sq.	116	28.50		144.50	fiberglass board insulation 1-3/8"
Demolish and Install	Sq.	116	62		178	thick, R-5.30 and C-0.19 values.
Minimum Charge	Job		157		157	

Insulation

Roof Insulation Board		Unit	Material	Labor	Equip.	Total	Specification
1-5/8" Fiberglass							
	Demolish	Sq.		33.50		33.50	Includes material and labor to install
	Install	Sq.	118	30		148	fiberglass board insulation 1-5/8"
	Demolish and Install	Sq.	118	63.50		181.50	thick, R-6.70 and C-0.15 values.
	Minimum Charge	Job		157		157	
2-1/4" Fiberglass							
	Demolish	Sq.		33.50		33.50	Includes material and labor to install
	Install	Sq.	141	33		174	fiberglass board insulation 2-1/4"
	Demolish and Install	Sq.	141	66.50		207.50	thick, R-8.30 and C-0.12 values.
	Minimum Charge	Job		157		157	
Perlite							
1" Perlite							
	Demolish	Sq.		33.50		33.50	Includes material and labor to install
	Install	Sq.	33	26		59	perlite board insulation 1" thick,
	Demolish and Install	Sq.	33	59.50		92.50	R-2.80 and C-0.36 values.
	Minimum Charge	Job		157		157	
4" Perlite							
	Demolish	Sq.		33.50		33.50	Includes material and labor to install
	Install	Sq.	132	50		182	perlite board insulation 4" thick,
	Demolish and Install	Sq.	132	83.50		215.50	R-10.0 and C-0.10 values.
	Minimum Charge	Job		157		157	
1-1/2" Perlite							
	Demolish	Sq.		33.50		33.50	Includes material and labor to install
	Install	Sq.	43	28.50		71.50	perlite board insulation 1-1/2" thick,
	Demolish and Install	Sq.	43	62		105	R-4.20 and C-0.24 values.
	Minimum Charge	Job		157		157	
2" Perlite							
	Demolish	Sq.		33.50		33.50	Includes material and labor to install
	Install	Sq.	67	35		102	perlite board insulation 2" thick,
	Demolish and Install	Sq.	67	68.50		135.50	R-5.30 and C-0.19 values.
	Minimum Charge	Job		157		157	
2-1/2" Perlite							
	Demolish	Sq.		33.50		33.50	Includes material and labor to install
	Install	Sq.	84.50	39		123.50	perlite board insulation 2-1/2" thick,
	Demolish and Install	Sq.	84.50	72.50		157	R-6.70 and C-0.15 values.
	Minimum Charge	Job		157		157	
3" Perlite							
	Demolish	Sq.		33.50		33.50	Includes material and labor to install
	Install	Sq.	99	42		141	perlite board insulation 3" thick,
	Demolish and Install	Sq.	99	75.50		174.50	R-8.30 and C-0.12 values.
	Minimum Charge	Job		157		157	
Polystyrene							
2" Polystyrene							
	Demolish	Sq.		33.50		33.50	Includes material and labor to install
	Install	Sq.	29.50	33		62.50	extruded polystyrene board, R5.0/ 2"
	Demolish and Install	Sq.	29.50	66.50		96	in thickness @25PSI compressive
	Minimum Charge	Job		157		157	strength.
1" Polystyrene							
	Demolish	Sq.		33.50		33.50	Includes material and labor to install
	Install	Sq.	28.50	22.50		51	extruded polystyrene board, R5.0/ 1"
	Demolish and Install	Sq.	28.50	56		84.50	in thickness @25PSI compressive
	Minimum Charge	Job		157		157	strength.
1-1/2" Polystyrene							
	Demolish	Sq.		33.50		33.50	Includes material and labor to install
	Install	Sq.	29	26		55	extruded polystyrene board, R5.0/
	Demolish and Install	Sq.	29	59.50		88.50	1-1/2" in thickness @25PSI
	Minimum Charge	Job		157		157	compressive strength.

Insulation

Roof Insulation Board		Unit	Material	Labor	Equip.	Total	Specification
Urethane							
3/4" Urethane							
	Demolish	Sq.		33.50		33.50	Includes material and labor to install
	Install	Sq.	35	19.60		54.60	urethane board insulation 3/4" thick,
	Demolish and Install	Sq.	35	53.10		88.10	R-5.30 and C-0.19 values.
	Minimum Charge	Job		157		157	
1" Urethane							
	Demolish	Sq.		33.50		33.50	Includes material and labor to install
	Install	Sq.	38.50	21		59.50	urethane board insulation 1" thick,
	Demolish and Install	Sq.	38.50	54.50		93	R-6.70 and C-0.15 values.
	Minimum Charge	Job		157		157	
1-1/2" Urethane							
	Demolish	Sq.		33.50		33.50	Includes material and labor to install
	Install	Sq.	40.50	24		64.50	urethane board insulation 1-1/2"
	Demolish and Install	Sq.	40.50	57.50		98	thick, R-10.1 and C-0.09 values.
	Minimum Charge	Job		157		157	
2" Urethane							
	Demolish	Sq.		33.50		33.50	Includes material and labor to install
	Install	Sq.	51	28.50		79.50	urethane board insulation 2" thick,
	Demolish and Install	Sq.	51	62		113	R-13.4 and C-0.06 values.
	Minimum Charge	Job		157		157	

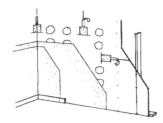

Gypsum Wallboard

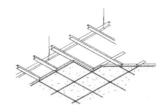

Tile Ceiling

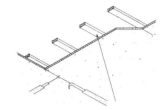

Gypsum Wallboard Ceiling

Wall Plaster	Unit	Material	Labor	Equip.	Total	Specification
Interior						
Cement						
Demolish	S.Y.		4.25		4.25	Includes material and labor to install
Install	S.Y.	13.75	17.75	1.34	32.84	cement plaster on interior wood
Demolish and Install	S.Y.	13.75	22	1.34	37.09	framing with metal lath.
Clean	S.Y.		2.09		2.09	
Paint	S.Y.	1.60	3.68		5.28	
Minimum Charge	Job		139		139	
Acoustic						
Demolish	S.Y.		4.25		4.25	Includes material and labor to install
Install	S.Y.	7.60	21.50	1.34	30.44	acoustical cement plaster on interior
Demolish and Install	S.Y.	7.60	25.75	1.34	34.69	wood framing with metal lath.
Clean	S.Y.		2.09		2.09	
Paint	S.Y.	1.60	3.68		5.28	
Minimum Charge	Job		139		139	
Gypsum						
Demolish	S.Y.		4.25		4.25	Includes material and labor to install
Install	S.Y.	6.55	18.30	1.14	25.99	gypsum wall or ceiling plaster on
Demolish and Install	S.Y.	6.55	22.55	1.14	30.24	interior wood framing with metal lath.
Clean	S.Y.		2.09		2.09	
Paint	S.Y.	1.60	3.68		5.28	
Minimum Charge	Job		139		139	
Stucco						
Demolish	S.Y.		10.20		10.20	Includes material and labor to install
Install	S.Y.	4.35	25	1.57	30.92	stucco on wire mesh over wood
Demolish and Install	S.Y.	4.35	35.20	1.57	41.12	framing.
Clean	S.Y.		2.09		2.09	
Paint	S.Y.	1.60	3.68		5.28	
Minimum Charge	Job		139		139	
Thin Coat						
Veneer						
Demolish	S.Y.		4.25		4.25	Includes materials and labor to install
Install	S.Y.	.59	3.28	.25	4.12	thin coat veneer plaster.
Demolish and Install	S.Y.	.59	7.53	.25	8.37	
Clean	S.Y.		2.09		2.09	
Paint	S.Y.	1.60	3.68		5.28	
Minimum Charge	Job		655	49.50	704.50	

Walls / Ceilings

Wall Plaster		Unit	Material	Labor	Equip.	Total	Specification
Lath							
Metal							
	Demolish	S.Y.		1		1	Includes material and labor to install
	Install	S.Y.	2.66	3.70		6.36	wire lath on steel framed walls.
	Demolish and Install	S.Y.	2.66	4.70		7.36	
	Clean	S.Y.		2.09		2.09	
	Minimum Charge	Job		65		65	
Gypsum							
	Demolish	S.Y.		1		1	Includes material and labor to install
	Install	S.Y.	3.86	3.47		7.33	gypsum lath on walls.
	Demolish and Install	S.Y.	3.86	4.47		8.33	
	Clean	S.Y.		2.09		2.09	
	Minimum Charge	Job		139		139	
Repairs							
Fill in Small Cracks							
	Install	L.F.	.06	2.22		2.28	Includes labor and material to fill-in
	Minimum Charge	Job		139		139	hairline cracks up to 1/8" with filler,
							sand and prep for paint.
Repair up to 1" Crack							
	Install	L.F.	.24	2.78		3.02	Includes labor and material to fill-in
	Minimum Charge	Job		139		139	hairline cracks up to 1" with filler,
							sand and prep for paint.
Patch Small Hole (2")							
	Install	Ea.	.06	2.93		2.99	Includes labor and material to patch
	Minimum Charge	Job		139		139	up to a 2" diameter hole in plaster.
Patch Large Hole							
	Install	Ea.	7.30	46.50		53.80	Includes labor and material to patch
	Minimum Charge	Job		139		139	up to a 20" diameter hole in plaster.
Patch Plaster Section							
	Install	Ea.	2.01	2.78		4.79	Includes labor and material to repair
	Minimum Charge	Job		139		139	section of plaster wall, area less than
							20 S.F.
Repair Inside Corner							
	Install	L.F.	2.93	6.95		9.88	Includes material and labor to patch
	Minimum Charge	Job		139		139	plaster corners.

Ceiling Plaster		Unit	Material	Labor	Equip.	Total	Specification
Interior							
Cement							
	Demolish	S.Y.		4.25		4.25	Includes material and labor to install
	Install	S.Y.	13.75	17.75	1.34	32.84	cement plaster on interior wood
	Demolish and Install	S.Y.	13.75	22	1.34	37.09	framing with metal lath.
	Clean	S.Y.		2.19		2.19	
	Paint	S.Y.	1.60	3.68		5.28	
	Minimum Charge	Job		139		139	
Acoustic							
	Demolish	S.Y.		4.25		4.25	Includes material and labor to install
	Install	S.Y.	7.60	21.50	1.34	30.44	acoustical cement plaster on interior
	Demolish and Install	S.Y.	7.60	25.75	1.34	34.69	wood framing with metal lath.
	Clean	S.Y.		2.19		2.19	
	Paint	S.Y.	1.60	3.68		5.28	
	Minimum Charge	Job		139		139	

Ceiling Plaster

	Unit	Material	Labor	Equip.	Total	Specification
Gypsum						
Demolish	S.Y.		4.25		4.25	Includes material and labor to install
Install	S.Y.	6.55	18.30	1.14	25.99	gypsum wall or ceiling plaster on
Demolish and Install	S.Y.	6.55	22.55	1.14	30.24	interior wood framing with metal lath.
Clean	S.Y.		2.19		2.19	
Paint	S.Y.	1.60	3.68		5.28	
Minimum Charge	Job		139		139	
Stucco						
Demolish	S.Y.		10.20		10.20	Includes material and labor to install
Install	S.Y.	4.35	25	1.57	30.92	stucco on wire mesh over wood
Demolish and Install	S.Y.	4.35	35.20	1.57	41.12	framing.
Clean	S.Y.		2.19		2.19	
Paint	S.Y.	1.60	3.68		5.28	
Minimum Charge	Job		139		139	
Thin Coat						
Veneer						
Demolish	S.Y.		4.25		4.25	Includes materials and labor to install
Install	S.Y.	.59	3.28	.25	4.12	thin coat veneer plaster.
Demolish and Install	S.Y.	.59	7.53	.25	8.37	
Clean	S.Y.		2.19		2.19	
Paint	S.Y.	1.60	3.68		5.28	
Minimum Charge	Job		139		139	
Lath						
Metal						
Demolish	S.Y.		1		1	Includes material and labor to install
Install	S.Y.	2.66	4.62		7.28	wire lath on steel framed ceilings.
Demolish and Install	S.Y.	2.66	5.62		8.28	
Minimum Charge	Job		139		139	
Gypsum						
Demolish	S.Y.		1		1	Includes material and labor to install
Install	S.Y.	3.86	2.89		6.75	gypsum lath on ceilings.
Demolish and Install	S.Y.	3.86	3.89		7.75	
Minimum Charge	Job		139		139	
Repairs						
Fill in Small Cracks						
Install	S.F.	.06	.51		.57	Includes labor and material to fill in
Minimum Charge	Job		278		278	hairline cracks up to 1/8" with filler, sand and prep for paint.

Gypsum Wallboard

	Unit	Material	Labor	Equip.	Total	Specification
Finished						
1/4"						
Demolish	S.F.		.26		.26	Cost includes material and labor to
Install	S.F.	.20	.94		1.14	install 1/4" drywall, including joint
Demolish and Install	S.F.	.20	1.20		1.40	taping, premixed compound and
Clean	S.F.		.21		.21	screws.
Paint	S.F.	.18	.24		.42	
Minimum Charge	Job		157		157	
3/8"						
Demolish	S.F.		.26		.26	Cost includes material and labor to
Install	S.F.	.20	.95		1.15	install 3/8" drywall including joint
Demolish and Install	S.F.	.20	1.21		1.41	tape, premixed compound and
Clean	S.F.		.21		.21	screws.
Paint	S.F.	.18	.24		.42	
Minimum Charge	Job		157		157	

Walls / Ceilings

Gypsum Wallboard		Unit	Material	Labor	Equip.	Total	Specification
1/2"							
	Demolish	S.F.		.26		.26	Cost includes material and labor to
	Install	S.F.	.19	.96		1.15	install 1/2" drywall including joint
	Demolish and Install	S.F.	.19	1.22		1.41	tape, premixed compound and
	Clean	S.F.		.21		.21	screws.
	Paint	S.F.	.18	.24		.42	
	Minimum Charge	Job		157		157	
1/2" Water Resistant							
	Demolish	S.F.		.26		.26	Cost includes material and labor to
	Install	S.F.	.18	.96		1.14	install 1/2" WR drywall including joint
	Demolish and Install	S.F.	.18	1.22		1.40	tape, premixed compound and
	Clean	S.F.		.21		.21	screws.
	Paint	S.F.	.18	.24		.42	
	Minimum Charge	Job		157		157	
1/2" Fire-rated							
	Demolish	S.F.		.26		.26	Cost includes material and labor to
	Install	S.F.	.20	.96		1.16	install 1/2" FR drywall including
	Demolish and Install	S.F.	.20	1.22		1.42	screws. Taping and finishing not
	Clean	S.F.		.21		.21	included.
	Paint	S.F.	.18	.24		.42	
	Minimum Charge	Job		157		157	
5/8"							
	Demolish	S.F.		.26		.26	Cost includes material and labor to
	Install	S.F.	.24	1.09		1.33	install 5/8" drywall including joint
	Demolish and Install	S.F.	.24	1.35		1.59	tape, premixed compound and
	Clean	S.F.		.21		.21	screws.
	Paint	S.F.	.18	.24		.42	
	Minimum Charge	Job		157		157	
5/8" Fire-rated							
	Demolish	S.F.		.26		.26	Cost includes material and labor to
	Install	S.F.	.25	1.09		1.34	install 5/8" FR drywall, including joint
	Demolish and Install	S.F.	.25	1.35		1.60	taping, premixed compound and
	Clean	S.F.		.21		.21	screws.
	Paint	S.F.	.18	.24		.42	
	Minimum Charge	Job		157		157	
2-hour fire-rated							
	Demolish	S.F.		.64		.64	Cost includes material and labor to
	Install	S.F.	.53	2.17		2.70	install 2 layers of 5/8" type X drywall
	Demolish and Install	S.F.	.53	2.81		3.34	including joint taping, premixed
	Clean	S.F.		.21		.21	compound and screws.
	Paint	S.F.	.18	.24		.42	
	Minimum Charge	Job		157		157	
Unfinished							
1/4"							
	Demolish	S.F.		.26		.26	Cost includes material and labor to
	Install	S.F.	.17	.33		.50	install 1/4" drywall including screws.
	Demolish and Install	S.F.	.17	.59		.76	Taping and finishing not included.
	Clean	S.F.		.21		.21	
	Minimum Charge	Job		157		157	
3/8"							
	Demolish	S.F.		.26		.26	Cost includes material and labor to
	Install	S.F.	.17	.33		.50	install 3/8" drywall including screws.
	Demolish and Install	S.F.	.17	.59		.76	Tape, finish or texture is not included.
	Clean	S.F.		.21		.21	
	Minimum Charge	Job		157		157	

Gypsum Wallboard		Unit	Material	Labor	Equip.	Total	Specification
1/2"							Cost includes material and labor to
	Demolish	S.F.		.26		.26	install 1/2" drywall including screws.
	Install	S.F.	.16	.33		.49	Taping and finishing not included.
	Demolish and Install	S.F.	.16	.59		.75	
	Clean	S.F.		.21		.21	
	Minimum Charge	Job		157		157	
1/2" Water Resistant							Cost includes material and labor to
	Demolish	S.F.		.26		.26	install 1/2" WR drywall including
	Install	S.F.	.15	.33		.48	screws. Taping and finishing is not
	Demolish and Install	S.F.	.15	.59		.74	included.
	Clean	S.F.		.21		.21	
	Minimum Charge	Job		157		157	
1/2" Fire-rated							Cost includes labor and material to
	Demolish	S.F.		.26		.26	install 1/2" FR drywall including
	Install	S.F.	.16	.23		.39	screws. Tape, finish or texture is not
	Demolish and Install	S.F.	.16	.49		.65	included.
	Clean	S.F.		.21		.21	
	Minimum Charge	Job		157		157	
5/8"							Cost includes material and labor to
	Demolish	S.F.		.26		.26	install 5/8" drywall including screws.
	Install	S.F.	.20	.37		.57	Taping and finishing not included.
	Demolish and Install	S.F.	.20	.63		.83	
	Clean	S.F.		.21		.21	
	Minimum Charge	Job		157		157	
5/8" Fire-rated							Cost includes material and labor to
	Demolish	S.F.		.26		.26	install 5/8" FR drywall including
	Install	S.F.	.21	.37		.58	screws. Taping and finishing not
	Demolish and Install	S.F.	.21	.63		.84	included.
	Clean	S.F.		.21		.21	
	Minimum Charge	Job		157		157	
2-hour Fire-rated							Cost includes material and labor to
	Demolish	S.F.		.64		.64	install 2 layers of 5/8" type X drywall
	Install	S.F.	.53	2.17		2.70	including joint taping, premixed
	Demolish and Install	S.F.	.53	2.81		3.34	compound and screws.
	Clean	S.F.		.21		.21	
	Minimum Charge	Job		157		157	
Accessories							
Tape & Finish							
	Install	S.F.	.04	.31		.35	Includes labor and material for taping,
	Minimum Charge	Job		157		157	sanding and finish.
Texture							
Sprayed							
	Install	S.F.	.04	.35		.39	Includes labor and material to install
	Minimum Charge	Job		157		157	by spray, texture finish.
Trowel							
	Install	S.F.		.18		.18	Includes labor and material to apply
	Minimum Charge	Job		157		157	hand troweled texture.
Patch Hole in Gypsum Wallboard							
	Install	S.F.	.31	41.50		41.81	Includes labor and material to patch a
	Minimum Charge	Job		157		157	one square foot area of damaged
							drywall.
Furring Strips							
	Demolish	S.F.		.06		.06	Cost includes material and labor to
	Install	S.F.	.15	.70		.85	install 1" x 2" furring strips, 16" O.C.
	Demolish and Install	S.F.	.15	.76		.91	per S.F. of surface area to be covered.
	Minimum Charge	Job		78.50		78.50	

Walls / Ceilings

Ceramic Wall Tile	Unit	Material	Labor	Equip.	Total	Specification
Thin Set						
2″ x 2″						
Demolish	S.F.		.85		.85	Cost includes material and labor to
Install	S.F.	6.05	4.73		10.78	install 2″ x 2″ ceramic tile in organic
Demolish and Install	S.F.	6.05	5.58		11.63	adhesive including cutting and grout.
Clean	S.F.		.44		.44	
Minimum Charge	Job		142		142	
4-1/4″ x 4-1/4″						
Demolish	S.F.		.85		.85	Cost includes material and labor to
Install	S.F.	2.30	2.65		4.95	install 4-1/4″ x 4-1/4″ ceramic tile in
Demolish and Install	S.F.	2.30	3.50		5.80	organic adhesive including grout.
Clean	S.F.		.44		.44	
Minimum Charge	Job		142		142	
6″ x 6″						
Demolish	S.F.		.85		.85	Cost includes material and labor to
Install	S.F.	2.89	2.52		5.41	install 6″ x 6″ ceramic tile in organic
Demolish and Install	S.F.	2.89	3.37		6.26	adhesive including and grout.
Clean	S.F.		.44		.44	
Minimum Charge	Job		142		142	
8″ x 8″						
Demolish	S.F.		.85		.85	Cost includes material and labor to
Install	S.F.	3.48	2.24		5.72	install 8″ x 8″ ceramic tile in organic
Demolish and Install	S.F.	3.48	3.09		6.57	adhesive including cutting and grout.
Clean	S.F.		.44		.44	
Minimum Charge	Job		142		142	
Thick Set						
2″ x 2″						
Demolish	S.F.		1.02		1.02	Cost includes material and labor to
Install	S.F.	4.92	5.70		10.62	install 2″ x 2″ ceramic tile in mortar
Demolish and Install	S.F.	4.92	6.72		11.64	bed including cutting and grout.
Clean	S.F.		.44		.44	
Minimum Charge	Job		142		142	
4-1/4″ x 4-1/4″						
Demolish	S.F.		1.02		1.02	Cost includes material and labor to
Install	S.F.	6.65	5.70		12.35	install 4-1/4″ x 4-1/4″ ceramic tile in
Demolish and Install	S.F.	6.65	6.72		13.37	mortar bed including cutting and
Clean	S.F.		.44		.44	grout.
Minimum Charge	Job		142		142	
6″ x 6″						
Demolish	S.F.		1.02		1.02	Cost includes material and labor to
Install	S.F.	7.45	5.15		12.60	install 6″ x 6″ ceramic tile in mortar
Demolish and Install	S.F.	7.45	6.17		13.62	bed including cutting and grout.
Clean	S.F.		.44		.44	
Minimum Charge	Job		142		142	
8″ x 8″						
Demolish	S.F.		1.02		1.02	Cost includes material and labor to
Install	S.F.	9.75	4.73		14.48	install 8″ x 8″ ceramic tile in mortar
Demolish and Install	S.F.	9.75	5.75		15.50	bed including cutting and grout.
Clean	S.F.		.44		.44	
Minimum Charge	Job		142		142	
Re-grout						
Install	S.F.	.14	2.84		2.98	Includes material and labor to regrout
Minimum Charge	Job		142		142	ceramic tile walls.

Walls / Ceilings

Wall Tile	Unit	Material	Labor	Equip.	Total	Specification
Mirror						
Custom Cut Wall						
Demolish	S.F.		.49		.49	Includes material and labor to install
Install	S.F.	1.75	3.08		4.83	distortion-free float glass mirror, 1/8"
Demolish and Install	S.F.	1.75	3.57		5.32	to 3/16" thick with cut and polished
Clean	S.F.	.02	.13		.15	edges.
Minimum Charge	Job		150		150	
Beveled Wall						
Demolish	S.F.		.49		.49	Includes material and labor to install
Install	S.F.	4.37	5.45		9.82	distortion-free float glass mirror, 1/8"
Demolish and Install	S.F.	4.37	5.94		10.31	to 3/16" thick with beveled edges.
Clean	S.F.	.02	.13		.15	
Minimum Charge	Job		150		150	
Smoked Wall						
Demolish	S.F.		.49		.49	Includes material and labor to install
Install	S.F.	6.95	5.45		12.40	distortion-free smoked float glass
Demolish and Install	S.F.	6.95	5.94		12.89	mirror, 1/8" to 3/16" thick with cut
Clean	S.F.	.02	.13		.15	and polished edges.
Minimum Charge	Job		150		150	
Marble						
Good Grade						
Demolish	S.F.		1.59		1.59	Cost includes material and labor to
Install	S.F.	7.60	12.35		19.95	install 3/8" x 12" x 12" marble tile in
Demolish and Install	S.F.	7.60	13.94		21.54	mortar bed with grout.
Clean	S.F.		.44		.44	
Minimum Charge	Job		142		142	
Better Grade						
Demolish	S.F.		1.59		1.59	Cost includes material and labor to
Install	S.F.	9.80	12.35		22.15	install 3/8" x 12" x 12" marble, in
Demolish and Install	S.F.	9.80	13.94		23.74	mortar bed with grout.
Clean	S.F.		.44		.44	
Minimum Charge	Job		142		142	
Premium Grade						
Demolish	S.F.		1.59		1.59	Cost includes material and labor to
Install	S.F.	18.90	12.35		31.25	install 3/8" x 12" x 12" marble, in
Demolish and Install	S.F.	18.90	13.94		32.84	mortar bed with grout.
Clean	S.F.		.44		.44	
Minimum Charge	Job		142		142	
Glass Block						
Demolish	S.F.		1.28		1.28	Cost includes material and labor to
Install	S.F.	16.55	12.45		29	install 3-7/8" thick, 6" x 6" smooth
Demolish and Install	S.F.	16.55	13.73		30.28	face glass block including mortar and
Clean	S.F.		.44		.44	cleaning.
Minimum Charge	Job		279		279	
Plastic						
Demolish	S.F.		.39		.39	Cost includes material and labor to
Install	S.F.	.92	1.62		2.54	install 4" plastic tile.
Demolish and Install	S.F.	.92	2.01		2.93	
Clean	S.F.		.44		.44	
Minimum Charge	Job		142		142	
Cork						
Demolish	S.F.		.20		.20	Cost includes material and labor to
Install	S.F.	2.86	1.14		4	install 3/16" thick cork tile and
Demolish and Install	S.F.	2.86	1.34		4.20	installation.
Clean	S.F.		.44		.44	
Minimum Charge	Job		137		137	

Walls / Ceilings

Wall Tile

	Unit	Material	Labor	Equip.	Total	Specification
Re-grout						
Install	S.F.	.14	2.84		2.98	Includes material and labor to regrout
Minimum Charge	Job		142		142	ceramic tile walls.

Wall Paneling

	Unit	Material	Labor	Equip.	Total	Specification
Plywood						
Good Grade						
Demolish	S.F.		.26		.26	Cost includes material and labor to
Install	S.F.	.87	1.25		2.12	install 4' x 8' x 1/4" paneling,
Demolish and Install	S.F.	.87	1.51		2.38	including nails and adhesive.
Clean	S.F.		.21		.21	
Paint	S.F.	.17	.37		.54	
Minimum Charge	Job		157		157	
Better Grade						
Demolish	S.F.		.26		.26	Cost includes material and labor to
Install	S.F.	1.33	1.49		2.82	install 4' x 8' x 1/4" plywood wall
Demolish and Install	S.F.	1.33	1.75		3.08	panel, including nails and adhesive.
Clean	S.F.		.21		.21	
Paint	S.F.	.17	.37		.54	
Minimum Charge	Job		157		157	
Premium Grade						
Demolish	S.F.		.26		.26	Cost includes material and labor to
Install	S.F.	1.94	1.79		3.73	install 4' x 8' x 1/4" plywood wall
Demolish and Install	S.F.	1.94	2.05		3.99	panel, including nails and adhesive.
Clean	S.F.		.21		.21	
Paint	S.F.	.17	.37		.54	
Minimum Charge	Job		157		157	
Hardboard						
Embossed Paneling						
Demolish	S.F.		.26		.26	Includes material and labor to install
Install	S.F.	.59	1.25		1.84	hardboard embossed paneling
Demolish and Install	S.F.	.59	1.51		2.10	including adhesives.
Clean	S.F.		.21		.21	
Minimum Charge	Job		157		157	
Plastic Molding						
Demolish	L.F.		.26		.26	Includes material and labor to install
Install	L.F.	.54	.78		1.32	plastic trim for paneling.
Demolish and Install	L.F.	.54	1.04		1.58	
Clean	L.F.	.02	.14		.16	
Minimum Charge	Job		157		157	
Hardboard						
Embossed Brick						
Demolish	S.F.		.26		.26	Cost includes material and labor to
Install	S.F.	1.97	1.39		3.36	install 4' x 8' non-ceramic mineral
Demolish and Install	S.F.	1.97	1.65		3.62	brick-like veneer wall panels including
Clean	S.F.		.21		.21	adhesive.
Paint	S.F.	.18	.42		.60	
Minimum Charge	Job		157		157	
Embossed Ceramic Tile						
Demolish	S.F.		.26		.26	Includes material and labor to install
Install	S.F.	.59	1.25		1.84	hardboard embossed paneling
Demolish and Install	S.F.	.59	1.51		2.10	including adhesives.
Clean	S.F.		.21		.21	
Paint	S.F.	.18	.42		.60	
Minimum Charge	Job		157		157	

Walls / Ceilings

Wall Paneling		Unit	Material	Labor	Equip.	Total	Specification
Pegboard							
	Demolish	S.F.		.26		.26	Cost includes material and labor to
	Install	S.F.	.41	1.25		1.66	install 4' x 8' x 1/4" pegboard panel,
	Demolish and Install	S.F.	.41	1.51		1.92	including adhesive.
	Clean	S.F.		.21		.21	
	Paint	S.F.	.18	.42		.60	
	Minimum Charge	Job		157		157	
Furring Strips							
	Demolish	S.F.		.06		.06	Cost includes material and labor to
	Install	S.F.	.15	.70		.85	install 1" x 2" furring strips, 16" O.C.
	Demolish and Install	S.F.	.15	.76		.91	per S.F. of surface area to be covered.
	Minimum Charge	Job		157		157	
Molding							
Unfinished Base							
	Demolish	L.F.		.43		.43	Cost includes material and labor to
	Install	L.F.	.49	.97		1.46	install 1/2" x 2-1/2" pine or birch
	Demolish and Install	L.F.	.49	1.40		1.89	base molding.
	Reinstall	L.F.		.77		.77	
	Clean	L.F.	.02	.13		.15	
	Paint	L.F.	.03	.30		.33	
	Minimum Charge	Job		157		157	
Unfinished Corner							
	Demolish	L.F.		.43		.43	Includes material and labor to install
	Install	L.F.	1.13	1.31		2.44	unfinished pine corner molding.
	Demolish and Install	L.F.	1.13	1.74		2.87	
	Reinstall	L.F.		1.05		1.05	
	Clean	Ea.	.73	20.50		21.23	
	Paint	L.F.	.12	.44		.56	
	Minimum Charge	Job		157		157	
Unfinished Cove							
	Demolish	L.F.		.43		.43	Cost includes material and labor to
	Install	L.F.	1.56	1.16		2.72	install 3/4" x 3-1/2" unfinished pine
	Demolish and Install	L.F.	1.56	1.59		3.15	cove molding.
	Reinstall	L.F.		.93		.93	
	Clean	Ea.	.73	20.50		21.23	
	Paint	L.F.	.12	.44		.56	
	Minimum Charge	Job		157		157	
Unfinished Crown							
	Demolish	L.F.		.51		.51	Includes material and labor to install
	Install	L.F.	1.84	1.25		3.09	unfinished pine crown molding.
	Demolish and Install	L.F.	1.84	1.76		3.60	
	Reinstall	L.F.		1		1	
	Clean	L.F.	.03	.18		.21	
	Paint	L.F.	.04	.30		.34	
	Minimum Charge	Job		157		157	
Prefinished Base							
	Demolish	L.F.		.43		.43	Cost includes material and labor to
	Install	L.F.	1.69	1.01		2.70	install 9/16" x 3-5/16" prefinished
	Demolish and Install	L.F.	1.69	1.44		3.13	base molding.
	Reinstall	L.F.		.80		.80	
	Clean	L.F.	.02	.14		.16	
	Paint	L.F.	.03	.30		.33	
	Minimum Charge	Job		157		157	

Walls / Ceilings

Wall Paneling

Wall Paneling	Unit	Material	Labor	Equip.	Total	Specification
Prefinished Corner						
Demolish	L.F.		.43		.43	Includes material and labor to install
Install	L.F.	1.98	1.31		3.29	inside or outside prefinished corner
Demolish and Install	L.F.	1.98	1.74		3.72	molding.
Reinstall	L.F.		1.05		1.05	
Clean	L.F.		.09		.09	
Paint	L.F.	.12	.44		.56	
Minimum Charge	Job		157		157	
Prefinished Crown						
Demolish	L.F.		.51		.51	Cost includes material and labor to
Install	L.F.	2.24	1.25		3.49	install 3/4" x 3-13/16" prefinished
Demolish and Install	L.F.	2.24	1.76		4	crown molding.
Reinstall	L.F.		1		1	
Clean	L.F.	.03	.18		.21	
Paint	L.F.	.04	.30		.34	
Minimum Charge	Job		157		157	

Wallcovering

Wallcovering	Unit	Material	Labor	Equip.	Total	Specification
Surface Prep						
Patch Walls						
Install	S.F.	.10	2.78		2.88	Includes labor and material to repair
Minimum Charge	Job		137		137	section of plaster wall, area less than
						20 S.F.
Sizing						
Install	S.F.	.07	.14		.21	Includes labor and materials to apply
Minimum Charge	Job		137		137	wall sizing.
Good Grade						
Walls						
Demolish	S.F.		.51		.51	Includes material and labor to install
Install	S.F.	.80	.57		1.37	pre-pasted, vinyl-coated wallcovering.
Demolish and Install	S.F.	.80	1.08		1.88	
Clean	S.F.	.01	.05		.06	
Minimum Charge	Job		137		137	
Ceiling						
Demolish	S.F.		.51		.51	Includes material and labor to install
Install	S.F.	.80	.57		1.37	pre-pasted, vinyl-coated wallcovering.
Demolish and Install	S.F.	.80	1.08		1.88	
Clean	S.F.	.02	.13		.15	
Minimum Charge	Job		137		137	
Better Grade						
Walls						
Demolish	S.F.		.51		.51	Includes material and labor to install
Install	S.F.	.90	.57		1.47	pre-pasted, vinyl-coated wallcovering.
Demolish and Install	S.F.	.90	1.08		1.98	
Clean	S.F.	.01	.05		.06	
Minimum Charge	Job		137		137	
Ceiling						
Demolish	S.F.		.51		.51	Includes material and labor to install
Install	S.F.	.90	.57		1.47	pre-pasted, vinyl-coated wallcovering.
Demolish and Install	S.F.	.90	1.08		1.98	
Clean	S.F.	.02	.13		.15	
Minimum Charge	Job		137		137	

Walls / Ceilings

Wallcovering

	Unit	Material	Labor	Equip.	Total	Specification
Premium Grade						
Walls						
Demolish	S.F.		.63		.63	Includes material and labor to install
Install	S.F.	1.14	.57		1.71	pre-pasted, vinyl-coated wallcovering.
Demolish and Install	S.F.	1.14	1.20		2.34	
Clean	S.F.	.01	.05		.06	
Minimum Charge	Job		137		137	
Ceiling						
Demolish	S.F.		.63		.63	Includes material and labor to install
Install	S.F.	1.14	.57		1.71	pre-pasted, vinyl-coated wallcovering.
Demolish and Install	S.F.	1.14	1.20		2.34	
Clean	S.F.	.02	.13		.15	
Minimum Charge	Job		137		137	
Mural						
Pre-printed						
Demolish	Ea.		79.50		79.50	Includes material and labor to install
Install	Ea.	213	206		419	wallcovering with a pre-printed mural.
Demolish and Install	Ea.	213	285.50		498.50	
Clean	Ea.		5.75		5.75	
Minimum Charge	Job		137		137	
Hand-painted						
Demolish	Ea.		79.50		79.50	Includes material and labor to install
Install	Ea.	430	415		845	wallcovering with a hand-painted
Demolish and Install	Ea.	430	494.50		924.50	mural.
Clean	Ea.		5.75		5.75	
Minimum Charge	Job		137		137	
Vinyl-coated						
Good Grade						
Demolish	Ea.		14.15		14.15	Includes material and labor to install
Install	Ea.	17.75	19.15		36.90	pre-pasted, vinyl-coated wallcovering.
Demolish and Install	Ea.	17.75	33.30		51.05	Based upon 30 S.F. rolls.
Clean	Ea.		5.75		5.75	
Minimum Charge	Job		137		137	
Better Grade						
Demolish	Ea.		14.15		14.15	Includes material and labor to install
Install	Ea.	23.50	19.15		42.65	pre-pasted, vinyl-coated wallcovering.
Demolish and Install	Ea.	23.50	33.30		56.80	Based upon 30 S.F. rolls.
Clean	Ea.		5.75		5.75	
Minimum Charge	Job		137		137	
Premium Grade						
Demolish	Ea.		14.15		14.15	Includes material and labor to install
Install	Ea.	41.50	19.15		60.65	pre-pasted, vinyl-coated wallcovering.
Demolish and Install	Ea.	41.50	33.30		74.80	Based upon 30 S.F. rolls.
Clean	Ea.		5.75		5.75	
Minimum Charge	Job		137		137	
Blank Underliner						
Demolish	Ea.		14.15		14.15	Includes material and labor to install
Install	Ea.	10.50	19.55		30.05	blank stock liner paper or bridging
Demolish and Install	Ea.	10.50	33.70		44.20	paper. Based upon 30 S.F. rolls.
Clean	Ea.		5.75		5.75	
Minimum Charge	Job		137		137	
Solid Vinyl						
Good Grade						
Demolish	Roll		16.90		16.90	Includes material and labor to install
Install	Ea.	23.50	19.15		42.65	pre-pasted, vinyl wallcovering. Based
Demolish and Install	Ea.	23.50	36.05		59.55	upon 30 S.F. rolls.
Clean	Ea.		5.75		5.75	
Minimum Charge	Job		137		137	

Walls / Ceilings

Wallcovering		Unit	Material	Labor	Equip.	Total	Specification
Better Grade							
	Demolish	Roll		16.90		16.90	Includes material and labor to install
	Install	Ea.	33	19.15		52.15	pre-pasted, vinyl wallcovering. Based
	Demolish and Install	Ea.	33	36.05		69.05	upon 30 S.F. rolls.
	Clean	Ea.		5.75		5.75	
	Minimum Charge	Job		137		137	
Premium Grade							
	Demolish	Roll		16.90		16.90	Includes material and labor to install
	Install	Ea.	42.50	19.15		61.65	pre-pasted, vinyl wallcovering. Based
	Demolish and Install	Ea.	42.50	36.05		78.55	upon 30 S.F. rolls.
	Clean	Ea.		5.75		5.75	
	Minimum Charge	Job		137		137	
Blank Underliner							
	Demolish	Ea.		14.15		14.15	Includes material and labor to install
	Install	Ea.	10.50	19.55		30.05	blank stock liner paper or bridging
	Demolish and Install	Ea.	10.50	33.70		44.20	paper. Based upon 30 S.F. rolls.
	Clean	Ea.		5.75		5.75	
	Minimum Charge	Job		137		137	
Grasscloth / String							
Good Grade							
	Demolish	Ea.		14.15		14.15	Includes material and labor to install
	Install	Ea.	33.50	25		58.50	grasscloth or string wallcovering.
	Demolish and Install	Ea.	33.50	39.15		72.65	Based upon 30 S.F. rolls.
	Clean	Ea.		7.65		7.65	
	Minimum Charge	Job		137		137	
Better Grade							
	Demolish	Ea.		14.15		14.15	Includes material and labor to install
	Install	Ea.	43.50	25		68.50	grasscloth or string wallcovering.
	Demolish and Install	Ea.	43.50	39.15		82.65	Based upon 30 S.F. rolls.
	Clean	Ea.		7.65		7.65	
	Minimum Charge	Job		137		137	
Premium Grade							
	Demolish	Ea.		14.15		14.15	Includes material and labor to install
	Install	Ea.	55.50	25		80.50	grasscloth or string wallcovering.
	Demolish and Install	Ea.	55.50	39.15		94.65	Based upon 30 S.F. rolls.
	Clean	Ea.		7.65		7.65	
	Minimum Charge	Job		137		137	
Blank Underliner							
	Demolish	Ea.		14.15		14.15	Includes material and labor to install
	Install	Ea.	10.50	19.55		30.05	blank stock liner paper or bridging
	Demolish and Install	Ea.	10.50	33.70		44.20	paper. Based upon 30 S.F. rolls.
	Clean	Ea.		5.75		5.75	
	Minimum Charge	Job		137		137	
Designer							
Premium Grade							
	Demolish	S.F.		.63		.63	Includes material and labor to install
	Install	Ea.	103	29		132	paperback vinyl, untrimmed and
	Demolish and Install	Ea.	103	29.63		132.63	hand-painted. Based upon 30 S.F.
	Clean	Ea.		5.75		5.75	rolls.
	Minimum Charge	Job		137		137	
Blank Underliner							
	Demolish	Ea.		14.15		14.15	Includes material and labor to install
	Install	Ea.	10.50	19.55		30.05	blank stock liner paper or bridging
	Demolish and Install	Ea.	10.50	33.70		44.20	paper. Based upon 30 S.F. rolls.
	Clean	Ea.		5.75		5.75	
	Minimum Charge	Job		137		137	

Walls / Ceilings

Wallcovering		Unit	Material	Labor	Equip.	Total	Specification
Foil							
	Demolish	Ea.		14.15		14.15	Includes material and labor to install
	Install	Ea.	36	25		61	foil coated wallcovering. Based upon
	Demolish and Install	Ea.	36	39.15		75.15	30 S.F. rolls.
	Clean	Ea.		5.75		5.75	
	Minimum Charge	Job		137		137	
Wallcovering Border							
Good Grade							
	Demolish	L.F.		.93		.93	Includes material and labor install
	Install	L.F.	1.77	1.05		2.82	vinyl-coated border 4" to 6" wide.
	Demolish and Install	L.F.	1.77	1.98		3.75	
	Clean	L.F.		.09		.09	
	Minimum Charge	Job		137		137	
Better Grade							
	Demolish	L.F.		.93		.93	Includes material and labor install
	Install	L.F.	2.37	1.05		3.42	vinyl-coated border 4" to 6" wide.
	Demolish and Install	L.F.	2.37	1.98		4.35	
	Clean	L.F.		.09		.09	
	Minimum Charge	Job		137		137	
Premium Grade							
	Demolish	L.F.		.93		.93	Includes material and labor install
	Install	L.F.	2.96	1.05		4.01	vinyl-coated border 4" to 6" wide.
	Demolish and Install	L.F.	2.96	1.98		4.94	
	Clean	L.F.		.09		.09	
	Minimum Charge	Job		137		137	

Acoustical Ceiling		Unit	Material	Labor	Equip.	Total	Specification
Complete System							
2' x 2' Square Edge							
	Demolish	S.F.		.43		.43	Includes material and labor to install
	Install	S.F.	1.09	.97		2.06	T-bar grid acoustical ceiling system,
	Demolish and Install	S.F.	1.09	1.40		2.49	random pinhole 5/8" thick square
	Reinstall	S.F.		.77		.77	edge tile.
	Clean	S.F.	.03	.17		.20	
	Minimum Charge	Job		157		157	
2' x 2' Reveal Edge							
	Demolish	S.F.		.43		.43	Includes material and labor to install
	Install	S.F.	1.85	1.25		3.10	T-bar grid acoustical ceiling system,
	Demolish and Install	S.F.	1.85	1.68		3.53	random pinhole 5/8" thick reveal
	Reinstall	S.F.		1		1	edge tile.
	Clean	S.F.	.03	.17		.20	
	Minimum Charge	Job		157		157	
2' x 2' Fire-rated							
	Demolish	S.F.		.43		.43	Includes material and labor to install
	Install	S.F.	2.23	.97		3.20	T-bar grid FR acoustical ceiling system,
	Demolish and Install	S.F.	2.23	1.40		3.63	random pinhole 5/8" thick square
	Reinstall	S.F.		.77		.77	edge tile.
	Clean	S.F.	.03	.17		.20	
	Minimum Charge	Job		157		157	
2' x 4' Square Edge							
	Demolish	S.F.		.43		.43	Cost includes material and labor to
	Install	S.F.	.92	.83		1.75	install 2' x 4' T-bar grid acoustical
	Demolish and Install	S.F.	.92	1.26		2.18	ceiling system, random pinhole 5/8"
	Reinstall	S.F.		.66		.66	thick square edge tile.
	Clean	S.F.	.03	.17		.20	
	Minimum Charge	Job		157		157	

Walls / Ceilings

Acoustical Ceiling		Unit	Material	Labor	Equip.	Total	Specification
2' x 4' Reveal Edge							
	Demolish	S.F.		.43		.43	Includes material and labor to install
	Install	S.F.	1.85	1.14		2.99	T-bar grid acoustical ceiling system,
	Demolish and Install	S.F.	1.85	1.57		3.42	random pinhole 5/8" thick reveal
	Reinstall	S.F.		.91		.91	edge tile.
	Clean	S.F.	.03	.17		.20	
	Minimum Charge	Job		157		157	
2' x 4' Fire-rated							
	Demolish	S.F.		.43		.43	Includes material and labor to install
	Install	S.F.	2.23	.97		3.20	T-bar grid FR acoustical ceiling system,
	Demolish and Install	S.F.	2.23	1.40		3.63	random pinhole 5/8" thick square
	Reinstall	S.F.		.77		.77	edge tile.
	Clean	S.F.	.03	.17		.20	
	Minimum Charge	Job		157		157	
T-bar Grid							
2' x 2'							
	Demolish	S.F.		.50		.50	Includes material and labor to install
	Install	S.F.	.56	.48		1.04	T-bar grid acoustical ceiling
	Demolish and Install	S.F.	.56	.98		1.54	suspension system.
	Reinstall	S.F.		.39		.39	
	Clean	S.F.	.03	.17		.20	
	Minimum Charge	Job		139		139	
2' x 4'							
	Demolish	S.F.		.50		.50	Includes material and labor to install
	Install	S.F.	.45	.39		.84	T-bar grid acoustical ceiling
	Demolish and Install	S.F.	.45	.89		1.34	suspension system.
	Reinstall	S.F.		.31		.31	
	Clean	S.F.	.03	.17		.20	
	Minimum Charge	Job		139		139	
Tile Panels							
2' x 2' Square Edge							
	Demolish	S.F.		.14		.14	Includes material and labor to install
	Install	S.F.	.47	.50		.97	square edge panels 5/8" thick with
	Demolish and Install	S.F.	.47	.64		1.11	pinhole pattern. T-bar grid not
	Reinstall	S.F.		.40		.40	included.
	Clean	S.F.	.03	.17		.20	
	Minimum Charge	Job		157		157	
2' x 2' Reveal Edge							
	Demolish	S.F.		.14		.14	Includes material and labor to install
	Install	S.F.	1.14	.67		1.81	reveal edge panels 5/8" thick with
	Demolish and Install	S.F.	1.14	.81		1.95	pinhole pattern. T-bar grid not
	Reinstall	S.F.		.53		.53	included.
	Clean	S.F.	.03	.17		.20	
	Minimum Charge	Job		157		157	
2' x 2' Fire-rated							
	Demolish	S.F.		.14		.14	Includes material and labor to install
	Install	S.F.	.92	.46		1.38	FR square edge panels 5/8" thick with
	Demolish and Install	S.F.	.92	.60		1.52	pinhole pattern. T-bar grid not
	Reinstall	S.F.		.37		.37	included.
	Clean	S.F.	.03	.17		.20	
	Minimum Charge	Job		157		157	
2' x 4' Square Edge							
	Demolish	S.F.		.14		.14	Includes material and labor to install
	Install	S.F.	.47	.50		.97	square edge panels 5/8" thick with
	Demolish and Install	S.F.	.47	.64		1.11	pinhole pattern. T-bar grid not
	Reinstall	S.F.		.40		.40	included.
	Clean	S.F.	.03	.17		.20	
	Minimum Charge	Job		157		157	

Walls / Ceilings

Acoustical Ceiling		Unit	Material	Labor	Equip.	Total	Specification
2' x 4' Reveal Edge							Includes material and labor to install
	Demolish	S.F.		.14		.14	reveal edge panels 5/8" thick with
	Install	S.F.	1.14	.67		1.81	pinhole pattern. T-bar grid not
	Demolish and Install	S.F.	1.14	.81		1.95	included.
	Reinstall	S.F.		.53		.53	
	Clean	S.F.	.03	.17		.20	
	Minimum Charge	Job		157		157	
2' x 4' Fire-rated							Includes material and labor to install
	Demolish	S.F.		.14		.14	FR square edge panels 5/8" thick with
	Install	S.F.	.92	.46		1.38	pinhole pattern. T-bar grid not
	Demolish and Install	S.F.	.92	.60		1.52	included.
	Reinstall	S.F.		.37		.37	
	Clean	S.F.	.03	.17		.20	
	Minimum Charge	Job		157		157	
Bleach Out Stains							Includes labor and material to apply
	Install	S.F.	.09	.23		.32	laundry type chlorine bleach to
	Minimum Charge	Job		157		157	remove light staining of acoustic tile
							material.
Adhesive Tile							
Good Grade							Cost includes material and labor to
	Demolish	S.F.		.57		.57	install 12" x 12" tile with embossed
	Install	S.F.	.85	1.05		1.90	texture pattern. Furring strips not
	Demolish and Install	S.F.	.85	1.62		2.47	included.
	Clean	S.F.	.02	.13		.15	
	Minimum Charge	Job		157		157	
Better Grade							Cost includes material and labor to
	Demolish	S.F.		.57		.57	install 12" x 12" tile with embossed
	Install	S.F.	1.13	1.05		2.18	texture pattern. Furring strips not
	Demolish and Install	S.F.	1.13	1.62		2.75	included.
	Clean	S.F.	.02	.13		.15	
	Minimum Charge	Job		157		157	
Premium Grade							Cost includes material and labor to
	Demolish	S.F.		.57		.57	install 12" x 12" tile with embossed
	Install	S.F.	1.75	1.05		2.80	texture pattern. Furring strips not
	Demolish and Install	S.F.	1.75	1.62		3.37	included.
	Clean	S.F.	.02	.13		.15	
	Minimum Charge	Job		157		157	
Bleach Out Stains							Includes labor and material to apply
	Install	S.F.	.09	.23		.32	laundry type chlorine bleach to
	Minimum Charge	Job		157		157	remove light staining of acoustic tile
							material.
Stapled Tile							
Good Grade							Cost includes material and labor to
	Demolish	S.F.		.34		.34	install 12" x 12" tile with embossed
	Install	S.F.	.85	1.05		1.90	texture pattern. Furring strips not
	Demolish and Install	S.F.	.85	1.39		2.24	included.
	Clean	S.F.	.02	.13		.15	
	Minimum Charge	Job		157		157	
Better Grade							Cost includes material and labor to
	Demolish	S.F.		.34		.34	install 12" x 12" tile with embossed
	Install	S.F.	1.13	1.05		2.18	texture pattern. Furring strips not
	Demolish and Install	S.F.	1.13	1.39		2.52	included.
	Clean	S.F.	.02	.13		.15	
	Minimum Charge	Job		157		157	

Walls / Ceilings

Acoustical Ceiling	Unit	Material	Labor	Equip.	Total	Specification
Premium Grade						
Demolish	S.F.		.34		.34	Cost includes material and labor to
Install	S.F.	1.75	1.05		2.80	install 12" x 12" tile with embossed
Demolish and Install	S.F.	1.75	1.39		3.14	texture pattern. Furring strips not
Clean	S.F.	.02	.13		.15	included.
Minimum Charge	Job		157		157	
Bleach Out Stains						
Install	S.F.	.09	.23		.32	Includes labor and material to apply
Minimum Charge	Job		157		157	laundry type chlorine bleach to
						remove light staining of acoustic tile
						material.
Blown						
Scrape Off Existing						
Install	S.F.		.47		.47	Includes labor and material to remove
Minimum Charge	Job		157		157	existing blown acoustical ceiling
						material.
Seal Ceiling Gypsum Wallboard						
Install	S.F.	.04	.24		.28	Includes labor and material to seal
Minimum Charge	Job		138		138	existing drywall with shellac-based
						material.
Reblow Existing						
Install	S.F.	.04	.35		.39	Includes labor and material to install
Minimum Charge	Job		375	61	436	by spray, texture finish.
Bleach Out Stains						
Install	S.F.	.09	.23		.32	Includes labor and material to apply
Minimum Charge	Job		157		157	laundry type chlorine bleach to
						remove light staining of acoustic tile
						material.
New Blown						
Install	S.F.	.04	.35		.39	Includes labor and material to install
Minimum Charge	Job		138		138	by spray, texture finish.
Apply Glitter To						
Install	S.F.	.09	.12		.21	Includes labor and material to apply
Minimum Charge	Job		138		138	glitter material to existing or new
						drywall.
Furring Strips						
Demolish	S.F.		.06		.06	Cost includes material and labor to
Install	S.F.	.15	.70		.85	install 1" x 2" furring strips, 16" O.C.
Demolish and Install	S.F.	.15	.76		.91	per S.F. of surface area to be covered.
Minimum Charge	Job		78.50		78.50	

Finished Stair

Bi-Fold Door

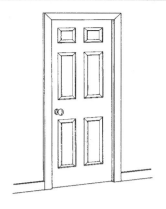

Raised Panel Door

Hollow Core Door	Unit	Material	Labor	Equip.	Total	Specification
Pre-hung Wood						
1' 6" x 6' 8"						
Demolish	Ea.		29		29	Cost includes material and labor to
Install	Ea.	142	14.95		156.95	install 1' 6" x 6' 8" pre-hung 1-3/8"
Demolish and Install	Ea.	142	43.95		185.95	lauan door with split pine jamb
Reinstall	Ea.		14.93		14.93	including casing, stops, and hinges.
Clean	Ea.	2.88	9.55		12.43	
Paint	Ea.	7.60	39.50		47.10	
Minimum Charge	Job		157		157	
2' x 6' 8"						
Demolish	Ea.		29		29	Cost includes material and labor to
Install	Ea.	146	15.70		161.70	install 2' x 6' 8" pre-hung 1-3/8"
Demolish and Install	Ea.	146	44.70		190.70	lauan door with split pine jamb
Reinstall	Ea.		15.68		15.68	including casing, stops, and hinges.
Clean	Ea.	2.88	9.55		12.43	
Paint	Ea.	7.60	39.50		47.10	
Minimum Charge	Job		157		157	
2' 4" x 6' 8"						
Demolish	Ea.		29		29	Cost includes material and labor to
Install	Ea.	153	15.70		168.70	install 2' 4" x 6' 8" pre-hung 1-3/8"
Demolish and Install	Ea.	153	44.70		197.70	lauan door with split pine jamb
Reinstall	Ea.		15.68		15.68	including casing, stops, and hinges.
Clean	Ea.	2.88	9.55		12.43	
Paint	Ea.	7.60	39.50		47.10	
Minimum Charge	Job		157		157	
2' 6" x 6' 8"						
Demolish	Ea.		29		29	Cost includes material and labor to
Install	Ea.	153	15.70		168.70	install 2' 6" x 6' 8" pre-hung 1-3/8"
Demolish and Install	Ea.	153	44.70		197.70	lauan door with split pine jamb
Reinstall	Ea.		15.68		15.68	including casing, stops, and hinges.
Clean	Ea.	2.88	9.55		12.43	
Paint	Ea.	7.60	39.50		47.10	
Minimum Charge	Job		157		157	
2' 8" x 6' 8"						
Demolish	Ea.		29		29	Cost includes material and labor to
Install	Ea.	156	15.70		171.70	install 2' 8" x 6' 8" pre-hung 1-3/8"
Demolish and Install	Ea.	156	44.70		200.70	lauan door with split pine jamb,
Reinstall	Ea.		15.68		15.68	including casing, stop and hinges.
Clean	Ea.	2.88	9.55		12.43	
Paint	Ea.	7.60	39.50		47.10	
Minimum Charge	Job		157		157	

Finish Carpentry

Hollow Core Door	Unit	Material	Labor	Equip.	Total	Specification
3' x 6' 8"						
Demolish	Ea.		29		29	Cost includes material and labor to
Install	Ea.	162	16.50		178.50	install 3' x 6' 8" pre-hung 1-3/8"
Demolish and Install	Ea.	162	45.50		207.50	lauan door with split pine jamb
Reinstall	Ea.		16.51		16.51	including casing, stops, and hinges.
Clean	Ea.	2.88	9.55		12.43	
Paint	Ea.	7.60	39.50		47.10	
Minimum Charge	Job		157		157	
Pocket Type						
Demolish	Ea.		43.50		43.50	Cost includes material and labor to
Install	Ea.	33	35		68	install 2' 8" x 6' 8" lauan door.
Demolish and Install	Ea.	33	78.50		111.50	
Reinstall	Ea.		34.84		34.84	
Clean	Ea.	2.88	9.55		12.43	
Paint	Ea.	7.60	39.50		47.10	
Minimum Charge	Job		157		157	
Stain						
Install	Ea.	3.98	23		26.98	Includes labor and material to stain
Minimum Charge	Job		138		138	single door and trim on both sides.
Pre-hung Masonite						
1' 6" x 6' 8"						
Demolish	Ea.		29		29	Cost includes material and labor to
Install	Ea.	185	14.95		199.95	install 1' 6" x 6' 8" pre-hung 1-3/8"
Demolish and Install	Ea.	185	43.95		228.95	masonite door with split pine jamb
Reinstall	Ea.		14.93		14.93	including casing, stops, and hinges.
Clean	Ea.	2.88	9.55		12.43	
Paint	Ea.	7.60	39.50		47.10	
Minimum Charge	Job		157		157	
2' x 6' 8"						
Demolish	Ea.		29		29	Cost includes material and labor to
Install	Ea.	193	15.70		208.70	install 2' x 6' 8" pre-hung 1-3/8"
Demolish and Install	Ea.	193	44.70		237.70	masonite door with split pine jamb
Reinstall	Ea.		15.68		15.68	including casing, stops, and hinges.
Clean	Ea.	2.88	9.55		12.43	
Paint	Ea.	7.60	39.50		47.10	
Minimum Charge	Job		157		157	
2' 4" x 6' 8"						
Demolish	Ea.		29		29	Cost includes material and labor to
Install	Ea.	194	15.70		209.70	install 2' 4" x 6' 8" pre-hung 1-3/8"
Demolish and Install	Ea.	194	44.70		238.70	masonite door with split pine jamb
Reinstall	Ea.		15.68		15.68	including casing, stops, and hinges.
Clean	Ea.	2.88	9.55		12.43	
Paint	Ea.	7.60	39.50		47.10	
Minimum Charge	Job		157		157	
2' 8" x 6' 8"						
Demolish	Ea.		29		29	Cost includes material and labor to
Install	Ea.	196	15.70		211.70	install 2' 8" x 6' 8" pre-hung 1-3/8"
Demolish and Install	Ea.	196	44.70		240.70	masonite door with split pine jamb
Reinstall	Ea.		15.68		15.68	including casing, stops, and hinges.
Clean	Ea.	2.88	9.55		12.43	
Paint	Ea.	7.60	39.50		47.10	
Minimum Charge	Job		157		157	
3' x 6' 8"						
Demolish	Ea.		29		29	Cost includes material and labor to
Install	Ea.	200	16.50		216.50	install 3' x 6' 8" pre-hung 1-3/8"
Demolish and Install	Ea.	200	45.50		245.50	masonite door with split pine jamb
Reinstall	Ea.		16.51		16.51	including casing, stops, and hinges.
Clean	Ea.	2.88	9.55		12.43	
Paint	Ea.	7.60	39.50		47.10	
Minimum Charge	Job		157		157	

Hollow Core Door		Unit	Material	Labor	Equip.	Total	Specification
Pocket Type							
	Demolish	Ea.		43.50		43.50	Cost includes material and labor to
	Install	Ea.	33	35		68	install 2' 8" x 6' 8" lauan door.
	Demolish and Install	Ea.	33	78.50		111.50	
	Reinstall	Ea.		34.84		34.84	
	Clean	Ea.	2.88	9.55		12.43	
	Paint	Ea.	7.60	39.50		47.10	
	Minimum Charge	Job		157		157	
Stain							
	Install	Ea.	3.98	23		26.98	Includes labor and material to stain
	Minimum Charge	Job		138		138	single door and trim on both sides.
Hollow Core Door Only							
1' 6" x 6' 8'							
	Demolish	Ea.		6.20		6.20	Cost includes material and labor to
	Install	Ea.	32.50	16.50		49	install 1' 6" x 6' 8" lauan hollow core
	Demolish and Install	Ea.	32.50	22.70		55.20	door slab.
	Reinstall	Ea.		16.51		16.51	
	Clean	Ea.	1.20	7.20		8.40	
	Paint	Ea.	2.37	34.50		36.87	
	Minimum Charge	Job		157		157	
2' x 6' 8'							
	Demolish	Ea.		6.20		6.20	Cost includes material and labor to
	Install	Ea.	34.50	17.40		51.90	install 2' x 6' 8" lauan hollow core
	Demolish and Install	Ea.	34.50	23.60		58.10	door slab.
	Reinstall	Ea.		17.42		17.42	
	Clean	Ea.	1.20	7.20		8.40	
	Paint	Ea.	2.37	34.50		36.87	
	Minimum Charge	Job		157		157	
2' 4" x 6' 8'							
	Demolish	Ea.		6.20		6.20	Cost includes material and labor to
	Install	Ea.	38	17.40		55.40	install 2' 4" x 6' 8" lauan hollow core
	Demolish and Install	Ea.	38	23.60		61.60	door slab.
	Reinstall	Ea.		17.42		17.42	
	Clean	Ea.	1.20	7.20		8.40	
	Paint	Ea.	2.37	34.50		36.87	
	Minimum Charge	Job		157		157	
2' 8" x 6' 8'							
	Demolish	Ea.		6.20		6.20	Cost includes material and labor to
	Install	Ea.	39.50	17.40		56.90	install 2' 8" x 6' 8" lauan hollow core
	Demolish and Install	Ea.	39.50	23.60		63.10	door slab.
	Reinstall	Ea.		17.42		17.42	
	Clean	Ea.	1.20	7.20		8.40	
	Paint	Ea.	2.37	34.50		36.87	
	Minimum Charge	Job		157		157	
3' x 6' 8'							
	Demolish	Ea.		6.20		6.20	Cost includes material and labor to
	Install	Ea.	41.50	18.45		59.95	install 3' x 6' 8" lauan hollow core
	Demolish and Install	Ea.	41.50	24.65		66.15	door slab.
	Reinstall	Ea.		18.45		18.45	
	Clean	Ea.	1.20	7.20		8.40	
	Paint	Ea.	2.37	34.50		36.87	
	Minimum Charge	Job		157		157	
Pocket Type							
	Demolish	Ea.		6.20		6.20	Cost includes material and labor to
	Install	Ea.	33	35		68	install 2' 8" x 6' 8" lauan door.
	Demolish and Install	Ea.	33	41.20		74.20	
	Reinstall	Ea.		34.84		34.84	
	Clean	Ea.	1.20	7.20		8.40	
	Paint	Ea.	2.37	34.50		36.87	
	Minimum Charge	Job		157		157	

Finish Carpentry

Hollow Core Door		Unit	Material	Labor	Equip.	Total	Specification
Stain							
	Install	Ea.	2.20	30.50		32.70	Includes labor and materials to stain a
	Minimum Charge	Job		138		138	flush style door on both sides by brush.
Casing Trim							
Single Width							
	Demolish	Ea.		5.55		5.55	Cost includes material and labor to
	Install	Opng.	15.90	53		68.90	install 11/16" x 2-1/2" pine ranch
	Demolish and Install	Opng.	15.90	58.55		74.45	style casing for one side of a standard
	Reinstall	Opng.		42.52		42.52	door opening.
	Clean	Opng.	.41	2.30		2.71	
	Paint	Ea.	3.08	5.50		8.58	
	Minimum Charge	Job		157		157	
Double Width							
	Demolish	Ea.		6.55		6.55	Cost includes material and labor to
	Install	Opng.	18.70	62.50		81.20	install 11/16" x 2-1/2" pine ranch
	Demolish and Install	Opng.	18.70	69.05		87.75	style door casing for one side of a
	Reinstall	Opng.		50.18		50.18	double door opening.
	Clean	Opng.	.48	4.59		5.07	
	Paint	Ea.	3.08	6.25		9.33	
	Minimum Charge	Job		157		157	
Hardware							
Doorknob w / Lock							
	Demolish	Ea.		11.60		11.60	Includes material and labor to install
	Install	Ea.	20	19.60		39.60	privacy lockset.
	Demolish and Install	Ea.	20	31.20		51.20	
	Reinstall	Ea.		15.68		15.68	
	Clean	Ea.	.29	7.20		7.49	
	Minimum Charge	Job		157		157	
Doorknob							
	Demolish	Ea.		11.60		11.60	Includes material and labor to install
	Install	Ea.	17.65	19.60		37.25	residential passage lockset, keyless
	Demolish and Install	Ea.	17.65	31.20		48.85	bored type.
	Reinstall	Ea.		15.68		15.68	
	Clean	Ea.	.29	7.20		7.49	
	Minimum Charge	Job		157		157	
Deadbolt							
	Demolish	Ea.		10.90		10.90	Includes material and labor to install a
	Install	Ea.	37.50	22.50		60	deadbolt.
	Demolish and Install	Ea.	37.50	33.40		70.90	
	Reinstall	Ea.		17.92		17.92	
	Clean	Ea.	.29	7.20		7.49	
	Minimum Charge	Job		157		157	
Lever Handle							
	Demolish	Ea.		11.60		11.60	Includes material and labor to install
	Install	Ea.	27	31.50		58.50	residential passage lockset, keyless
	Demolish and Install	Ea.	27	43.10		70.10	bored type with lever handle.
	Reinstall	Ea.		25.09		25.09	
	Clean	Ea.	.29	7.20		7.49	
	Minimum Charge	Job		157		157	
Closer							
	Demolish	Ea.		6.55		6.55	Includes material and labor to install
	Install	Ea.	93	48.50		141.50	pneumatic light duty closer for interior
	Demolish and Install	Ea.	93	55.05		148.05	type doors.
	Reinstall	Ea.		38.60		38.60	
	Clean	Ea.	.14	9.55		9.69	
	Minimum Charge	Job		157		157	

Finish Carpentry

Hollow Core Door

Hollow Core Door	Unit	Material	Labor	Equip.	Total	Specification
Push Plate						
Demolish	Ea.		9.70		9.70	Includes material and labor to install
Install	Ea.	17.25	26		43.25	bronze push-pull plate.
Demolish and Install	Ea.	17.25	35.70		52.95	
Reinstall	Ea.		20.91		20.91	
Clean	Ea.	.29	7.20		7.49	
Minimum Charge	Job		157		157	
Kickplate						
Demolish	Ea.		9.70		9.70	Includes material and labor to install
Install	Ea.	19.80	21		40.80	aluminum kickplate, 10" x 28".
Demolish and Install	Ea.	19.80	30.70		50.50	
Reinstall	Ea.		16.73		16.73	
Clean	Ea.	.29	7.20		7.49	
Minimum Charge	Job		157		157	
Exit Sign						
Demolish	Ea.		17.35		17.35	Includes material and labor to install a
Install	Ea.	44	43		87	wall mounted interior electric exit sign.
Demolish and Install	Ea.	44	60.35		104.35	
Reinstall	Ea.		34.36		34.36	
Clean	Ea.	.02	5.15		5.17	
Minimum Charge	Job		157		157	
Jamb						
Demolish	Ea.		22		22	Includes material and labor to install
Install	Ea.	80	8.50		88.50	flat pine jamb with square cut heads
Demolish and Install	Ea.	80	30.50		110.50	and rabbeted sides for 6' 8" high and
Clean	Ea.	.09	5.75		5.84	3-9/16" door including trim sets for
Paint	Ea.	5.60	18.40		24	both sides.
Minimum Charge	Job		157		157	
Shave & Refit						
Install	Ea.		26		26	Includes labor to shave and rework
Minimum Charge	Job		157		157	door to fit opening at the job site.

Solid Core Door

Solid Core Door	Unit	Material	Labor	Equip.	Total	Specification
Pre-hung Wood						
2' 4" x 6' 8"						
Demolish	Ea.		29		29	Cost includes material and labor to
Install	Ea.	174	15.70		189.70	install 2' 4" x 6' 8" pre-hung 1-3/4"
Demolish and Install	Ea.	174	44.70		218.70	door, 4-1/2" split jamb, including
Reinstall	Ea.		15.68		15.68	casing and stop, hinges, aluminum sill,
Clean	Ea.	2.88	9.55		12.43	weatherstripping.
Paint	Ea.	8.95	39.50		48.45	
Minimum Charge	Job		157		157	
2' 6" x 6' 8"						
Demolish	Ea.		29		29	Cost includes material and labor to
Install	Ea.	178	15.70		193.70	install 2' 6" x 6' 8" pre-hung 1-3/4"
Demolish and Install	Ea.	178	44.70		222.70	door, 4-1/2" split jamb, including
Reinstall	Ea.		15.68		15.68	casing and stop, hinges, aluminum sill,
Clean	Ea.	2.88	9.55		12.43	weatherstripping.
Paint	Ea.	8.95	39.50		48.45	
Minimum Charge	Job		157		157	

Finish Carpentry

Solid Core Door

Solid Core Door	Unit	Material	Labor	Equip.	Total	Specification
2' 8" x 6' 8"						
Demolish	Ea.		29		29	Cost includes material and labor to
Install	Ea.	202	15.70		217.70	install 2' 8" x 6' 8" pre-hung 1-3/4"
Demolish and Install	Ea.	202	44.70		246.70	door, 4-1/2" split jamb, including
Reinstall	Ea.		15.68		15.68	casing and stop, hinges, aluminum sill,
Clean	Ea.	2.88	9.55		12.43	weatherstripping.
Paint	Ea.	8.95	39.50		48.45	
Minimum Charge	Job		157		157	
3' x 6' 8"						
Demolish	Ea.		29		29	Cost includes material and labor to
Install	Ea.	210	16.50		226.50	install 3' x 6' 8" pre-hung 1-3/4"
Demolish and Install	Ea.	210	45.50		255.50	door, including casing and stop,
Reinstall	Ea.		16.51		16.51	hinges, jamb, aluminum sill,
Clean	Ea.	2.88	9.55		12.43	weatherstripped.
Paint	Ea.	8.95	39.50		48.45	
Minimum Charge	Job		157		157	
3' 6" x 6' 8"						
Demolish	Ea.		29		29	Cost includes material and labor to
Install	Ea.	330	17.40		347.40	install 3' 6" x 6' 8" pre-hung 1-3/4"
Demolish and Install	Ea.	330	46.40		376.40	door, 4-1/2" split jamb, including
Reinstall	Ea.		17.42		17.42	casing and stop, hinges, aluminum sill,
Clean	Ea.	2.88	9.55		12.43	weatherstripping.
Paint	Ea.	8.95	39.50		48.45	
Minimum Charge	Job		157		157	
Raised Panel						
Demolish	Ea.		29		29	Cost includes material and labor to
Install	Ea.	370	17.40		387.40	install 3' x 6' 8" pre-hung 1-3/4"
Demolish and Install	Ea.	370	46.40		416.40	door, jamb, including casing and
Reinstall	Ea.		17.42		17.42	stop, hinges, aluminum sill,
Clean	Ea.	2.88	9.55		12.43	weatherstripping.
Paint	Ea.	8.95	92		100.95	
Minimum Charge	Job		157		157	
Stain						
Install	Ea.	3.98	23		26.98	Includes labor and material to stain
Minimum Charge	Job		138		138	single door and trim on both sides.

Bifold Door

Bifold Door	Unit	Material	Labor	Equip.	Total	Specification
Wood, Single						
3'						
Demolish	Ea.		22		22	Cost includes material and labor to
Install	Ea.	54.50	48.50		103	install 3' wide solid wood bi-fold door
Demolish and Install	Ea.	54.50	70.50		125	not including jamb, trim, track, and
Reinstall	Ea.		48.25		48.25	hardware.
Clean	Ea.	2.88	9.55		12.43	
Paint	Ea.	7.60	39.50		47.10	
Minimum Charge	Job		157		157	
5'						
Demolish	Ea.		22		22	Includes material and labor to install
Install	Ea.	100	57		157	pair of 2' 6" wide solid wood bi-fold
Demolish and Install	Ea.	100	79		179	doors not including jamb, trim, track,
Reinstall	Ea.		57.02		57.02	and hardware.
Clean	Ea.	1.27	16.40		17.67	
Paint	Ea.	16.45	69		85.45	
Minimum Charge	Job		157		157	

Finish Carpentry

Bifold Door		Unit	Material	Labor	Equip.	Total	Specification
6'							
	Demolish	Ea.		22		22	Includes material and labor to install
	Install	Ea.	108	62.50		170.50	pair of 3' wide solid wood bi-fold
	Demolish and Install	Ea.	108	84.50		192.50	doors not including jamb, trim, track,
	Reinstall	Ea.		62.72		62.72	and hardware.
	Clean	Ea.	1.44	19.15		20.59	
	Paint	Ea.	16.45	69		85.45	
	Minimum Charge	Job		157		157	
8'							
	Demolish	Ea.		22		22	Includes material and labor to install
	Install	Ea.	190	35		225	pair of 4' wide solid wood bi-fold
	Demolish and Install	Ea.	190	57		247	doors not including jamb, trim, track,
	Reinstall	Ea.		34.84		34.84	and hardware.
	Clean	Ea.	1.98	23		24.98	
	Paint	Ea.	8.60	69		77.60	
	Minimum Charge	Job		157		157	
Mirrored							
5'							
	Demolish	Ea.		22		22	Includes material and labor to install
	Install	Ea.	375	28.50		403.50	pair of 2' 6" wide mirrored bi-fold
	Demolish and Install	Ea.	375	50.50		425.50	doors including jamb, trim, track, and
	Reinstall	Ea.		28.51		28.51	hardware.
	Clean	Ea.	1.27	8.20		9.47	
	Minimum Charge	Job		157		157	
6'							
	Demolish	Ea.		22		22	Includes material and labor to install
	Install	Ea.	415	35		450	pair of 3' wide mirrored bi-fold doors
	Demolish and Install	Ea.	415	57		472	including jamb, trim, track, and
	Reinstall	Ea.		34.84		34.84	hardware.
	Clean	Ea.	1.44	9.55		10.99	
	Minimum Charge	Job		157		157	
8'							
	Demolish	Ea.		22		22	Includes material and labor to install
	Install	Ea.	665	52.50		717.50	pair of 4' wide mirrored bi-fold doors
	Demolish and Install	Ea.	665	74.50		739.50	including jamb, trim, track, and
	Reinstall	Ea.		52.27		52.27	hardware.
	Clean	Ea.	1.98	11.50		13.48	
	Minimum Charge	Job		157		157	
Louvered							
3'							
	Demolish	Ea.		22		22	Cost includes material and labor to
	Install	Ea.	162	48.50		210.50	install 3' wide louvered bi-fold door
	Demolish and Install	Ea.	162	70.50		232.50	not including jamb, trim, track, and
	Reinstall	Ea.		48.25		48.25	hardware.
	Clean	Ea.	2.88	9.55		12.43	
	Paint	Ea.	7.60	39.50		47.10	
	Minimum Charge	Job		157		157	
5'							
	Demolish	Ea.		22		22	Includes material and labor to install
	Install	Ea.	204	57		261	pair of 2' 6" wide louvered wood
	Demolish and Install	Ea.	204	79		283	bi-fold doors not including jamb, trim,
	Reinstall	Ea.		57.02		57.02	track, and hardware.
	Clean	Ea.	1.27	16.40		17.67	
	Paint	Ea.	16.45	69		85.45	
	Minimum Charge	Job		157		157	

Finish Carpentry

Bifold Door	Unit	Material	Labor	Equip.	Total	Specification
6'						
Demolish	Ea.		22		22	Includes material and labor to install
Install	Ea.	224	62.50		286.50	pair of 3' wide louvered wood bi-fold
Demolish and Install	Ea.	224	84.50		308.50	doors not including jamb, trim, track,
Reinstall	Ea.		62.72		62.72	and hardware.
Clean	Ea.	1.44	19.15		20.59	
Paint	Ea.	16.45	69		85.45	
Minimum Charge	Job		157		157	
8'						
Demolish	Ea.		22		22	Includes material and labor to install
Install	Ea.	395	31.50		426.50	pair of 4' wide solid wood bi-fold
Demolish and Install	Ea.	395	53.50		448.50	doors not including jamb, trim, track,
Reinstall	Ea.		31.36		31.36	and hardware.
Clean	Ea.	6.90	16.40		23.30	
Paint	Ea.	8.60	69		77.60	
Minimum Charge	Job		157		157	
Accordion Type						
Demolish	S.F.		.75		.75	Includes material and labor to install
Install	S.F.	1.77	1.57		3.34	vinyl folding accordion closet doors,
Demolish and Install	S.F.	1.77	2.32		4.09	including track and frame.
Reinstall	S.F.		1.57		1.57	
Clean	S.F.		.27		.27	
Minimum Charge	Job		157		157	
Casing Trim						
Single Width						
Demolish	Ea.		5.55		5.55	Cost includes material and labor to
Install	Opng.	15.90	53		68.90	install 11/16" x 2-1/2" pine ranch
Demolish and Install	Opng.	15.90	58.55		74.45	style casing for one side of a standard
Reinstall	Opng.		42.52		42.52	door opening.
Clean	Opng.	.41	2.30		2.71	
Paint	Ea.	3.08	5.50		8.58	
Minimum Charge	Job		157		157	
Double Width						
Demolish	Ea.		6.55		6.55	Cost includes material and labor to
Install	Opng.	18.70	62.50		81.20	install 11/16" x 2-1/2" pine ranch
Demolish and Install	Opng.	18.70	69.05		87.75	style door casing for one side of a
Reinstall	Opng.		50.18		50.18	double door opening.
Clean	Opng.	.48	4.59		5.07	
Paint	Ea.	3.08	6.25		9.33	
Minimum Charge	Job		157		157	
Jamb						
Demolish	Ea.		22		22	Includes material and labor to install
Install	Ea.	80	8.50		88.50	flat pine jamb with square cut heads
Demolish and Install	Ea.	80	30.50		110.50	and rabbeted sides for 6' 8" high and
Clean	Ea.	.09	5.75		5.84	3-9/16" door including trim sets for
Paint	Ea.	5.60	18.40		24	both sides.
Minimum Charge	Job		78.50		78.50	
Stain						
Install	Ea.	3.98	23		26.98	Includes labor and material to stain
Minimum Charge	Job		138		138	single door and trim on both sides.
Shave & Refit						
Install	Ea.		26		26	Includes labor to shave and rework
Minimum Charge	Job		78.50		78.50	door to fit opening at the job site.

Finish Carpentry

Louver Door

	Unit	Material	Labor	Equip.	Total	Specification
Full						
2' x 6' 8"						
Demolish	Ea.		6.20		6.20	Cost includes material and labor to
Install	Ea.	166	31.50		197.50	install 2' x 6' 8" pine full louver door,
Demolish and Install	Ea.	166	37.70		203.70	jamb, hinges.
Reinstall	Ea.		31.36		31.36	
Clean	Ea.	5.35	28.50		33.85	
Paint	Ea.	7.60	39.50		47.10	
Minimum Charge	Job		157		157	
2' 8" x 6' 8"						
Demolish	Ea.		6.20		6.20	Cost includes material and labor to
Install	Ea.	185	31.50		216.50	install 2' 8" x 6' 8" pine full louver
Demolish and Install	Ea.	185	37.70		222.70	door, jamb, hinges.
Reinstall	Ea.		31.36		31.36	
Clean	Ea.	5.35	28.50		33.85	
Paint	Ea.	7.60	39.50		47.10	
Minimum Charge	Job		157		157	
3' x 6' 8"						
Demolish	Ea.		6.20		6.20	Cost includes material and labor to
Install	Ea.	196	33		229	install 3' x 6' 8" pine full louver door,
Demolish and Install	Ea.	196	39.20		235.20	jamb, hinges.
Reinstall	Ea.		33.01		33.01	
Clean	Ea.	5.35	28.50		33.85	
Paint	Ea.	7.60	39.50		47.10	
Minimum Charge	Job		157		157	
Half						
2' x 6' 8"						
Demolish	Ea.		6.20		6.20	Cost includes material and labor to
Install	Ea.	277	31.50		308.50	install 2' x 6' 8" pine half louver door,
Demolish and Install	Ea.	277	37.70		314.70	jamb, hinges.
Reinstall	Ea.		31.36		31.36	
Clean	Ea.	5.35	28.50		33.85	
Paint	Ea.	7.60	39.50		47.10	
Minimum Charge	Job		157		157	
2' 8" x 6' 8"						
Demolish	Ea.		6.20		6.20	Cost includes material and labor to
Install	Ea.	300	31.50		331.50	install 2' 8" x 6' 8" pine half louver
Demolish and Install	Ea.	300	37.70		337.70	door, jamb, hinges.
Reinstall	Ea.		31.36		31.36	
Clean	Ea.	5.35	28.50		33.85	
Paint	Ea.	7.60	39.50		47.10	
Minimum Charge	Job		157		157	
3' x 6' 8"						
Demolish	Ea.		6.20		6.20	Cost includes material and labor to
Install	Ea.	315	33		348	install 3' x 6' 8" pine half louver door,
Demolish and Install	Ea.	315	39.20		354.20	jamb, hinges.
Reinstall	Ea.		33.01		33.01	
Clean	Ea.	5.35	28.50		33.85	
Paint	Ea.	7.60	39.50		47.10	
Minimum Charge	Job		157		157	
Jamb						
Demolish	Ea.		22		22	Includes material and labor to install
Install	Ea.	80	8.50		88.50	flat pine jamb with square cut heads
Demolish and Install	Ea.	80	30.50		110.50	and rabbeted sides for 6' 8" high and
Clean	Ea.	.09	5.75		5.84	3-9/16" door including trim sets for
Paint	Ea.	5.60	18.40		24	both sides.
Minimum Charge	Job		78.50		78.50	

Finish Carpentry

Louver Door		Unit	Material	Labor	Equip.	Total	Specification
Stain							
	Install	Ea.	3.98	23		26.98	Includes labor and material to stain
	Minimum Charge	Job		138		138	single door and trim on both sides.
Shave & Refit							
	Install	Ea.		26		26	Includes labor to shave and rework
	Minimum Charge	Job		138		138	door to fit opening at the job site.

Bypass Sliding Door		Unit	Material	Labor	Equip.	Total	Specification
Wood							
5'							
	Demolish	Ea.		29		29	Includes material and labor to install
	Install	Opng.	188	57		245	pair of 2' 6" wide lauan hollow core
	Demolish and Install	Opng.	188	86		274	wood bi-pass doors with jamb, track
	Reinstall	Opng.		57.02		57.02	hardware.
	Clean	Ea.	1.27	16.40		17.67	
	Paint	Ea.	16.45	69		85.45	
	Minimum Charge	Job		157		157	
6'							
	Demolish	Ea.		29		29	Includes material and labor to install
	Install	Opng.	201	62.50		263.50	pair of 3' wide lauan hollow core
	Demolish and Install	Opng.	201	91.50		292.50	wood bi-pass doors with jamb, track
	Reinstall	Opng.		62.72		62.72	and hardware.
	Clean	Ea.	1.44	19.15		20.59	
	Paint	Ea.	16.45	69		85.45	
	Minimum Charge	Job		157		157	
8'							
	Demolish	Ea.		29		29	Includes material and labor to install
	Install	Opng.	310	78.50		388.50	pair of 4' wide lauan hollow core
	Demolish and Install	Opng.	310	107.50		417.50	wood bi-pass doors with jamb, track
	Reinstall	Opng.		78.40		78.40	and hardware.
	Clean	Ea.	1.98	23		24.98	
	Paint	Ea.	8.60	69		77.60	
	Minimum Charge	Job		157		157	
Mirrored							
5'							
	Demolish	Ea.		29		29	Includes material and labor to install
	Install	Opng.	223	62.50		285.50	pair of 2' 6 " wide mirrored bi-pass
	Demolish and Install	Opng.	223	91.50		314.50	doors with jamb, trim, track hardware.
	Reinstall	Opng.		62.72		62.72	
	Clean	Ea.	1.27	8.20		9.47	
	Minimum Charge	Job		157		157	
6'							
	Demolish	Ea.		29		29	Includes material and labor to install
	Install	Opng.	254	62.50		316.50	pair of 3' wide mirrored bi-pass doors
	Demolish and Install	Opng.	254	91.50		345.50	with jamb, trim, track hardware.
	Reinstall	Opng.		62.72		62.72	
	Clean	Ea.	1.44	9.55		10.99	
	Minimum Charge	Job		157		157	
8'							
	Demolish	Ea.		29		29	Includes material and labor to install
	Install	Opng.	530	69.50		599.50	pair of 4' wide mirrored bi-pass doors
	Demolish and Install	Opng.	530	98.50		628.50	with jamb, trim, track hardware.
	Reinstall	Opng.		69.69		69.69	
	Clean	Ea.	1.98	11.50		13.48	
	Minimum Charge	Job		157		157	

Finish Carpentry

Bypass Sliding Door

Bypass Sliding Door	Unit	Material	Labor	Equip.	Total	Specification
Casing Trim						
Single Width						Cost includes material and labor to
Demolish	Ea.		5.55		5.55	install 11/16″ x 2-1/2″ pine ranch
Install	Opng.	15.90	53		68.90	style casing for one side of a standard
Demolish and Install	Opng.	15.90	58.55		74.45	door opening.
Reinstall	Opng.		42.52		42.52	
Clean	Opng.	.41	2.30		2.71	
Paint	Ea.	3.08	5.50		8.58	
Minimum Charge	Job		157		157	
Double Width						Cost includes material and labor to
Demolish	Ea.		6.55		6.55	install 11/16″ x 2-1/2″ pine ranch
Install	Opng.	18.70	62.50		81.20	style door casing for one side of a
Demolish and Install	Opng.	18.70	69.05		87.75	double door opening.
Reinstall	Opng.		50.18		50.18	
Clean	Opng.	.48	4.59		5.07	
Paint	Ea.	3.08	6.25		9.33	
Minimum Charge	Job		157		157	
Jamb						Includes material and labor to install
Demolish	Ea.		22		22	flat pine jamb with square cut heads
Install	Ea.	80	8.50		88.50	and rabbeted sides for 6′ 8″ high and
Demolish and Install	Ea.	80	30.50		110.50	3-9/16″ door including trim sets for
Clean	Ea.	.09	5.75		5.84	both sides.
Paint	Ea.	5.60	18.40		24	
Minimum Charge	Job		78.50		78.50	
Stain						Includes labor and material to stain
Install	Ea.	3.98	23		26.98	single door and trim on both sides.
Minimum Charge	Job		138		138	
Shave & Refit						Includes labor to shave and rework
Install	Ea.		26		26	door to fit opening at the job site.
Minimum Charge	Job		78.50		78.50	

Base Molding

Base Molding	Unit	Material	Labor	Equip.	Total	Specification
One Piece						
1-5/8″						Includes material and labor to install
Demolish	L.F.		.43		.43	pine or birch base molding.
Install	L.F.	.23	.93		1.16	
Demolish and Install	L.F.	.23	1.36		1.59	
Reinstall	L.F.		.74		.74	
Clean	L.F.	.02	.13		.15	
Paint	L.F.	.03	.30		.33	
Minimum Charge	Job		157		157	
2-1/2″						Cost includes material and labor to
Demolish	L.F.		.43		.43	install 1/2″ x 2-1/2″ pine or birch
Install	L.F.	.49	.97		1.46	base molding.
Demolish and Install	L.F.	.49	1.40		1.89	
Reinstall	L.F.		.77		.77	
Clean	L.F.	.02	.13		.15	
Paint	L.F.	.03	.30		.33	
Minimum Charge	Job		157		157	

Finish Carpentry

Base Molding		Unit	Material	Labor	Equip.	Total	Specification
3-1/2"							
	Demolish	L.F.		.43		.43	Cost includes material and labor to
	Install	L.F.	.94	1.01		1.95	install 9/16" x 3-1/2" pine or birch
	Demolish and Install	L.F.	.94	1.44		2.38	base molding.
	Reinstall	L.F.		.80		.80	
	Clean	L.F.	.02	.13		.15	
	Paint	L.F.	.03	.33		.36	
	Minimum Charge	Job		157		157	
6"							
	Demolish	L.F.		.43		.43	Cost includes material and labor to
	Install	L.F.	.75	1.10		1.85	install 1" x 6" pine or birch base
	Demolish and Install	L.F.	.75	1.53		2.28	molding.
	Reinstall	L.F.		.88		.88	
	Clean	L.F.	.02	.19		.21	
	Paint	L.F.	.04	.42		.46	
	Minimum Charge	Job		157		157	
8"							
	Demolish	L.F.		.51		.51	Cost includes material and labor to
	Install	L.F.	1.16	1.21		2.37	install 1" x 8" pine or birch base
	Demolish and Install	L.F.	1.16	1.72		2.88	molding.
	Reinstall	L.F.		.96		.96	
	Clean	L.F.	.03	.23		.26	
	Paint	L.F.	.06	.46		.52	
	Minimum Charge	Job		157		157	
Premium / Custom Grade							
	Demolish	L.F.		.51		.51	Includes material and labor to install
	Install	L.F.	3.83	3.02		6.85	custom 2 or 3 piece oak or other
	Demolish and Install	L.F.	3.83	3.53		7.36	hardwood, stain grade trim.
	Reinstall	L.F.		2.41		2.41	
	Clean	L.F.	.03	.23		.26	
	Paint	L.F.	.06	.46		.52	
	Minimum Charge	Job		157		157	
One Piece Oak							
1-5/8"							
	Demolish	L.F.		.43		.43	Cost includes material and labor to
	Install	L.F.	.70	.93		1.63	install 1/2" x 1-5/8" oak base
	Demolish and Install	L.F.	.70	1.36		2.06	molding.
	Reinstall	L.F.		.74		.74	
	Clean	L.F.	.02	.13		.15	
	Paint	L.F.	.08	.41		.49	
	Minimum Charge	Job		157		157	
2-1/2"							
	Demolish	L.F.		.43		.43	Cost includes material and labor to
	Install	L.F.	1.07	.95		2.02	install 1/2" x 2-1/2" oak base
	Demolish and Install	L.F.	1.07	1.38		2.45	molding.
	Reinstall	L.F.		.76		.76	
	Clean	L.F.	.02	.13		.15	
	Paint	L.F.	.08	.42		.50	
	Minimum Charge	Job		157		157	
3-1/2"							
	Demolish	L.F.		.43		.43	Cost includes material and labor to
	Install	L.F.	2.39	.97		3.36	install 1/2" x 3-1/2" oak base
	Demolish and Install	L.F.	2.39	1.40		3.79	molding.
	Reinstall	L.F.		.77		.77	
	Clean	L.F.	.02	.13		.15	
	Paint	L.F.	.08	.42		.50	
	Minimum Charge	Job		157		157	

Finish Carpentry

Base Molding

		Unit	Material	Labor	Equip.	Total	Specification
6″							
	Demolish	L.F.		.43		.43	Cost includes material and labor to
	Install	L.F.	5.55	1.10		6.65	install 1″ x 6″ oak base molding.
	Demolish and Install	L.F.	5.55	1.53		7.08	
	Reinstall	L.F.		.88		.88	
	Clean	L.F.	.03	.23		.26	
	Paint	L.F.	.08	.44		.52	
	Minimum Charge	Job		157		157	
8″							
	Demolish	L.F.		.51		.51	Cost includes material and labor to
	Install	L.F.	6.45	1.21		7.66	install 1″ x 8″ oak base molding.
	Demolish and Install	L.F.	6.45	1.72		8.17	
	Reinstall	L.F.		.96		.96	
	Clean	L.F.	.03	.23		.26	
	Paint	L.F.	.08	.46		.54	
	Minimum Charge	Job		157		157	

Vinyl Base Molding

		Unit	Material	Labor	Equip.	Total	Specification
2-1/2″							
	Demolish	L.F.		.43		.43	Cost includes material and labor to
	Install	L.F.	.48	.90		1.38	install 2-1/2″ vinyl or rubber cove
	Demolish and Install	L.F.	.48	1.33		1.81	base including adhesive.
	Reinstall	L.F.		.90		.90	
	Clean	L.F.		.21		.21	
	Minimum Charge	Job		143		143	
4″							
	Demolish	L.F.		.43		.43	Cost includes material and labor to
	Install	L.F.	.54	.90		1.44	install 4″ vinyl or rubber cove base
	Demolish and Install	L.F.	.54	1.33		1.87	including adhesive.
	Reinstall	L.F.		.90		.90	
	Clean	L.F.		.21		.21	
	Minimum Charge	Job		143		143	
6″							
	Demolish	L.F.		.43		.43	Cost includes material and labor to
	Install	L.F.	.88	.90		1.78	install 6″ vinyl or rubber cove base
	Demolish and Install	L.F.	.88	1.33		2.21	including adhesive.
	Reinstall	L.F.		.90		.90	
	Clean	L.F.		.21		.21	
	Minimum Charge	Job		143		143	

Prefinished

		Unit	Material	Labor	Equip.	Total	Specification
	Demolish	L.F.		.43		.43	Cost includes material and labor to
	Install	L.F.	1.69	1.01		2.70	install 9/16″ x 3-5/16″ prefinished
	Demolish and Install	L.F.	1.69	1.44		3.13	base molding.
	Reinstall	L.F.		.80		.80	
	Clean	L.F.	.02	.14		.16	
	Paint	L.F.	.03	.33		.36	
	Minimum Charge	Job		157		157	

Shoe

		Unit	Material	Labor	Equip.	Total	Specification
	Demolish	L.F.		.43		.43	Includes material and labor to install
	Install	L.F.	.23	.93		1.16	pine or birch base molding.
	Demolish and Install	L.F.	.23	1.36		1.59	
	Reinstall	L.F.		.74		.74	
	Clean	L.F.	.02	.13		.15	
	Paint	L.F.	.12	.44		.56	
	Minimum Charge	Job		157		157	

Finish Carpentry

Base Molding

Base Molding	Unit	Material	Labor	Equip.	Total	Specification
Carpeted						
Demolish	L.F.		.17		.17	Includes material and labor to install
Install	L.F.	1.20	1.72		2.92	carpet baseboard with adhesive.
Demolish and Install	L.F.	1.20	1.89		3.09	
Reinstall	L.F.		1.38		1.38	
Clean	L.F.		.09		.09	
Minimum Charge	Job		142		142	
Ceramic Tile						
Demolish	L.F.		.93		.93	Includes material and labor to install
Install	L.F.	3.40	3.94		7.34	thin set ceramic tile cove base
Demolish and Install	L.F.	3.40	4.87		8.27	including grout.
Clean	L.F.		.44		.44	
Minimum Charge	Job		142		142	
Quarry Tile						
Demolish	L.F.		.93		.93	Includes material and labor to install
Install	L.F.	4.24	4.58		8.82	quarry tile base including grout.
Demolish and Install	L.F.	4.24	5.51		9.75	
Clean	L.F.		.44		.44	
Minimum Charge	Job		142		142	
Caulk and Renail						
Install	L.F.	.01	.49		.50	Includes labor and material to renail
Minimum Charge	Job		157		157	base and caulk where necessary.

Molding

Molding	Unit	Material	Labor	Equip.	Total	Specification
Oak Base						
Demolish	L.F.		.43		.43	Cost includes material and labor to
Install	L.F.	1.07	.95		2.02	install 1/2" x 2-1/2" oak base
Demolish and Install	L.F.	1.07	1.38		2.45	molding.
Reinstall	L.F.		.76		.76	
Clean	L.F.	.02	.13		.15	
Paint	L.F.	.08	.42		.50	
Minimum Charge	Job		157		157	
Shoe						
Demolish	L.F.		.43		.43	Includes material and labor to install
Install	L.F.	.23	.93		1.16	pine or birch base molding.
Demolish and Install	L.F.	.23	1.36		1.59	
Reinstall	L.F.		.74		.74	
Clean	L.F.	.02	.13		.15	
Paint	L.F.	.12	.44		.56	
Minimum Charge	Job		157		157	
Quarter Round						
Demolish	L.F.		.43		.43	Includes material and labor to install
Install	L.F.	.40	1.23		1.63	pine base quarter round molding.
Demolish and Install	L.F.	.40	1.66		2.06	
Reinstall	L.F.		.98		.98	
Clean	L.F.	.02	.13		.15	
Paint	L.F.	.06	.69		.75	
Minimum Charge	Job		157		157	
Chair Rail						
Demolish	L.F.		.43		.43	Cost includes material and labor to
Install	L.F.	.94	1.16		2.10	install 5/8" x 2-1/2" oak chair rail
Demolish and Install	L.F.	.94	1.59		2.53	molding.
Reinstall	L.F.		.93		.93	
Clean	L.F.	.02	.13		.15	
Paint	L.F.	.06	.69		.75	
Minimum Charge	Job		157		157	

Finish Carpentry

Molding

Molding		Unit	Material	Labor	Equip.	Total	Specification
Crown							
	Demolish	L.F.		.43		.43	Includes material and labor to install
	Install	L.F.	1.84	1.25		3.09	unfinished pine crown molding.
	Demolish and Install	L.F.	1.84	1.68		3.52	
	Reinstall	L.F.		1		1	
	Clean	L.F.	.03	.18		.21	
	Paint	L.F.	.06	.69		.75	
	Minimum Charge	Job		157		157	
Cove							
	Demolish	L.F.		.43		.43	Cost includes material and labor to
	Install	L.F.	1.38	1.23		2.61	install 1/2" x 2-3/4" pine cove
	Demolish and Install	L.F.	1.38	1.66		3.04	molding.
	Reinstall	L.F.		.98		.98	
	Clean	L.F.	.02	.13		.15	
	Paint	L.F.	.06	.69		.75	
	Minimum Charge	Job		157		157	
Corner							
	Demolish	L.F.		.43		.43	Includes material and labor to install
	Install	L.F.	.56	.49		1.05	inside or outside pine corner molding.
	Demolish and Install	L.F.	.56	.92		1.48	
	Reinstall	L.F.		.39		.39	
	Clean	L.F.	.02	.13		.15	
	Paint	L.F.	.06	.69		.75	
	Minimum Charge	Job		157		157	
Picture							
	Demolish	L.F.		.43		.43	Cost includes material and labor to
	Install	L.F.	.94	1.31		2.25	install 9/16" x 2-1/2" pine casing.
	Demolish and Install	L.F.	.94	1.74		2.68	
	Reinstall	L.F.		1.05		1.05	
	Clean	L.F.	.02	.13		.15	
	Paint	L.F.	.06	.69		.75	
	Minimum Charge	Job		157		157	
Single Cased Opening							
	Demolish	Ea.		22		22	Includes material and labor to install
	Install	Ea.	13.95	19.60		33.55	flat pine jamb with square cut heads
	Demolish and Install	Ea.	13.95	41.60		55.55	and rabbeted sides for 6' 8" high
	Clean	Ea.	.20	4.78		4.98	opening including trim sets for both
	Paint	Ea.	1	6.90		7.90	sides.
	Minimum Charge	Job		157		157	
Double Cased Opening							
	Demolish	Ea.		22		22	Includes material and labor to install
	Install	Ea.	28	22.50		50.50	flat pine jamb with square cut heads
	Demolish and Install	Ea.	28	44.50		72.50	and rabbeted sides for 6' 8" high
	Clean	Ea.	.23	5.60		5.83	opening including trim sets for both
	Paint	Ea.	1.17	8.10		9.27	sides.
	Minimum Charge	Job		157		157	

Window Trim Set

Window Trim Set		Unit	Material	Labor	Equip.	Total	Specification
Single							
	Demolish	Ea.		5.65		5.65	Cost includes material and labor to
	Install	Opng.	26	31.50		57.50	install 11/16" x 2-1/2" pine ranch
	Demolish and Install	Opng.	26	37.15		63.15	style window casing.
	Reinstall	Opng.		25.09		25.09	
	Clean	Ea.	.29	3.28		3.57	
	Paint	Ea.	.95	6.55		7.50	
	Minimum Charge	Job		157		157	

Finish Carpentry

Window Trim Set

	Unit	Material	Labor	Equip.	Total	Specification
Double						
Demolish	Ea.		6.40		6.40	Cost includes material and labor to
Install	Ea.	35	39		74	install 11/16" x 2-1/2" pine ranch
Demolish and Install	Ea.	35	45.40		80.40	style trim for casing one side of a
Reinstall	Ea.		31.36		31.36	double window opening.
Clean	Ea.	.52	3.83		4.35	
Paint	Ea.	1.58	11.05		12.63	
Minimum Charge	Job		157		157	
Triple						
Demolish	Ea.		7.30		7.30	Cost includes material and labor to
Install	Opng.	60.50	52.50		113	install 11/16" x 2-1/2" pine ranch
Demolish and Install	Opng.	60.50	59.80		120.30	style trim for casing one side of a
Reinstall	Opng.		41.81		41.81	triple window opening.
Clean	Ea.	.73	4.59		5.32	
Paint	Ea.	2.21	15.35		17.56	
Minimum Charge	Job		157		157	
Window Casing Per L.F.						
Demolish	L.F.		.43		.43	Cost includes material and labor to
Install	L.F.	.94	1.31		2.25	install 11/16" x 2-1/2" pine ranch
Demolish and Install	L.F.	.94	1.74		2.68	style trim for casing.
Reinstall	L.F.		1.05		1.05	
Clean	L.F.	.02	.13		.15	
Paint	L.F.	.06	.69		.75	
Minimum Charge	Job		157		157	

Window Sill

	Unit	Material	Labor	Equip.	Total	Specification
Wood						
Demolish	L.F.		.43		.43	Includes material and labor to install
Install	L.F.	1.68	1.57		3.25	flat, wood window stool.
Demolish and Install	L.F.	1.68	2		3.68	
Reinstall	L.F.		1.25		1.25	
Clean	L.F.	.02	.13		.15	
Paint	L.F.	.06	.69		.75	
Minimum Charge	Job		157		157	
Marble						
Demolish	L.F.		.64		.64	Includes material and labor to install
Install	L.F.	7.80	6.55		14.35	marble window sills, 6" x 3/4" thick.
Demolish and Install	L.F.	7.80	7.19		14.99	
Reinstall	L.F.		5.24		5.24	
Clean	L.F.	.02	.13		.15	
Minimum Charge	Job		142		142	

Shelving

	Unit	Material	Labor	Equip.	Total	Specification
Pine						
18"						
Demolish	L.F.		.73		.73	Cost includes material and labor to
Install	L.F.	4.91	3.30		8.21	install 1" thick custom shelving board
Demolish and Install	L.F.	4.91	4.03		8.94	18" wide.
Reinstall	L.F.		2.64		2.64	
Clean	L.F.	.04	.38		.42	
Paint	L.F.	.73	.43		1.16	
Minimum Charge	Job		157		157	

Finish Carpentry

Shelving

Shelving		Unit	Material	Labor	Equip.	Total	Specification
24"							
	Demolish	L.F.		.73		.73	Cost includes material and labor to
	Install	L.F.	6.55	3.69		10.24	install 1" thick custom shelving board.
	Demolish and Install	L.F.	6.55	4.42		10.97	
	Reinstall	L.F.		2.95		2.95	
	Clean	L.F.	.06	.57		.63	
	Paint	L.F.	.99	.58		1.57	
	Minimum Charge	Job		157		157	
Particle Board							
	Demolish	L.F.		.67		.67	Cost includes material and labor to
	Install	L.F.	1.03	1.65		2.68	install 3/4" thick particle shelving
	Demolish and Install	L.F.	1.03	2.32		3.35	board 18" wide.
	Reinstall	L.F.		1.32		1.32	
	Clean	L.F.	.04	.38		.42	
	Paint	S.F.	.10	.64		.74	
	Minimum Charge	Job		157		157	
Plywood							
	Demolish	L.F.		.67		.67	Cost includes material and labor to
	Install	L.F.	1.02	4.82		5.84	install 3/4" thick plywood shelving
	Demolish and Install	L.F.	1.02	5.49		6.51	board.
	Reinstall	L.F.		3.86		3.86	
	Clean	L.F.	.04	.38		.42	
	Paint	S.F.	.10	.64		.74	
	Minimum Charge	Job		157		157	
Hardwood (oak)							
	Demolish	L.F.		.67		.67	Includes material and labor to install
	Install	L.F.	8.25	2.16		10.41	oak shelving, 1" x 12", incl. cleats and
	Demolish and Install	L.F.	8.25	2.83		11.08	bracing.
	Reinstall	L.F.		1.73		1.73	
	Clean	L.F.	.04	.38		.42	
	Paint	S.F.	.10	.64		.74	
	Minimum Charge	Job		157		157	
Custom Bookcase							
	Demolish	L.F.		8.50		8.50	Includes material and labor to install
	Install	L.F.	21.88	31.50		53.38	custom modular bookcase unit with
	Demolish and Install	L.F.	21.88	40		61.88	clear pine faced frames, shelves 12"
	Reinstall	L.F.		25.09		25.09	O.C., 7' high, 8" deep.
	Clean	L.F.	1.16	8.20		9.36	
	Paint	L.F.	.95	21		21.95	
	Minimum Charge	Job		157		157	
Glass							
	Demolish	L.F.		.67		.67	Includes material and labor to install
	Install	L.F.	23.50	2.16		25.66	glass shelving, 1" x 12", incl. cleats
	Demolish and Install	L.F.	23.50	2.83		26.33	and bracing.
	Reinstall	L.F.		1.73		1.73	
	Clean	L.F.	.06	.35		.41	
	Minimum Charge	Job		150		150	

Closet Shelving

Closet Shelving		Unit	Material	Labor	Equip.	Total	Specification
Closet Shelf and Rod							
	Demolish	L.F.		.67		.67	Cost includes material and labor to
	Install	L.F.	2.33	5.25		7.58	install 1" thick custom shelving board
	Demolish and Install	L.F.	2.33	5.92		8.25	18" wide and 1" diameter clothes pole
	Reinstall	L.F.		4.18		4.18	with brackets 3' O.C.
	Clean	L.F.	.03	.29		.32	
	Paint	L.F.	.11	.69		.80	
	Minimum Charge	Job		157		157	

Finish Carpentry

Closet Shelving

	Unit	Material	Labor	Equip.	Total	Specification
Clothing Rod						Includes material and labor to install
Demolish	L.F.		.44		.44	fir closet pole, 1-5/8" diameter.
Install	L.F.	1.31	1.57		2.88	
Demolish and Install	L.F.	1.31	2.01		3.32	
Reinstall	L.F.		1.25		1.25	
Clean	L.F.	.03	.39		.42	
Paint	L.F.	.12	.42		.54	
Minimum Charge	Job		157		157	

Stair Assemblies

	Unit	Material	Labor	Equip.	Total	Specification
Straight Hardwood						Includes material and labor to install
Demolish	Ea.		300		300	factory cut and assembled straight
Install	Flight	955	209		1164	closed box stairs with oak treads and
Demolish and Install	Flight	955	509		1464	prefinished stair rail with balusters.
Clean	Flight	2.88	19.55		22.43	
Paint	Flight	10.10	37		47.10	
Minimum Charge	Job		157		157	
Clear Oak Tread						Includes material and labor to install
Demolish	Ea.		7.95		7.95	1-1/4" thick clear oak treads per riser.
Install	Ea.	81.50	17.40		98.90	
Demolish and Install	Ea.	81.50	25.35		106.85	
Clean	Ea.	.09	1.95		2.04	
Paint	Ea.	.34	2.04		2.38	
Minimum Charge	Job		157		157	
Landing						Includes material and labor to install
Demolish	S.F.		2.66		2.66	clear oak landing.
Install	S.F.	9.45	17.40		26.85	
Demolish and Install	S.F.	9.45	20.06		29.51	
Clean	S.F.		.29		.29	
Paint	S.F.	.26	.68		.94	
Minimum Charge	Job		157		157	
Refinish / Stain						Includes labor and material to sand
Install	S.F.	1.14	1.77		2.91	and finish (3 passes) with two coats of
Minimum Charge	Job		138		138	urethane on a new floor.
Spiral Hardwood						Cost includes material and labor to
Demolish	Ea.		300		300	install 4' - 6' diameter spiral stairs with
Install	Flight	4850	420		5270	oak treads, factory cut and assembled
Demolish and Install	Flight	4850	720		5570	with double handrails.
Clean	Flight	2.88	19.55		22.43	
Paint	Flight	10.10	37		47.10	
Minimum Charge	Job		157		157	

Stairs

	Unit	Material	Labor	Equip.	Total	Specification
Disappearing						Includes material and labor to install a
Demolish	Ea.		15.95		15.95	folding staircase.
Install	Ea.	110	89.50		199.50	
Demolish and Install	Ea.	110	105.45		215.45	
Reinstall	Ea.		71.68		71.68	
Clean	Ea.	.43	14.35		14.78	
Minimum Charge	Job		157		157	

Finish Carpentry

Stair Components

	Unit	Material	Labor	Equip.	Total	Specification
Handrail w / Balusters						
Demolish	L.F.		2.13		2.13	Includes material and labor to install
Install	L.F.	29.50	6.55		36.05	prefinished assembled stair rail with
Demolish and Install	L.F.	29.50	8.68		38.18	brackets, turned balusters and newel.
Clean	L.F.	.57	1.58		2.15	
Paint	L.F.	1.34	3.25		4.59	
Minimum Charge	Job		157		157	
Bannister						
Demolish	L.F.		1.28		1.28	Includes material and labor to install
Install	L.F.	33	5.25		38.25	built-up oak railings.
Demolish and Install	L.F.	33	6.53		39.53	
Clean	L.F.	.57	1.44		2.01	
Paint	L.F.	1.34	2.51		3.85	
Minimum Charge	Job		157		157	

Cabinets and Countertops

Wall and Base Cabinets

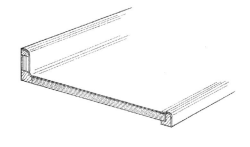

Countertop

Wall Cabinets	Unit	Material	Labor	Equip.	Total	Specification
Good Grade Laminated						
12″ Wide, 1 Door						
Demolish	Ea.		6.40		6.40	Includes material and labor to install
Install	Ea.	101	28.50		129.50	good modular unit with melamine
Demolish and Install	Ea.	101	34.90		135.90	laminated to particle board, in
Reinstall	Ea.		22.81		22.81	textured colors or wood grain print
Clean	Ea.	.43	4.42		4.85	finish, hinges and pulls.
Paint	Ea.	.41	9.85		10.26	
Minimum Charge	Job		157		157	
15″ Wide, 1 Door						
Demolish	Ea.		7.95		7.95	Includes material and labor to install
Install	Ea.	114	29.50		143.50	good modular unit with melamine
Demolish and Install	Ea.	114	37.45		151.45	laminated to particle board, in
Reinstall	Ea.		23.45		23.45	textured colors or wood grain print
Clean	Ea.	.54	5.45		5.99	finish, hinges and pulls.
Paint	Ea.	.51	12.55		13.06	
Minimum Charge	Job		157		157	
18″ Wide, 1 Door						
Demolish	Ea.		9.65		9.65	Includes material and labor to install
Install	Ea.	124	30		154	good modular unit with melamine
Demolish and Install	Ea.	124	39.65		163.65	laminated to particle board, in
Reinstall	Ea.		24.01		24.01	textured colors or wood grain print
Clean	Ea.	.65	6.55		7.20	finish, hinges and pulls.
Paint	Ea.	.61	14.55		15.16	
Minimum Charge	Job		157		157	
21″ Wide, 1 Door						
Demolish	Ea.		11.10		11.10	Includes material and labor to install
Install	Ea.	146	30.50		176.50	good modular unit with melamine
Demolish and Install	Ea.	146	41.60		187.60	laminated to particle board, in
Reinstall	Ea.		24.36		24.36	textured colors or wood grain print
Clean	Ea.	.76	7.65		8.41	finish, hinges and pulls.
Paint	Ea.	.70	17.25		17.95	
Minimum Charge	Job		157		157	
24″ w / Blind Corner						
Demolish	Ea.		12.75		12.75	Includes material and labor to install
Install	Ea.	141	31		172	good modular unit with melamine
Demolish and Install	Ea.	141	43.75		184.75	laminated to particle board, in
Reinstall	Ea.		24.72		24.72	textured colors or wood grain print
Clean	Ea.	.87	8.85		9.72	finish, hinges and pulls.
Paint	Ea.	.80	19.70		20.50	
Minimum Charge	Job		157		157	

Cabinets and Countertops

Wall Cabinets	Unit	Material	Labor	Equip.	Total	Specification
27" Wide, 2 Door						
Demolish	Ea.		14.15		14.15	Includes material and labor to install
Install	Ea.	174	31.50		205.50	good modular unit with melamine
Demolish and Install	Ea.	174	45.65		219.65	laminated to particle board, in
Reinstall	Ea.		25.34		25.34	textured colors or wood grain print
Clean	Ea.	.98	10		10.98	finish, hinges and pulls.
Paint	Ea.	.91	23		23.91	
Minimum Charge	Job		157		157	
30" Wide, 2 Door						
Demolish	Ea.		15.95		15.95	Includes material and labor to install
Install	Ea.	168	32.50		200.50	good modular unit with melamine
Demolish and Install	Ea.	168	48.45		216.45	laminated to particle board, in
Reinstall	Ea.		26		26	textured colors or wood grain print
Clean	Ea.	1.08	10.95		12.03	finish, hinges and pulls.
Paint	Ea.	1.01	25		26.01	
Minimum Charge	Job		157		157	
36" Wide, 2 Door						
Demolish	Ea.		18.90		18.90	Includes material and labor to install
Install	Ea.	193	33.50		226.50	good modular unit with melamine
Demolish and Install	Ea.	193	52.40		245.40	laminated to particle board, in
Reinstall	Ea.		26.69		26.69	textured colors or wood grain print
Clean	Ea.	1.30	13.50		14.80	finish, hinges and pulls.
Paint	Ea.	1.22	30.50		31.72	
Minimum Charge	Job		157		157	
48" Wide, 2 Door						
Demolish	Ea.		25.50		25.50	Includes material and labor to install
Install	Ea.	235	34		269	good modular unit with melamine
Demolish and Install	Ea.	235	59.50		294.50	laminated to particle board, in
Reinstall	Ea.		27.27		27.27	textured colors or wood grain print
Clean	Ea.	1.73	17.65		19.38	finish, hinges and pulls.
Paint	Ea.	1.62	39.50		41.12	
Minimum Charge	Job		157		157	
Above-Appliance						
Demolish	Ea.		15.95		15.95	Includes material and labor to install
Install	Ea.	111	25.50		136.50	good modular unit with melamine
Demolish and Install	Ea.	111	41.45		152.45	laminated to particle board, in
Reinstall	Ea.		20.23		20.23	textured colors or wood grain print
Clean	Ea.	1.08	10.95		12.03	finish, hinges and pulls.
Paint	Ea.	1.01	25		26.01	
Minimum Charge	Job		157		157	
Better Grade Wood						
12" Wide, 1 Door						
Demolish	Ea.		6.40		6.40	Includes material and labor to install
Install	Ea.	134	28.50		162.50	custom modular unit with solid
Demolish and Install	Ea.	134	34.90		168.90	hardwood faced frames, hardwood
Reinstall	Ea.		22.81		22.81	door frames and drawer fronts,
Clean	Ea.	.43	4.42		4.85	hardwood veneer on raised door
Paint	Ea.	.41	10.60		11.01	panels, hinges and pulls.
Minimum Charge	Job		157		157	
15" Wide, 1 Door						
Demolish	Ea.		7.95		7.95	Includes material and labor to install
Install	Ea.	152	29.50		181.50	custom modular unit with solid
Demolish and Install	Ea.	152	37.45		189.45	hardwood faced frames, hardwood
Reinstall	Ea.		23.45		23.45	door frames and drawer fronts,
Clean	Ea.	.54	5.45		5.99	hardwood veneer on raised door
Paint	Ea.	.51	13.15		13.66	panels, hinges and pulls.
Minimum Charge	Job		157		157	

Cabinets and Countertops

Wall Cabinets		Unit	Material	Labor	Equip.	Total	Specification
18" Wide, 1 Door							
	Demolish	Ea.		9.65		9.65	Includes material and labor to install
	Install	Ea.	166	30		196	custom modular unit with solid
	Demolish and Install	Ea.	166	39.65		205.65	hardwood faced frames, hardwood
	Reinstall	Ea.		24.01		24.01	door frames and drawer fronts,
	Clean	Ea.	.65	6.55		7.20	hardwood veneer on raised door
	Paint	Ea.	.61	16.25		16.86	panels, hinges and pulls.
	Minimum Charge	Job		157		157	
21" Wide, 1 Door							
	Demolish	Ea.		11.10		11.10	Includes material and labor to install
	Install	Ea.	189	30.50		219.50	custom modular unit with solid
	Demolish and Install	Ea.	189	41.60		230.60	hardwood faced frames, hardwood
	Reinstall	Ea.		24.36		24.36	door frames and drawer fronts,
	Clean	Ea.	.76	7.65		8.41	hardwood veneer on raised door
	Paint	Ea.	.70	18.40		19.10	panels, hinges and pulls.
	Minimum Charge	Job		157		157	
24" w / Blind Corner							
	Demolish	Ea.		12.75		12.75	Includes material and labor to install
	Install	Ea.	187	31		218	custom modular unit with solid
	Demolish and Install	Ea.	187	43.75		230.75	hardwood faced frames, hardwood
	Reinstall	Ea.		24.72		24.72	door frames and drawer fronts,
	Clean	Ea.	.87	8.85		9.72	hardwood veneer on raised door
	Paint	Ea.	.80	21		21.80	panels, hinges and pulls.
	Minimum Charge	Job		157		157	
27" Wide, 2 Door							
	Demolish	Ea.		14.15		14.15	Includes material and labor to install
	Install	Ea.	211	31.50		242.50	custom modular unit with solid
	Demolish and Install	Ea.	211	45.65		256.65	hardwood faced frames, hardwood
	Reinstall	Ea.		25.34		25.34	door frames and drawer fronts,
	Clean	Ea.	.98	10		10.98	hardwood veneer on raised door
	Paint	Ea.	.91	23		23.91	panels, hinges and pulls.
	Minimum Charge	Job		157		157	
30" Wide, 2 Door							
	Demolish	Ea.		15.95		15.95	Includes material and labor to install
	Install	Ea.	224	32.50		256.50	custom modular unit with solid
	Demolish and Install	Ea.	224	48.45		272.45	hardwood faced frames, hardwood
	Reinstall	Ea.		26		26	door frames and drawer fronts,
	Clean	Ea.	1.08	10.95		12.03	hardwood veneer on raised door
	Paint	Ea.	1.01	27.50		28.51	panels, hinges and pulls.
	Minimum Charge	Job		157		157	
36" Wide, 2 Door							
	Demolish	Ea.		18.90		18.90	Includes material and labor to install
	Install	Ea.	256	33.50		289.50	custom modular unit with solid
	Demolish and Install	Ea.	256	52.40		308.40	hardwood faced frames, hardwood
	Reinstall	Ea.		26.69		26.69	door frames and drawer fronts,
	Clean	Ea.	1.30	13.50		14.80	hardwood veneer on raised door
	Paint	Ea.	1.22	30.50		31.72	panels, hinges and pulls.
	Minimum Charge	Job		157		157	
48" Wide, 2 Door							
	Demolish	Ea.		25.50		25.50	Includes material and labor to install
	Install	Ea.	315	34		349	custom modular unit with solid
	Demolish and Install	Ea.	315	59.50		374.50	hardwood faced frames, hardwood
	Reinstall	Ea.		27.27		27.27	door frames and drawer fronts,
	Clean	Ea.	1.73	17.65		19.38	hardwood veneer on raised door
	Paint	Ea.	1.62	39.50		41.12	panels, hinges and pulls.
	Minimum Charge	Job		157		157	

Cabinets and Countertops

Wall Cabinets		Unit	Material	Labor	Equip.	Total	Specification
Above-appliance							
	Demolish	Ea.		15.95		15.95	Includes material and labor to install
	Install	Ea.	153	25.50		178.50	custom modular unit with solid
	Demolish and Install	Ea.	153	41.45		194.45	hardwood faced frames, hardwood
	Reinstall	Ea.		20.23		20.23	door frames and drawer fronts,
	Clean	Ea.	1.08	10.95		12.03	hardwood veneer on raised door
	Paint	Ea.	1.01	27.50		28.51	panels, hinges and pulls.
	Minimum Charge	Job		157		157	
Premium Grade Wood							
12" Wide, 1 Door							
	Demolish	Ea.		6.40		6.40	Includes material and labor to install
	Install	Ea.	174	28.50		202.50	premium modular unit with solid
	Demolish and Install	Ea.	174	34.90		208.90	hardwood faced frames, hardwood
	Reinstall	Ea.		22.81		22.81	door frames and drawer fronts,
	Clean	Ea.	.43	4.42		4.85	hardwood veneer on raised door
	Paint	Ea.	.41	11.50		11.91	panels, hinges and pulls.
	Minimum Charge	Job		157		157	
15" Wide, 1 Door							
	Demolish	Ea.		7.95		7.95	Includes material and labor to install
	Install	Ea.	202	29.50		231.50	premium modular unit with solid
	Demolish and Install	Ea.	202	37.45		239.45	hardwood faced frames, hardwood
	Reinstall	Ea.		23.45		23.45	door frames and drawer fronts,
	Clean	Ea.	.54	5.45		5.99	hardwood veneer on raised door
	Paint	Ea.	.51	14.55		15.06	panels, hinges and pulls.
	Minimum Charge	Job		157		157	
18" Wide, 1 Door							
	Demolish	Ea.		9.65		9.65	Includes material and labor to install
	Install	Ea.	213	30		243	premium modular unit with solid
	Demolish and Install	Ea.	213	39.65		252.65	hardwood faced frames, hardwood
	Reinstall	Ea.		24.01		24.01	door frames and drawer fronts,
	Clean	Ea.	.65	6.55		7.20	hardwood veneer on raised door
	Paint	Ea.	.61	17.25		17.86	panels, hinges and pulls.
	Minimum Charge	Job		157		157	
21" Wide, 1 Door							
	Demolish	Ea.		11.10		11.10	Includes material and labor to install
	Install	Ea.	227	30.50		257.50	premium modular unit with solid
	Demolish and Install	Ea.	227	41.60		268.60	hardwood faced frames, hardwood
	Reinstall	Ea.		24.36		24.36	door frames and drawer fronts,
	Clean	Ea.	.76	7.65		8.41	hardwood veneer on raised door
	Paint	Ea.	.70	19.70		20.40	panels, hinges and pulls.
	Minimum Charge	Job		157		157	
24" w / Blind Corner							
	Demolish	Ea.		12.75		12.75	Includes material and labor to install
	Install	Ea.	249	31		280	premium modular unit with solid
	Demolish and Install	Ea.	249	43.75		292.75	hardwood faced frames, hardwood
	Reinstall	Ea.		24.72		24.72	door frames and drawer fronts,
	Clean	Ea.	.87	8.85		9.72	hardwood veneer on raised door
	Paint	Ea.	.80	23		23.80	panels, hinges and pulls.
	Minimum Charge	Job		157		157	
27" Wide, 2 Door							
	Demolish	Ea.		14.15		14.15	Includes material and labor to install
	Install	Ea.	232	31.50		263.50	premium modular unit with solid
	Demolish and Install	Ea.	232	45.65		277.65	hardwood faced frames, hardwood
	Reinstall	Ea.		25.34		25.34	door frames and drawer fronts,
	Clean	Ea.	.98	10		10.98	hardwood veneer on raised door
	Paint	Ea.	.91	25		25.91	panels, hinges and pulls.
	Minimum Charge	Job		157		157	

Cabinets and Countertops

Wall Cabinets		Unit	Material	Labor	Equip.	Total	Specification
30" Wide, 2 Door							
	Demolish	Ea.		15.95		15.95	Includes material and labor to install
	Install	Ea.	283	32.50		315.50	premium modular unit with solid
	Demolish and Install	Ea.	283	48.45		331.45	hardwood faced frames, hardwood
	Reinstall	Ea.		26		26	door frames and drawer fronts,
	Clean	Ea.	1.08	10.95		12.03	hardwood veneer on raised door
	Paint	Ea.	1.01	27.50		28.51	panels, hinges and pulls.
	Minimum Charge	Job		157		157	
36" Wide, 2 Door							
	Demolish	Ea.		18.90		18.90	Includes material and labor to install
	Install	Ea.	325	33.50		358.50	premium modular unit with solid
	Demolish and Install	Ea.	325	52.40		377.40	hardwood faced frames, hardwood
	Reinstall	Ea.		26.69		26.69	door frames and drawer fronts,
	Clean	Ea.	1.30	13.50		14.80	hardwood veneer on raised door
	Paint	Ea.	1.22	34.50		35.72	panels, hinges and pulls.
	Minimum Charge	Job		157		157	
48" Wide, 2 Door							
	Demolish	Ea.		25.50		25.50	Includes material and labor to install
	Install	Ea.	400	34		434	premium modular unit with solid
	Demolish and Install	Ea.	400	59.50		459.50	hardwood faced frames, hardwood
	Reinstall	Ea.		27.27		27.27	door frames and drawer fronts,
	Clean	Ea.	1.73	17.65		19.38	hardwood veneer on raised door
	Paint	Ea.	1.62	46		47.62	panels, hinges and pulls.
	Minimum Charge	Job		157		157	
Above-appliance							
	Demolish	Ea.		15.95		15.95	Includes material and labor to install
	Install	Ea.	189	25.50		214.50	premium modular unit with solid
	Demolish and Install	Ea.	189	41.45		230.45	hardwood faced frames, hardwood
	Reinstall	Ea.		20.23		20.23	door frames and drawer fronts,
	Clean	Ea.	1.08	10.95		12.03	hardwood veneer on raised door
	Paint	Ea.	1.01	27.50		28.51	panels, hinges, pulls.
	Minimum Charge	Job		157		157	
Premium Hardwood							
12" Wide, 1 Door							
	Demolish	Ea.		6.40		6.40	Includes material and labor to install
	Install	Ea.	202	28.50		230.50	premium hardwood modular unit with
	Demolish and Install	Ea.	202	34.90		236.90	solid hardwood faced frames, drawer
	Reinstall	Ea.		22.81		22.81	fronts, and door panels, steel drawer
	Clean	Ea.	.43	4.42		4.85	guides, hinges, pulls, laminated
	Paint	Ea.	.41	12.55		12.96	interior.
	Minimum Charge	Job		157		157	
15" Wide, 1 Door							
	Demolish	Ea.		7.95		7.95	Includes material and labor to install
	Install	Ea.	266	29.50		295.50	premium hardwood modular unit with
	Demolish and Install	Ea.	266	37.45		303.45	solid hardwood faced frames, drawer
	Reinstall	Ea.		23.45		23.45	fronts, and door panels, steel drawer
	Clean	Ea.	.54	5.45		5.99	guides, hinges, pulls, laminated
	Paint	Ea.	.51	15.35		15.86	interior.
	Minimum Charge	Job		157		157	
18" Wide, 1 Door							
	Demolish	Ea.		9.65		9.65	Includes material and labor to install
	Install	Ea.	242	30		272	premium hardwood modular unit with
	Demolish and Install	Ea.	242	39.65		281.65	solid hardwood faced frames, drawer
	Reinstall	Ea.		24.01		24.01	fronts, and door panels, steel drawer
	Clean	Ea.	.65	6.55		7.20	guides, hinges, pulls, laminated
	Paint	Ea.	.61	18.40		19.01	interior.
	Minimum Charge	Job		157		157	

Cabinets and Countertops

Wall Cabinets	Unit	Material	Labor	Equip.	Total	Specification
21" Wide, 1 Door						
Demolish	Ea.		11.10		11.10	Includes material and labor to install
Install	Ea.	249	30.50		279.50	premium hardwood modular unit with
Demolish and Install	Ea.	249	41.60		290.60	solid hardwood faced frames, drawer
Reinstall	Ea.		24.36		24.36	fronts, and door panels, steel drawer
Clean	Ea.	.76	7.65		8.41	guides, hinges, pulls, laminated
Paint	Ea.	.70	21		21.70	interior.
Minimum Charge	Job		157		157	
24" w / Blind Corner						
Demolish	Ea.		12.75		12.75	Includes material and labor to install
Install	Ea.	277	31		308	premium hardwood modular unit with
Demolish and Install	Ea.	277	43.75		320.75	solid hardwood faced frames, drawer
Reinstall	Ea.		24.72		24.72	fronts, and door panels, steel drawer
Clean	Ea.	.87	8.85		9.72	guides, hinges, pulls, laminated
Paint	Ea.	.80	25		25.80	interior.
Minimum Charge	Job		157		157	
27" Wide, 2 Door						
Demolish	Ea.		14.15		14.15	Includes material and labor to install
Install	Ea.	470	31.50		501.50	premium hardwood modular unit with
Demolish and Install	Ea.	470	45.65		515.65	solid hardwood faced frames, drawer
Reinstall	Ea.		25.34		25.34	fronts, and door panels, steel drawer
Clean	Ea.	.98	10		10.98	guides, hinges, pulls, laminated
Paint	Ea.	.91	27.50		28.41	interior.
Minimum Charge	Job		157		157	
30" Wide, 2 Door						
Demolish	Ea.		15.95		15.95	Includes material and labor to install
Install	Ea.	310	32.50		342.50	premium hardwood modular unit with
Demolish and Install	Ea.	310	48.45		358.45	solid hardwood faced frames, drawer
Reinstall	Ea.		26		26	fronts, and door panels, steel drawer
Clean	Ea.	1.08	10.95		12.03	guides, hinges, pulls, laminated
Paint	Ea.	1.01	30.50		31.51	interior.
Minimum Charge	Job		157		157	
36" Wide, 2 Door						
Demolish	Ea.		18.90		18.90	Includes material and labor to install
Install	Ea.	350	33.50		383.50	premium hardwood modular unit with
Demolish and Install	Ea.	350	52.40		402.40	solid hardwood faced frames, drawer
Reinstall	Ea.		26.69		26.69	fronts, and door panels, steel drawer
Clean	Ea.	1.30	13.50		14.80	guides, hinges, pulls, laminated
Paint	Ea.	1.22	39.50		40.72	interior.
Minimum Charge	Job		157		157	
48" Wide, 2 Door						
Demolish	Ea.		25.50		25.50	Includes material and labor to install
Install	Ea.	430	34		464	premium hardwood modular unit with
Demolish and Install	Ea.	430	59.50		489.50	solid hardwood faced frames, drawer
Reinstall	Ea.		27.27		27.27	fronts, and door panels, steel drawer
Clean	Ea.	1.73	17.65		19.38	guides, hinges, pulls, laminated
Paint	Ea.	1.62	46		47.62	interior.
Minimum Charge	Job		157		157	
Above-appliance						
Demolish	Ea.		15.95		15.95	Includes material and labor to install
Install	Ea.	218	25.50		243.50	premium hardwood modular unit with
Demolish and Install	Ea.	218	41.45		259.45	solid hardwood faced frames, drawer
Reinstall	Ea.		20.23		20.23	fronts, and door panels, steel drawer
Clean	Ea.	1.08	10.95		12.03	guides, hinges, pulls, laminated
Paint	Ea.	1.01	30.50		31.51	interior.
Minimum Charge	Job		157		157	

Wall Cabinets	Unit	Material	Labor	Equip.	Total	Specification
Medicine Cabinet						
Good Grade						
Demolish	Ea.		10.65		10.65	Cost includes material and labor to
Install	Ea.	75	22.50		97.50	install 14" x 18" recessed medicine
Demolish and Install	Ea.	75	33.15		108.15	cabinet with stainless steel frame,
Reinstall	Ea.		17.92		17.92	mirror. Light not included.
Clean	Ea.	.02	10.30		10.32	
Paint	Ea.	.99	17.25		18.24	
Minimum Charge	Job		157		157	
Good Grade Lighted						
Demolish	Ea.		10.65		10.65	Cost includes material and labor to
Install	Ea.	234	26		260	install 14" x 18" recessed medicine
Demolish and Install	Ea.	234	36.65		270.65	cabinet with stainless steel frame,
Reinstall	Ea.		20.91		20.91	mirror and light.
Clean	Ea.	.02	10.30		10.32	
Paint	Ea.	.99	17.25		18.24	
Minimum Charge	Job		157		157	
Better Grade						
Demolish	Ea.		10.65		10.65	Includes material and labor to install
Install	Ea.	103	22.50		125.50	recessed medicine cabinet with
Demolish and Install	Ea.	103	33.15		136.15	polished chrome frame and mirror.
Reinstall	Ea.		17.92		17.92	Light not included.
Clean	Ea.	.02	10.30		10.32	
Paint	Ea.	.99	17.25		18.24	
Minimum Charge	Job		157		157	
Better Grade Lighted						
Demolish	Ea.		10.65		10.65	Includes material and labor to install
Install	Ea.	263	26		289	recessed medicine cabinet with
Demolish and Install	Ea.	263	36.65		299.65	polished chrome frame, mirror and
Reinstall	Ea.		20.91		20.91	light.
Clean	Ea.	.02	10.30		10.32	
Paint	Ea.	.99	17.25		18.24	
Minimum Charge	Job		157		157	
Premium Grade						
Demolish	Ea.		10.65		10.65	Cost includes material and labor to
Install	Ea.	132	22.50		154.50	install 14" x 24" recessed medicine
Demolish and Install	Ea.	132	33.15		165.15	cabinet with beveled mirror, frameless
Reinstall	Ea.		17.92		17.92	swing door. Light not included.
Clean	Ea.	.02	10.30		10.32	
Paint	Ea.	.99	17.25		18.24	
Minimum Charge	Job		157		157	
Premium Grade Light						
Demolish	Ea.		10.65		10.65	Cost includes material and labor to
Install	Ea.	292	26		318	install 36" x 30" surface mounted
Demolish and Install	Ea.	292	36.65		328.65	medicine cabinet with beveled tri-view
Reinstall	Ea.		20.91		20.91	mirror, frameless swing door and light.
Clean	Ea.	.02	10.30		10.32	
Paint	Ea.	.99	17.25		18.24	
Minimum Charge	Job		157		157	
Bath Mirror						
Demolish	Ea.		7.95		7.95	Cost includes material and labor to
Install	Ea.	64.50	15.70		80.20	install 18" x 24" surface mount mirror
Demolish and Install	Ea.	64.50	23.65		88.15	with stainless steel frame.
Reinstall	Ea.		12.54		12.54	
Clean	Ea.	.02	.69		.71	
Minimum Charge	Job		157		157	

Cabinets and Countertops

Wall Cabinets		Unit	Material	Labor	Equip.	Total	Specification
Garage							Includes material and labor to install
	Demolish	L.F.		6.40		6.40	good modular unit with melamine
	Install	L.F.	49	15.70		64.70	laminated to particle board, in
	Demolish and Install	L.F.	49	22.10		71.10	textured colors or wood grain print
	Reinstall	L.F.		12.54		12.54	finish, hinges and pulls.
	Clean	L.F.	.43	4.42		4.85	
	Paint	L.F.	.41	9.85		10.26	
	Minimum Charge	Job		157		157	
Strip and Refinish							
	Install	L.F.	1.08	5		6.08	Includes labor and material to strip,
	Minimum Charge	Job		138		138	prep and refinish exterior of cabinets.

Tall Cabinets		Unit	Material	Labor	Equip.	Total	Specification
Good Grade Laminated							Includes material and labor to install
	Demolish	L.F.		11.35		11.35	custom modular unit with solid
	Install	L.F.	199	177		376	hardwood faced frames, hardwood
	Demolish and Install	L.F.	199	188.35		387.35	door frames and drawer fronts,
	Reinstall	L.F.		141.34		141.34	hardwood veneer on raised door
	Clean	L.F.	1.16	8.20		9.36	panels, hinges and pulls.
	Paint	L.F.	.95	19.70		20.65	
	Minimum Charge	Job		157		157	
Better Grade Wood							Includes material and labor to install
	Demolish	L.F.		11.35		11.35	custom modular unit with solid
	Install	L.F.	265	177		442	hardwood faced frames, hardwood
	Demolish and Install	L.F.	265	188.35		453.35	door frames and drawer fronts,
	Reinstall	L.F.		141.34		141.34	hardwood veneer on raised door
	Clean	L.F.	1.16	8.20		9.36	panels, hinges and pulls.
	Paint	L.F.	.95	21		21.95	
	Minimum Charge	Job		157		157	
Premium Grade Wood							Includes material and labor to install
	Demolish	L.F.		11.35		11.35	premium modular unit with solid
	Install	L.F.	350	177		527	hardwood faced frames, hardwood
	Demolish and Install	L.F.	350	188.35		538.35	door frames and drawer fronts,
	Reinstall	L.F.		141.34		141.34	hardwood veneer on raised door
	Clean	L.F.	1.16	8.20		9.36	panels, hinges and pulls.
	Paint	L.F.	.95	23		23.95	
	Minimum Charge	Job		157		157	
Premium Hardwood							Includes material and labor to install
	Demolish	L.F.		11.35		11.35	premium hardwood modular unit with
	Install	L.F.	460	177		637	solid hardwood faced frames, drawer
	Demolish and Install	L.F.	460	188.35		648.35	fronts, and door panels, steel drawer
	Reinstall	L.F.		141.34		141.34	guides, hinges and pulls, laminated
	Clean	L.F.	1.16	8.20		9.36	interior.
	Paint	L.F.	.95	25		25.95	
	Minimum Charge	Job		157		157	

Cabinets and Countertops

Tall Cabinets	Unit	Material	Labor	Equip.	Total	Specification
Single Oven, 27" Wide						
Good Grade Laminated						
Demolish	Ea.		25.50		25.50	Includes material and labor to install
Install	Ea.	445	78.50		523.50	good modular unit with melamine
Demolish and Install	Ea.	445	104		549	laminated to particle board, in
Reinstall	Ea.		62.72		62.72	textured colors or wood grain print
Clean	Ea.	3.61	14.35		17.96	finish, hinges and pulls.
Paint	Ea.	1.58	44		45.58	
Minimum Charge	Job		157		157	
Better Grade Veneer						
Demolish	Ea.		25.50		25.50	Includes material and labor to install
Install	Ea.	790	78.50		868.50	custom modular unit with solid
Demolish and Install	Ea.	790	104		894	hardwood faced frames, hardwood
Reinstall	Ea.		62.72		62.72	door frames and drawer fronts,
Clean	Ea.	3.61	14.35		17.96	hardwood veneer on raised door
Paint	Ea.	1.58	47.50		49.08	panels, hinges and pulls.
Minimum Charge	Job		157		157	
Premium Grade Veneer						
Demolish	Ea.		25.50		25.50	Includes material and labor to install
Install	Ea.	705	78.50		783.50	premium modular unit with solid
Demolish and Install	Ea.	705	104		809	hardwood faced frames, hardwood
Reinstall	Ea.		62.72		62.72	door frames and drawer fronts,
Clean	Ea.	3.61	14.35		17.96	hardwood veneer on raised door
Paint	Ea.	1.58	52		53.58	panels, hinges and pulls.
Minimum Charge	Job		157		157	
Premium Hardwood						
Demolish	Ea.		25.50		25.50	Includes material and labor to install
Install	Ea.	900	78.50		978.50	premium hardwood modular unit with
Demolish and Install	Ea.	900	104		1004	solid hardwood faced frames, drawer
Reinstall	Ea.		62.72		62.72	fronts, and door panels, steel drawer
Clean	Ea.	3.61	14.35		17.96	guides, hinges and pulls, laminated
Paint	Ea.	1.58	56.50		58.08	interior.
Minimum Charge	Job		157		157	
Single Oven, 30" Wide						
Good Grade Laminated						
Demolish	Ea.		25.50		25.50	Includes material and labor to install
Install	Ea.	405	78.50		483.50	good modular unit with melamine
Demolish and Install	Ea.	405	104		509	laminated to particle board, in
Reinstall	Ea.		62.72		62.72	textured colors or wood grain print
Clean	Ea.	3.61	14.35		17.96	finish, hinges and pulls.
Paint	Ea.	1.58	44		45.58	
Minimum Charge	Job		157		157	
Better Grade Veneer						
Demolish	Ea.		25.50		25.50	Includes material and labor to install
Install	Ea.	595	78.50		673.50	custom modular unit with solid
Demolish and Install	Ea.	595	104		699	hardwood faced frames, hardwood
Reinstall	Ea.		62.72		62.72	door frames and drawer fronts,
Clean	Ea.	3.61	14.35		17.96	hardwood veneer on raised door
Paint	Ea.	1.58	44		45.58	panels, hinges and pulls.
Minimum Charge	Job		157		157	

Cabinets and Countertops

Tall Cabinets	Unit	Material	Labor	Equip.	Total	Specification
Premium Grade Veneer						
Demolish	Ea.		25.50		25.50	Includes material and labor to install
Install	Ea.	695	78.50		773.50	premium modular unit with solid
Demolish and Install	Ea.	695	104		799	hardwood faced frames, hardwood
Reinstall	Ea.		62.72		62.72	door frames and drawer fronts,
Clean	Ea.	3.61	14.35		17.96	hardwood veneer on raised door
Paint	Ea.	1.58	44		45.58	panels, hinges and pulls.
Minimum Charge	Job		157		157	
Premium Hardwood						
Demolish	Ea.		25.50		25.50	Includes material and labor to install
Install	Ea.	825	78.50		903.50	premium hardwood modular unit with
Demolish and Install	Ea.	825	104		929	solid hardwood faced frames, drawer
Reinstall	Ea.		62.72		62.72	fronts, and door panels, steel drawer
Clean	Ea.	3.61	14.35		17.96	guides, hinges and pulls, laminated
Paint	Ea.	1.58	44		45.58	interior.
Minimum Charge	Job		157		157	
Utility 18" W x 12" D						
Good Grade Laminated						
Demolish	Ea.		25.50		25.50	Includes material and labor to install
Install	Ea.	365	62.50		427.50	good modular unit with melamine
Demolish and Install	Ea.	365	88		453	laminated to particle board, in
Reinstall	Ea.		50.18		50.18	textured colors or wood grain print
Clean	Ea.	.85	6.95		7.80	finish, hinges and pulls.
Paint	Ea.	.61	17.25		17.86	
Minimum Charge	Job		157		157	
Better Grade Veneer						
Demolish	Ea.		25.50		25.50	Includes material and labor to install
Install	Ea.	395	62.50		457.50	custom modular unit with solid
Demolish and Install	Ea.	395	88		483	hardwood faced frames, hardwood
Reinstall	Ea.		50.18		50.18	door frames and drawer fronts,
Clean	Ea.	.85	6.95		7.80	hardwood veneer on raised door
Paint	Ea.	.61	18.40		19.01	panels, hinges and pulls.
Minimum Charge	Job		157		157	
Premium Grade Veneer						
Demolish	Ea.		25.50		25.50	Includes material and labor to install
Install	Ea.	505	62.50		567.50	premium modular unit with solid
Demolish and Install	Ea.	505	88		593	hardwood faced frames, hardwood
Reinstall	Ea.		50.18		50.18	door frames and drawer fronts,
Clean	Ea.	.85	6.95		7.80	hardwood veneer on raised door
Paint	Ea.	.61	21		21.61	panels, hinges and pulls.
Minimum Charge	Job		157		157	
Premium Hardwood						
Demolish	Ea.		25.50		25.50	Includes material and labor to install
Install	Ea.	645	62.50		707.50	premium hardwood modular unit with
Demolish and Install	Ea.	645	88		733	solid hardwood faced frames, drawer
Reinstall	Ea.		50.18		50.18	fronts, and door panels, steel drawer
Clean	Ea.	.85	6.95		7.80	guides, hinges and pulls, laminated
Paint	Ea.	.61	23		23.61	interior.
Minimum Charge	Job		157		157	

Cabinets and Countertops

Tall Cabinets	Unit	Material	Labor	Equip.	Total	Specification
Utility 24″ W x 24″ D						
Good Grade Laminated						
Demolish	Ea.		25.50		25.50	Includes material and labor to install
Install	Ea.	440	78.50		518.50	good modular unit with melamine
Demolish and Install	Ea.	440	104		544	laminated to particle board, in
Reinstall	Ea.		62.72		62.72	textured colors or wood grain print
Clean	Ea.	1.11	9.20		10.31	finish, hinges and pulls.
Paint	Ea.	.80	23		23.80	
Minimum Charge	Job		157		157	
Better Grade Veneer						
Demolish	Ea.		25.50		25.50	Includes material and labor to install
Install	Ea.	535	78.50		613.50	custom modular unit with solid
Demolish and Install	Ea.	535	104		639	hardwood faced frames, hardwood
Reinstall	Ea.		62.72		62.72	door frames and drawer fronts,
Clean	Ea.	1.11	9.20		10.31	hardwood veneer on raised door
Paint	Ea.	.80	25		25.80	panels, hinges and pulls.
Minimum Charge	Job		157		157	
Premium Grade Veneer						
Demolish	Ea.		25.50		25.50	Includes material and labor to install
Install	Ea.	675	78.50		753.50	premium modular unit with solid
Demolish and Install	Ea.	675	104		779	hardwood faced frames, hardwood
Reinstall	Ea.		62.72		62.72	door frames and drawer fronts,
Clean	Ea.	1.11	9.20		10.31	hardwood veneer on raised door
Paint	Ea.	.80	27.50		28.30	panels, hinges and pulls.
Minimum Charge	Job		157		157	
Premium Hardwood						
Demolish	Ea.		25.50		25.50	Includes material and labor to install
Install	Ea.	825	78.50		903.50	premium hardwood modular unit with
Demolish and Install	Ea.	825	104		929	solid hardwood faced frames, drawer
Reinstall	Ea.		62.72		62.72	fronts, and door panels, steel drawer
Clean	Ea.	1.11	9.20		10.31	guides, hinges and pulls, laminated
Paint	Ea.	.80	30.50		31.30	interior.
Minimum Charge	Job		157		157	
Plain Shelves						
Install	Ea.	15.31	19.60		34.91	Includes labor and material to install
Rotating Shelves						tall cabinet shelving, per set.
Install	Ea.	72.50	15.70		88.20	Includes labor and material to install
						lazy susan shelving for wall cabinet.
Strip and Refinish						
Install	L.F.	1.68	12.55		14.23	Includes labor and material to strip,
Minimum Charge	Job		138		138	prep and refinish exterior of cabinets.
Stain						
Install	L.F.	.15	2.51		2.66	Includes labor and material to replace
Minimum Charge	Job		138		138	normal prep and stain on doors and
						exterior cabinets.

Cabinets and Countertops

Base Cabinets	Unit	Material	Labor	Equip.	Total	Specification
Good Grade Laminated						
12" w, 1 Door, 1 Drawer						
Demolish	Ea.		6.40		6.40	Includes material and labor to install
Install	Ea.	105	25.50		130.50	good modular unit with melamine
Demolish and Install	Ea.	105	31.90		136.90	laminate, textured colors or wood
Reinstall	Ea.		20.23		20.23	grain print finish including hinges and
Clean	Ea.	.57	4.59		5.16	pulls.
Paint	Ea.	.41	11.50		11.91	
Minimum Charge	Job		157		157	
15" w, 1 Door, 1 Drawer						
Demolish	Ea.		7.95		7.95	Includes material and labor to install
Install	Ea.	143	26		169	good modular unit with melamine
Demolish and Install	Ea.	143	33.95		176.95	laminate, textured colors or wood
Reinstall	Ea.		20.91		20.91	grain print finish including hinges and
Clean	Ea.	.73	5.75		6.48	pulls.
Paint	Ea.	.51	14.55		15.06	
Minimum Charge	Job		157		157	
18" w, 1 Door, 1 Drawer						
Demolish	Ea.		9.45		9.45	Includes material and labor to install
Install	Ea.	155	27		182	good modular unit with melamine
Demolish and Install	Ea.	155	36.45		191.45	laminate, textured colors or wood
Reinstall	Ea.		21.53		21.53	grain print finish including hinges and
Clean	Ea.	.85	6.95		7.80	pulls.
Paint	Ea.	.61	17.25		17.86	
Minimum Charge	Job		157		157	
21" w, 1 Door, 1 Drawer						
Demolish	Ea.		11.10		11.10	Includes material and labor to install
Install	Ea.	162	27.50		189.50	good modular unit with melamine
Demolish and Install	Ea.	162	38.60		200.60	laminate, textured colors or wood
Reinstall	Ea.		22.10		22.10	grain print finish including hinges and
Clean	Ea.	1.03	8.20		9.23	pulls.
Paint	Ea.	.70	19.70		20.40	
Minimum Charge	Job		157		157	
24" w, 1 Door, 1 Drawer						
Demolish	Ea.		12.75		12.75	Includes material and labor to install
Install	Ea.	186	28		214	good modular unit with melamine
Demolish and Install	Ea.	186	40.75		226.75	laminate, textured colors or wood
Reinstall	Ea.		22.50		22.50	grain print finish including hinges and
Clean	Ea.	1.11	9.20		10.31	pulls.
Paint	Ea.	.80	23		23.80	
Minimum Charge	Job		157		157	
30" Wide Sink Base						
Demolish	Ea.		15.95		15.95	Includes material and labor to install
Install	Ea.	183	29.50		212.50	good modular unit with melamine
Demolish and Install	Ea.	183	45.45		228.45	laminate, textured colors or wood
Reinstall	Ea.		23.45		23.45	grain print finish including hinges and
Clean	Ea.	1.44	11.50		12.94	pulls.
Paint	Ea.	1.01	27.50		28.51	
Minimum Charge	Job		157		157	
36" Wide Sink Base						
Demolish	Ea.		19.60		19.60	Includes material and labor to install
Install	Ea.	205	31		236	good modular unit with melamine
Demolish and Install	Ea.	205	50.60		255.60	laminate, textured colors or wood
Reinstall	Ea.		24.72		24.72	grain print finish including hinges and
Clean	Ea.	1.61	13.50		15.11	pulls.
Paint	Ea.	1.22	34.50		35.72	
Minimum Charge	Job		157		157	

Cabinets and Countertops

Base Cabinets	Unit	Material	Labor	Equip.	Total	Specification
36″ Blind Corner						
Demolish	Ea.		19.60		19.60	Includes material and labor to install
Install	Ea.	300	35		335	good modular unit with melamine
Demolish and Install	Ea.	300	54.60		354.60	laminate, textured colors or wood
Reinstall	Ea.		27.88		27.88	grain print finish including hinges and
Clean	Ea.	1.61	13.50		15.11	pulls.
Paint	Ea.	1.22	34.50		35.72	
Minimum Charge	Job		157		157	
42″ w, 2 Door, 2 Drawer						
Demolish	Ea.		23		23	Includes material and labor to install
Install	Ea.	254	31.50		285.50	good modular unit with melamine
Demolish and Install	Ea.	254	54.50		308.50	laminate, textured colors or wood
Reinstall	Ea.		25.34		25.34	grain print finish including hinges and
Clean	Ea.	2.06	16.40		18.46	pulls.
Paint	Ea.	1.41	39.50		40.91	
Minimum Charge	Job		157		157	
Clean Interior						
Clean	L.F.	.57	3.48		4.05	Includes labor and materials to clean
Minimum Charge	Job		115		115	the interior of a 27″ cabinet.
Better Grade Wood						
12″ w, 1 Door, 1 Drawer						
Demolish	Ea.		6.40		6.40	Includes material and labor to install
Install	Ea.	140	25.50		165.50	custom modular unit including solid
Demolish and Install	Ea.	140	31.90		171.90	hardwood faced frames, hardwood
Reinstall	Ea.		20.23		20.23	door frames and drawer fronts,
Clean	Ea.	.57	4.59		5.16	hardwood veneer on raised door
Paint	Ea.	.41	12.55		12.96	panels, hinges and pulls.
Minimum Charge	Job		157		157	
15″ w, 1 Door, 1 Drawer						
Demolish	Ea.		7.95		7.95	Includes material and labor to install
Install	Ea.	190	26		216	custom modular unit including solid
Demolish and Install	Ea.	190	33.95		223.95	hardwood faced frames, hardwood
Reinstall	Ea.		20.91		20.91	door frames and drawer fronts,
Clean	Ea.	.73	5.75		6.48	hardwood veneer on raised door
Paint	Ea.	.51	16.25		16.76	panels, hinges and pulls.
Minimum Charge	Job		157		157	
18″ w, 1 Door, 1 Drawer						
Demolish	Ea.		9.45		9.45	Includes material and labor to install
Install	Ea.	207	27		234	custom modular unit including solid
Demolish and Install	Ea.	207	36.45		243.45	hardwood faced frames, hardwood
Reinstall	Ea.		21.53		21.53	door frames and drawer fronts,
Clean	Ea.	.85	6.95		7.80	hardwood veneer on raised door
Paint	Ea.	.61	18.40		19.01	panels, hinges and pulls.
Minimum Charge	Job		157		157	
21″ w, 1 Door, 1 Drawer						
Demolish	Ea.		11.10		11.10	Includes material and labor to install
Install	Ea.	216	27.50		243.50	custom modular unit including solid
Demolish and Install	Ea.	216	38.60		254.60	hardwood faced frames, hardwood
Reinstall	Ea.		22.10		22.10	door frames and drawer fronts,
Clean	Ea.	1.03	8.20		9.23	hardwood veneer on raised door
Paint	Ea.	.70	21		21.70	panels, hinges and pulls.
Minimum Charge	Job		157		157	

Cabinets and Countertops

Base Cabinets	Unit	Material	Labor	Equip.	Total	Specification
24" w, 1 Door, 1 Drawer						
Demolish	Ea.		12.75		12.75	Includes material and labor to install
Install	Ea.	248	28		276	custom modular unit including solid
Demolish and Install	Ea.	248	40.75		288.75	hardwood faced frames, hardwood
Reinstall	Ea.		22.50		22.50	door frames and drawer fronts,
Clean	Ea.	1.11	9.20		10.31	hardwood veneer on raised door
Paint	Ea.	.80	25		25.80	panels, hinges and pulls.
Minimum Charge	Job		157		157	
30" Wide Sink Base						
Demolish	Ea.		15.95		15.95	Includes material and labor to install
Install	Ea.	243	29.50		272.50	custom modular unit including solid
Demolish and Install	Ea.	243	45.45		288.45	hardwood faced frames, hardwood
Reinstall	Ea.		23.45		23.45	door frames and drawer fronts,
Clean	Ea.	1.44	11.50		12.94	hardwood veneer on raised door
Paint	Ea.	1.01	30.50		31.51	panels, hinges and pulls.
Minimum Charge	Job		157		157	
36" Wide Sink Base						
Demolish	Ea.		19.60		19.60	Includes material and labor to install
Install	Ea.	273	31		304	custom modular unit including solid
Demolish and Install	Ea.	273	50.60		323.60	hardwood faced frames, hardwood
Reinstall	Ea.		24.72		24.72	door frames and drawer fronts,
Clean	Ea.	1.61	13.50		15.11	hardwood veneer on raised door
Paint	Ea.	1.22	39.50		40.72	panels, hinges and pulls.
Minimum Charge	Job		157		157	
36" Blind Corner						
Demolish	Ea.		19.60		19.60	Includes material and labor to install
Install	Ea.	400	35		435	custom modular unit including solid
Demolish and Install	Ea.	400	54.60		454.60	hardwood faced frames, hardwood
Reinstall	Ea.		27.88		27.88	door frames and drawer fronts,
Clean	Ea.	1.61	13.50		15.11	hardwood veneer on raised door
Paint	Ea.	1.22	39.50		40.72	panels, hinges and pulls.
Minimum Charge	Job		157		157	
42" w, 2 Door, 2 Drawer						
Demolish	Ea.		23		23	Includes material and labor to install
Install	Ea.	340	31.50		371.50	modular unit including solid hardwood
Demolish and Install	Ea.	340	54.50		394.50	faced frames, hardwood door frames
Reinstall	Ea.		25.34		25.34	and drawer fronts, hardwood veneer
Clean	Ea.	2.06	16.40		18.46	on raised door panels, hinges and
Paint	Ea.	1.41	46		47.41	pulls.
Minimum Charge	Job		157		157	
Clean Interior						
Clean	L.F.	.57	3.48		4.05	Includes labor and materials to clean
Minimum Charge	Job		115		115	the interior of a 27" cabinet.
Premium Grade Wood						
12" w, 1 Door, 1 Drawer						
Demolish	Ea.		6.40		6.40	Includes material and labor to install
Install	Ea.	240	12.65		252.65	premium hardwood modular unit with
Demolish and Install	Ea.	240	19.05		259.05	solid hardwood faced frames, drawer
Reinstall	Ea.		10.12		10.12	fronts, and door panels, steel drawer
Clean	Ea.	.57	4.59		5.16	guides, hinges, pulls and laminated
Paint	Ea.	.41	13.80		14.21	interior.
Minimum Charge	Job		157		157	

Cabinets and Countertops

Base Cabinets	Unit	Material	Labor	Equip.	Total	Specification
15" w, 1 Door, 1 Drawer						
Demolish	Ea.		7.95		7.95	Includes material and labor to install
Install	Ea.	249	26		275	premium modular unit including solid
Demolish and Install	Ea.	249	33.95		282.95	hardwood faced frames, hardwood
Reinstall	Ea.		20.91		20.91	door frames and drawer fronts,
Clean	Ea.	.73	5.75		6.48	hardwood veneer on raised door
Paint	Ea.	.51	17.25		17.76	panels, hinges and pulls.
Minimum Charge	Job		157		157	
18" w, 1 Door, 1 Drawer						
Demolish	Ea.		9.45		9.45	Includes material and labor to install
Install	Ea.	268	27		295	premium modular unit including solid
Demolish and Install	Ea.	268	36.45		304.45	hardwood faced frames, hardwood
Reinstall	Ea.		21.53		21.53	door frames and drawer fronts,
Clean	Ea.	.85	6.95		7.80	hardwood veneer on raised door
Paint	Ea.	.61	21		21.61	panels, hinges and pulls.
Minimum Charge	Job		157		157	
21" w, 1 Door, 1 Drawer						
Demolish	Ea.		11.10		11.10	Includes material and labor to install
Install	Ea.	279	27.50		306.50	premium modular unit including solid
Demolish and Install	Ea.	279	38.60		317.60	hardwood faced frames, hardwood
Reinstall	Ea.		22.10		22.10	door frames and drawer fronts,
Clean	Ea.	1.03	8.20		9.23	hardwood veneer on raised door
Paint	Ea.	.70	23		23.70	panels, hinges and pulls.
Minimum Charge	Job		157		157	
24" w, 1 Door, 1 Drawer						
Demolish	Ea.		12.75		12.75	Includes material and labor to install
Install	Ea.	320	28		348	premium modular unit including solid
Demolish and Install	Ea.	320	40.75		360.75	hardwood faced frames, hardwood
Reinstall	Ea.		22.50		22.50	door frames and drawer fronts,
Clean	Ea.	1.11	9.20		10.31	hardwood veneer on raised door
Paint	Ea.	.80	27.50		28.30	panels, hinges and pulls.
Minimum Charge	Job		157		157	
30" Wide Sink Base						
Demolish	Ea.		15.95		15.95	Includes material and labor to install
Install	Ea.	283	29.50		312.50	premium modular unit including solid
Demolish and Install	Ea.	283	45.45		328.45	hardwood faced frames, hardwood
Reinstall	Ea.		23.45		23.45	door frames and drawer fronts,
Clean	Ea.	1.44	11.50		12.94	hardwood veneer on raised door
Paint	Ea.	1.01	34.50		35.51	panels, hinges and pulls.
Minimum Charge	Job		157		157	
36" Wide Sink Base						
Demolish	Ea.		19.60		19.60	Includes material and labor to install
Install	Ea.	320	31		351	premium modular unit including solid
Demolish and Install	Ea.	320	50.60		370.60	hardwood faced frames, hardwood
Reinstall	Ea.		24.72		24.72	door frames and drawer fronts,
Clean	Ea.	1.61	13.50		15.11	hardwood veneer on raised door
Paint	Ea.	1.22	39.50		40.72	panels, hinges and pulls.
Minimum Charge	Job		157		157	
36" Blind Corner						
Demolish	Ea.		19.60		19.60	Includes material and labor to install
Install	Ea.	665	35		700	premium modular unit including solid
Demolish and Install	Ea.	665	54.60		719.60	hardwood faced frames, hardwood
Reinstall	Ea.		27.88		27.88	door frames and drawer fronts,
Clean	Ea.	1.61	13.50		15.11	hardwood veneer on raised door
Paint	Ea.	1.22	39.50		40.72	panels, hinges and pulls.
Minimum Charge	Job		157		157	

Cabinets and Countertops

Base Cabinets	Unit	Material	Labor	Equip.	Total	Specification
42″ w, 2 Door, 2 Drawer						
Demolish	Ea.		23		23	Includes material and labor to install
Install	Ea.	450	31.50		481.50	premium modular unit including solid
Demolish and Install	Ea.	450	54.50		504.50	hardwood faced frames, hardwood
Reinstall	Ea.		25.34		25.34	door frames and drawer fronts,
Clean	Ea.	2.06	16.40		18.46	hardwood veneer on raised door
Paint	Ea.	1.41	46		47.41	panels, hinges and pulls.
Minimum Charge	Job		157		157	
Clean Interior						
Clean	L.F.	.57	3.48		4.05	Includes labor and materials to clean
Minimum Charge	Job		115		115	the interior of a 27″ cabinet.
Premium Hardwood						
12″ w, 1 Door, 1 Drawer						
Demolish	Ea.		6.40		6.40	Includes material and labor to install
Install	Ea.	294	25.50		319.50	premium hardwood modular unit with
Demolish and Install	Ea.	294	31.90		325.90	solid hardwood faced frames, drawer
Reinstall	Ea.		20.23		20.23	fronts, and door panels, steel drawer
Clean	Ea.	.57	4.59		5.16	guides, hinges, pulls and laminated
Paint	Ea.	.41	15.35		15.76	interior.
Minimum Charge	Job		157		157	
15″ w, 1 Door, 1 Drawer						
Demolish	Ea.		7.95		7.95	Includes material and labor to install
Install	Ea.	298	26		324	premium hardwood modular unit with
Demolish and Install	Ea.	298	33.95		331.95	solid hardwood faced frames, drawer
Reinstall	Ea.		20.91		20.91	fronts, and door panels, steel drawer
Clean	Ea.	.73	5.75		6.48	guides, hinges, pulls and laminated
Paint	Ea.	.51	19.70		20.21	interior.
Minimum Charge	Job		157		157	
18″ w, 1 Door, 1 Drawer						
Demolish	Ea.		9.45		9.45	Includes material and labor to install
Install	Ea.	325	27		352	premium hardwood modular unit with
Demolish and Install	Ea.	325	36.45		361.45	solid hardwood faced frames, drawer
Reinstall	Ea.		21.53		21.53	fronts, and door panels, steel drawer
Clean	Ea.	.85	6.95		7.80	guides, hinges, pulls and laminated
Paint	Ea.	.61	23		23.61	interior.
Minimum Charge	Job		157		157	
21″ w, 1 Door, 1 Drawer						
Demolish	Ea.		11.10		11.10	Includes material and labor to install
Install	Ea.	335	27.50		362.50	premium hardwood modular unit with
Demolish and Install	Ea.	335	38.60		373.60	solid hardwood faced frames, drawer
Reinstall	Ea.		22.10		22.10	fronts, and door panels, steel drawer
Clean	Ea.	1.03	8.20		9.23	guides, hinges, pulls and laminated
Paint	Ea.	.70	25		25.70	interior.
Minimum Charge	Job		157		157	
24″ w, 1 Door, 1 Drawer						
Demolish	Ea.		12.75		12.75	Includes material and labor to install
Install	Ea.	375	28		403	premium hardwood modular unit with
Demolish and Install	Ea.	375	40.75		415.75	solid hardwood faced frames, drawer
Reinstall	Ea.		22.50		22.50	fronts, and door panels, steel drawer
Clean	Ea.	1.11	9.20		10.31	guides, hinges and pulls, laminated
Paint	Ea.	.80	30.50		31.30	interior.
Minimum Charge	Job		157		157	

Cabinets and Countertops

Base Cabinets		Unit	Material	Labor	Equip.	Total	Specification
30" Wide Sink Base							
	Demolish	Ea.		15.95		15.95	Includes material and labor to install
	Install	Ea.	335	29.50		364.50	premium hardwood modular unit with
	Demolish and Install	Ea.	335	45.45		380.45	solid hardwood faced frames, drawer
	Reinstall	Ea.		23.45		23.45	fronts, and door panels, steel drawer
	Clean	Ea.	1.44	11.50		12.94	guides, hinges, pulls and laminated
	Paint	Ea.	1.01	34.50		35.51	interior.
	Minimum Charge	Job		157		157	
36" Wide Sink Base							
	Demolish	Ea.		19.60		19.60	Includes material and labor to install
	Install	Ea.	375	31		406	premium hardwood modular unit with
	Demolish and Install	Ea.	375	50.60		425.60	solid hardwood faced frames, drawer
	Reinstall	Ea.		24.72		24.72	fronts, and door panels, steel drawer
	Clean	Ea.	1.61	13.50		15.11	guides, hinges, pulls and laminated
	Paint	Ea.	1.22	46		47.22	interior.
	Minimum Charge	Job		157		157	
36" Blind Corner							
	Demolish	Ea.		19.60		19.60	Includes material and labor to install
	Install	Ea.	720	35		755	premium hardwood modular unit with
	Demolish and Install	Ea.	720	54.60		774.60	solid hardwood faced frames, drawer
	Reinstall	Ea.		27.88		27.88	fronts, and door panels, steel drawer
	Clean	Ea.	1.61	13.50		15.11	guides, hinges, pulls and laminated
	Paint	Ea.	1.22	46		47.22	interior.
	Minimum Charge	Job		157		157	
42" w, 2 Door, 2 Drawer							
	Demolish	Ea.		23		23	Includes material and labor to install
	Install	Ea.	595	31.50		626.50	premium hardwood modular unit with
	Demolish and Install	Ea.	595	54.50		649.50	solid hardwood faced frames, drawer
	Reinstall	Ea.		25.34		25.34	fronts, and door panels, steel drawer
	Clean	Ea.	2.06	16.40		18.46	guides, hinges, pulls and laminated
	Paint	Ea.	1.41	55		56.41	interior.
	Minimum Charge	Job		157		157	
Clean Interior							
	Clean	L.F.	.57	3.48		4.05	Includes labor and materials to clean
	Minimum Charge	Job		115		115	the interior of a 27" cabinet.
Island							
24" Wide							
	Demolish	Ea.		12.75		12.75	Includes material and labor to install
	Install	Ea.	350	26		376	custom modular unit with solid
	Demolish and Install	Ea.	350	38.75		388.75	hardwood faced frames, hardwood
	Reinstall	Ea.		20.91		20.91	door frames and drawer fronts,
	Clean	Ea.	1.61	13.50		15.11	hardwood veneer on raised door
	Paint	Ea.	1.21	39.50		40.71	panels, hinges and pulls.
	Minimum Charge	Job		157		157	
30" Wide							
	Demolish	Ea.		15.95		15.95	Includes material and labor to install
	Install	Ea.	415	32.50		447.50	custom modular unit with solid
	Demolish and Install	Ea.	415	48.45		463.45	hardwood faced frames, hardwood
	Reinstall	Ea.		26.13		26.13	door frames and drawer fronts,
	Clean	Ea.	2.06	16.40		18.46	hardwood veneer on raised door
	Paint	Ea.	1.52	46		47.52	panels, hinges and pulls.
	Minimum Charge	Job		157		157	

Cabinets and Countertops

Base Cabinets		Unit	Material	Labor	Equip.	Total	Specification
36" Wide							
	Demolish	Ea.		19.60		19.60	Includes material and labor to install
	Install	Ea.	475	39		514	custom modular unit with solid
	Demolish and Install	Ea.	475	58.60		533.60	hardwood faced frames, hardwood
	Reinstall	Ea.		31.36		31.36	door frames and drawer fronts,
	Clean	Ea.	2.41	21		23.41	hardwood veneer on raised door
	Paint	Ea.	1.84	55		56.84	panels, hinges and pulls.
	Minimum Charge	Job		157		157	
48" Wide							
	Demolish	Ea.		25.50		25.50	Includes material and labor to install
	Install	Ea.	705	52.50		757.50	custom modular unit with solid
	Demolish and Install	Ea.	705	78		783	hardwood faced frames, hardwood
	Reinstall	Ea.		41.81		41.81	door frames and drawer fronts,
	Clean	Ea.	2.88	25.50		28.38	hardwood veneer on raised door
	Paint	Ea.	2.10	69		71.10	panels, hinges and pulls.
	Minimum Charge	Job		157		157	
60" Wide							
	Demolish	Ea.		28.50		28.50	Includes material and labor to install
	Install	Ea.	830	65.50		895.50	custom modular unit with solid
	Demolish and Install	Ea.	830	94		924	hardwood faced frames, hardwood
	Reinstall	Ea.		52.27		52.27	door frames and drawer fronts,
	Clean	Ea.	3.61	33		36.61	hardwood veneer on raised door
	Paint	Ea.	2.10	92		94.10	panels, hinges and pulls.
	Minimum Charge	Job		157		157	
Clean Interior							
	Clean	L.F.	.57	3.48		4.05	Includes labor and materials to clean
	Minimum Charge	Job		115		115	the interior of a 27" cabinet.
Liquor Bar							
	Demolish	L.F.		6.40		6.40	Includes material and labor to install
	Install	L.F.	128	31.50		159.50	custom modular unit with solid
	Demolish and Install	L.F.	128	37.90		165.90	hardwood faced frames, hardwood
	Reinstall	L.F.		25.09		25.09	door frames and drawer fronts,
	Clean	L.F.	.57	4.78		5.35	hardwood veneer on raised door
	Paint	L.F.	.41	12.55		12.96	panels, hinges and pulls.
	Minimum Charge	Job		157		157	
Clean Interior							
	Clean	L.F.	.57	3.48		4.05	Includes labor and materials to clean
	Minimum Charge	Job		115		115	the interior of a 27" cabinet.
Seal Interior							
	Install	L.F.	.28	4.52		4.80	Includes labor and materials to seal a
	Minimum Charge	Job		138		138	cabinet interior.
Back Bar							
	Demolish	L.F.		6.40		6.40	Includes material and labor to install
	Install	L.F.	128	31.50		159.50	custom modular unit with solid
	Demolish and Install	L.F.	128	37.90		165.90	hardwood faced frames, hardwood
	Reinstall	L.F.		25.09		25.09	door frames and drawer fronts,
	Clean	L.F.	.57	4.78		5.35	hardwood veneer on raised door
	Paint	L.F.	.41	12.55		12.96	panels, hinges and pulls.
	Minimum Charge	Job		157		157	
Clean Interior							
	Clean	L.F.	.57	3.48		4.05	Includes labor and materials to clean
	Minimum Charge	Job		115		115	the interior of a 27" cabinet.

Cabinets and Countertops

Base Cabinets		Unit	Material	Labor	Equip.	Total	Specification
Seal Interior							
	Install	L.F.	.28	4.52		4.80	Includes labor and materials to seal a
	Minimum Charge	Job		138		138	cabinet interior.
Storage (garage)							
	Demolish	L.F.		6.40		6.40	Includes material and labor to install
	Install	L.F.	95.50	31.50		127	good modular unit with melamine
	Demolish and Install	L.F.	95.50	37.90		133.40	laminate, textured colors or wood
	Reinstall	L.F.		25.09		25.09	grain print finish including hinges and
	Clean	L.F.	.57	4.78		5.35	pulls.
	Paint	L.F.	.41	12.55		12.96	
	Minimum Charge	Job		157		157	
Clean Interior							
	Clean	L.F.	.57	3.48		4.05	Includes labor and materials to clean
	Minimum Charge	Job		115		115	the interior of a 27" cabinet.
Seal Interior							
	Install	L.F.	.28	4.52		4.80	Includes labor and materials to seal a
	Minimum Charge	Job		138		138	cabinet interior.
Built-in Desk							
	Demolish	L.F.		6.40		6.40	Includes material and labor to install
	Install	L.F.	128	31.50		159.50	custom modular unit with solid
	Demolish and Install	L.F.	128	37.90		165.90	hardwood faced frames, hardwood
	Reinstall	L.F.		25.09		25.09	door frames and drawer fronts,
	Clean	L.F.	.57	4.78		5.35	hardwood veneer on raised door
	Paint	L.F.	.41	12.55		12.96	panels, hinges and pulls.
	Minimum Charge	Job		157		157	
Built-in Bookcase							
	Demolish	L.F.		8.50		8.50	Includes material and labor to install
	Install	L.F.	21.88	31.50		53.38	custom modular bookcase unit with
	Demolish and Install	L.F.	21.88	40		61.88	clear pine faced frames, shelves 12"
	Reinstall	L.F.		25.09		25.09	O.C., 7' high, 8" deep.
	Clean	L.F.	1.16	8.20		9.36	
	Paint	L.F.	.95	21		21.95	
	Minimum Charge	Job		157		157	
Strip and Refinish							
	Install	L.F.	1.08	5		6.08	Includes labor and material to strip,
	Minimum Charge	Job		138		138	prep and refinish exterior of cabinets.
Stain							
	Install	L.F.	.35	1.28		1.63	Includes labor and material to stain
	Minimum Charge	Job		157		157	door and exterior of cabinets.

Countertop		Unit	Material	Labor	Equip.	Total	Specification
Laminated							
With Splash							
	Demolish	L.F.		4.25		4.25	Includes material and labor to install
	Install	L.F.	27.50	11.20		38.70	one piece laminated top with 4"
	Demolish and Install	L.F.	27.50	15.45		42.95	backsplash.
	Reinstall	L.F.		8.96		8.96	
	Clean	L.F.	.08	.57		.65	
	Minimum Charge	Job		157		157	

Cabinets and Countertops

Countertop		Unit	Material	Labor	Equip.	Total	Specification
Roll Top							
	Demolish	L.F.		4.25		4.25	Includes material and labor to install
	Install	L.F.	9.70	10.45		20.15	one piece laminated top with rolled
	Demolish and Install	L.F.	9.70	14.70		24.40	drip edge (post formed) and
	Reinstall	L.F.		8.36		8.36	backsplash.
	Clean	L.F.	.10	.87		.97	
	Minimum Charge	Job		157		157	
Ceramic Tile							
	Demolish	S.F.		2.13		2.13	Cost includes material and labor to
	Install	S.F.	14.05	6.25		20.30	install 4-1/4" x 4-1/4" to 6" x 6"
	Demolish and Install	S.F.	14.05	8.38		22.43	glazed tile set in mortar bed and grout
	Clean	S.F.	.06	.44		.50	on particle board substrate.
	Minimum Charge	Job		142		142	
Cultured Marble							
	Demolish	L.F.		4.25		4.25	Includes material and labor to install
	Install	L.F.	25	32		57	marble countertop, 24" wide, no
	Demolish and Install	L.F.	25	36.25		61.25	backsplash.
	Reinstall	L.F.		25.54		25.54	
	Clean	L.F.	.10	.87		.97	
	Minimum Charge	Job		142		142	
With Splash							
	Demolish	L.F.		4.25		4.25	Includes material and labor to install
	Install	L.F.	28	32		60	marble countertop, 24" wide, with
	Demolish and Install	L.F.	28	36.25		64.25	backsplash.
	Reinstall	L.F.		25.54		25.54	
	Clean	L.F.	.10	.87		.97	
	Minimum Charge	Job		142		142	
Quarry Tile							
	Demolish	S.F.		2.13		2.13	Includes material and labor to install
	Install	S.F.	7.70	6.25		13.95	quarry tile countertop with no
	Demolish and Install	S.F.	7.70	8.38		16.08	backsplash.
	Clean	S.F.	.06	.44		.50	
	Minimum Charge	Job		142		142	
With Splash							
	Demolish	S.F.		2.13		2.13	Includes material and labor to install
	Install	S.F.	9.80	6.95		16.75	quarry tile countertop with backsplash.
	Demolish and Install	S.F.	9.80	9.08		18.88	
	Clean	S.F.	.06	.44		.50	
	Minimum Charge	Job		142		142	
Butcher Block							
	Demolish	S.F.		2.13		2.13	Includes material and labor to install
	Install	S.F.	28.50	5.60		34.10	solid laminated maple countertop with
	Demolish and Install	S.F.	28.50	7.73		36.23	no backsplash.
	Reinstall	S.F.		4.48		4.48	
	Clean	S.F.	.02	.18		.20	
	Minimum Charge	Job		157		157	
Solid Surface							
	Demolish	L.F.		4.25		4.25	Includes material and labor to install
	Install	L.F.	48	22.50		70.50	solid surface counter top 22" deep.
	Demolish and Install	L.F.	48	26.75		74.75	
	Reinstall	L.F.		17.92		17.92	
	Clean	L.F.	.10	.87		.97	
	Minimum Charge	Job		157		157	

Cabinets and Countertops

Vanity Cabinets	Unit	Material	Labor	Equip.	Total	Specification
Good Grade Laminated						
24″ Wide, 2 Door						
Demolish	Ea.		12.75		12.75	Includes material and labor to install
Install	Ea.	140	31.50		171.50	modular unit with melamine laminated
Demolish and Install	Ea.	140	44.25		184.25	to particle board, textured colors or
Reinstall	Ea.		25.09		25.09	wood grain print finish, hinges and
Clean	Ea.	1.11	9.20		10.31	pulls.
Paint	Ea.	.80	23		23.80	
Minimum Charge	Job		157		157	
30″ Wide, 2 Door						
Demolish	Ea.		15.95		15.95	Includes material and labor to install
Install	Ea.	161	39		200	modular unit with melamine laminated
Demolish and Install	Ea.	161	54.95		215.95	to particle board, textured colors or
Reinstall	Ea.		31.36		31.36	wood grain print finish, hinges and
Clean	Ea.	1.44	11.50		12.94	pulls.
Paint	Ea.	1.01	27.50		28.51	
Minimum Charge	Job		157		157	
36″ Wide, 2 Door						
Demolish	Ea.		19.60		19.60	Includes material and labor to install
Install	Ea.	215	47		262	modular unit with melamine laminated
Demolish and Install	Ea.	215	66.60		281.60	to particle board, textured colors or
Reinstall	Ea.		37.64		37.64	wood grain print finish, hinges and
Clean	Ea.	1.61	13.50		15.11	pulls.
Paint	Ea.	1.22	34.50		35.72	
Minimum Charge	Job		157		157	
42″ w, 3 Doors / Drawer						
Demolish	Ea.		23		23	Includes material and labor to install
Install	Ea.	305	50.50		355.50	modular unit with melamine laminated
Demolish and Install	Ea.	305	73.50		378.50	to particle board, textured colors or
Reinstall	Ea.		40.53		40.53	wood grain print finish, hinges and
Clean	Ea.	2.06	16.40		18.46	pulls.
Paint	Ea.	1.41	39.50		40.91	
Minimum Charge	Job		157		157	
48″ w, 3 Doors / Drawer						
Demolish	Ea.		25.50		25.50	Includes material and labor to install
Install	Ea.	255	55		310	modular unit with melamine laminated
Demolish and Install	Ea.	255	80.50		335.50	to particle board, textured colors or
Reinstall	Ea.		43.90		43.90	wood grain print finish, hinges and
Clean	Ea.	2.88	25.50		28.38	pulls.
Paint	Ea.	2.10	69		71.10	
Minimum Charge	Job		157		157	
Clean Interior						
Install	L.F.	.57	4.78		5.35	Includes labor and materials to clean
Minimum Charge	Job		115		115	base cabinetry.
Better Grade Wood						
24″ Wide, 2 Door						
Demolish	Ea.		12.75		12.75	Includes material and labor to install
Install	Ea.	186	31.50		217.50	custom modular unit with solid
Demolish and Install	Ea.	186	44.25		230.25	hardwood faced frames, hardwood
Reinstall	Ea.		25.09		25.09	door frames and drawer fronts,
Clean	Ea.	1.11	9.20		10.31	hardwood veneer on raised door
Paint	Ea.	.80	25		25.80	panels, hinges and pulls.
Minimum Charge	Job		157		157	

Cabinets and Countertops

Vanity Cabinets		Unit	Material	Labor	Equip.	Total	Specification
30" Wide, 2 Door							
	Demolish	Ea.		15.95		15.95	Includes material and labor to install
	Install	Ea.	213	39		252	custom modular unit with solid
	Demolish and Install	Ea.	213	54.95		267.95	hardwood faced frames, hardwood
	Reinstall	Ea.		31.36		31.36	door frames and drawer fronts,
	Clean	Ea.	1.44	11.50		12.94	hardwood veneer on raised door
	Paint	Ea.	1.01	30.50		31.51	panels, hinges and pulls.
	Minimum Charge	Job		157		157	
36" Wide, 2 Door							
	Demolish	Ea.		19.60		19.60	Includes material and labor to install
	Install	Ea.	285	47		332	custom modular unit with solid
	Demolish and Install	Ea.	285	66.60		351.60	hardwood faced frames, hardwood
	Reinstall	Ea.		37.64		37.64	door frames and drawer fronts,
	Clean	Ea.	1.61	13.50		15.11	hardwood veneer on raised door
	Paint	Ea.	1.22	39.50		40.72	panels, hinges and pulls.
	Minimum Charge	Job		157		157	
42" w, 3 Doors / Drawer							
	Demolish	Ea.		23		23	Includes material and labor to install
	Install	Ea.	305	50.50		355.50	custom modular unit with solid
	Demolish and Install	Ea.	305	73.50		378.50	hardwood faced frames, hardwood
	Reinstall	Ea.		40.53		40.53	door frames and drawer fronts,
	Clean	Ea.	2.06	16.40		18.46	hardwood veneer on raised door
	Paint	Ea.	1.41	46		47.41	panels, hinges and pulls.
	Minimum Charge	Job		157		157	
48" w, 3 Doors / Drawer							
	Demolish	Ea.		25.50		25.50	Includes material and labor to install
	Install	Ea.	340	55		395	custom modular unit with solid
	Demolish and Install	Ea.	340	80.50		420.50	hardwood faced frames, hardwood
	Reinstall	Ea.		43.90		43.90	door frames and drawer fronts,
	Clean	Ea.	2.88	25.50		28.38	hardwood veneer on raised door
	Paint	Ea.	2.10	69		71.10	panels, hinges and pulls.
	Minimum Charge	Job		157		157	
Clean Interior							
	Install	L.F.	.57	4.78		5.35	Includes labor and materials to clean
	Minimum Charge	Job		115		115	base cabinetry.
Premium Grade Wood							
24" Wide, 2 Door							
	Demolish	Ea.		12.75		12.75	Includes material and labor to install
	Install	Ea.	221	31.50		252.50	premium modular unit with solid
	Demolish and Install	Ea.	221	44.25		265.25	hardwood faced frames, hardwood
	Reinstall	Ea.		25.09		25.09	door frames and drawer fronts,
	Clean	Ea.	1.11	9.20		10.31	hardwood veneer on raised door
	Paint	Ea.	.80	27.50		28.30	panels, hinges and pulls.
	Minimum Charge	Job		157		157	
30" Wide, 2 Door							
	Demolish	Ea.		15.95		15.95	Includes material and labor to install
	Install	Ea.	254	39		293	premium modular unit with solid
	Demolish and Install	Ea.	254	54.95		308.95	hardwood faced frames, hardwood
	Reinstall	Ea.		31.36		31.36	door frames and drawer fronts,
	Clean	Ea.	1.44	11.50		12.94	hardwood veneer on raised door
	Paint	Ea.	1.01	34.50		35.51	panels, hinges and pulls.
	Minimum Charge	Job		157		157	

Cabinets and Countertops

Vanity Cabinets

Vanity Cabinets	Unit	Material	Labor	Equip.	Total	Specification
36″ Wide, 2 Door						
Demolish	Ea.		19.60		19.60	Includes material and labor to install
Install	Ea.	395	47		442	premium modular unit with solid
Demolish and Install	Ea.	395	66.60		461.60	hardwood faced frames, hardwood
Reinstall	Ea.		37.64		37.64	door frames and drawer fronts,
Clean	Ea.	1.61	13.50		15.11	hardwood veneer on raised door
Paint	Ea.	1.22	39.50		40.72	panels, hinges and pulls.
Minimum Charge	Job		157		157	
42″ w, 3 Doors / Drawer						
Demolish	Ea.		23		23	Includes material and labor to install
Install	Ea.	380	50.50		430.50	premium modular unit with solid
Demolish and Install	Ea.	380	73.50		453.50	hardwood faced frames, hardwood
Reinstall	Ea.		40.53		40.53	door frames and drawer fronts,
Clean	Ea.	2.06	16.40		18.46	hardwood veneer on raised door
Paint	Ea.	1.41	46		47.41	panels, hinges and pulls.
Minimum Charge	Job		157		157	
48″ w, 3 Doors / Drawer						
Demolish	Ea.		25.50		25.50	Includes material and labor to install
Install	Ea.	460	55		515	premium modular unit with solid
Demolish and Install	Ea.	460	80.50		540.50	hardwood faced frames, hardwood
Reinstall	Ea.		43.90		43.90	door frames and drawer fronts,
Clean	Ea.	2.88	25.50		28.38	hardwood veneer on raised door
Paint	Ea.	2.10	69		71.10	panels, hinges and pulls.
Minimum Charge	Job		157		157	
Clean Interior						
Install	L.F.	.57	4.78		5.35	Includes labor and materials to clean
Minimum Charge	Job		115		115	base cabinetry.

Premium Hardwood

	Unit	Material	Labor	Equip.	Total	Specification
24″ Wide, 2 Door						
Demolish	Ea.		12.75		12.75	Includes material and labor to install
Install	Ea.	277	31.50		308.50	premium modular unit with solid
Demolish and Install	Ea.	277	44.25		321.25	hardwood faced frames, hardwood
Reinstall	Ea.		25.09		25.09	door frames and drawer fronts,
Clean	Ea.	1.11	9.20		10.31	hardwood veneer on raised door
Paint	Ea.	.80	30.50		31.30	panels, hinges and pulls.
Minimum Charge	Job		157		157	
30″ Wide, 2 Door						
Demolish	Ea.		15.95		15.95	Includes material and labor to install
Install	Ea.	310	39		349	premium modular unit with solid
Demolish and Install	Ea.	310	54.95		364.95	hardwood faced frames, hardwood
Reinstall	Ea.		31.36		31.36	door frames and drawer fronts,
Clean	Ea.	1.44	11.50		12.94	hardwood veneer on raised door
Paint	Ea.	1.01	34.50		35.51	panels, hinges and pulls.
Minimum Charge	Job		157		157	
36″ Wide, 2 Door						
Demolish	Ea.		19.60		19.60	Includes material and labor to install
Install	Ea.	455	47		502	premium modular unit with solid
Demolish and Install	Ea.	455	66.60		521.60	hardwood faced frames, hardwood
Reinstall	Ea.		37.64		37.64	door frames and drawer fronts,
Clean	Ea.	1.61	13.50		15.11	hardwood veneer on raised door
Paint	Ea.	1.22	46		47.22	panels, hinges and pulls.
Minimum Charge	Job		157		157	

Cabinets and Countertops

Vanity Cabinets	Unit	Material	Labor	Equip.	Total	Specification
42" w, 3 Doors / Drawer						Includes material and labor to install
Demolish	Ea.		23		23	premium modular unit with solid
Install	Ea.	425	50.50		475.50	hardwood faced frames, hardwood
Demolish and Install	Ea.	425	73.50		498.50	door frames and drawer fronts,
Reinstall	Ea.		40.53		40.53	hardwood veneer on raised door
Clean	Ea.	2.06	16.40		18.46	panels, hinges and pulls.
Paint	Ea.	1.41	55		56.41	
Minimum Charge	Job		157		157	
48" w, 3 Doors / Drawer						Includes material and labor to install
Demolish	Ea.		25.50		25.50	premium modular unit with solid
Install	Ea.	525	55		580	hardwood faced frames, hardwood
Demolish and Install	Ea.	525	80.50		605.50	door frames and drawer fronts,
Reinstall	Ea.		43.90		43.90	hardwood veneer on raised door
Clean	Ea.	2.88	25.50		28.38	panels, hinges and pulls.
Paint	Ea.	2.10	69		71.10	
Minimum Charge	Job		157		157	
Clean Interior						
Install	L.F.	.57	4.78		5.35	Includes labor and materials to clean
Minimum Charge	Job		115		115	base cabinetry.
Strip and Refinish						
Install	L.F.	1.08	5		6.08	Includes labor and material to strip,
Minimum Charge	Job		138		138	prep and refinish exterior of cabinets.

Vanity Cabinet Tops	Unit	Material	Labor	Equip.	Total	Specification
Cultured Marble Top						
Demolish	L.F.		4.25		4.25	Includes material and labor to install
Install	L.F.	34	18.80		52.80	cultured marble vanity top with
Demolish and Install	L.F.	34	23.05		57.05	integral sink, 22" deep.
Reinstall	L.F.		15.02		15.02	
Clean	L.F.	.10	.87		.97	
Minimum Charge	Job		142		142	
Laminated Top						
Demolish	L.F.		2.55		2.55	Includes material and labor to install
Install	L.F.	27.50	11.20		38.70	one piece laminated top with 4"
Demolish and Install	L.F.	27.50	13.75		41.25	backsplash.
Reinstall	L.F.		8.96		8.96	
Clean	L.F.	.08	.57		.65	
Minimum Charge	Job		157		157	
Solid Surface						
Demolish	L.F.		4.25		4.25	Includes material and labor to install
Install	L.F.	67	23		90	solid surface vanity top with integral
Demolish and Install	L.F.	67	27.25		94.25	sink installed, 22" deep.
Reinstall	L.F.		18.58		18.58	
Clean	L.F.	.10	.87		.97	
Minimum Charge	Job		157		157	

Painting

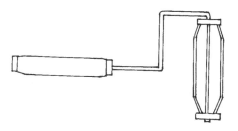

Roller Handle

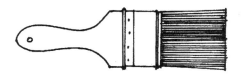

Paint Brush

Paint Preparation, Walls		Unit	Material	Labor	Equip.	Total	Specification
Cover / Protect Floors							
	Install	S.F.	.01	.11		.12	Includes material and labor to install
	Minimum Charge	Job		137		137	plastic masking sheet on large uninterrupted areas.
Cover / Protect Walls							
	Install	S.F.	.01	.11		.12	Includes material and labor to install
	Minimum Charge	Job		137		137	plastic masking sheet on walls.
Clean Walls							
Light							
	Clean	S.F.		.09		.09	Includes labor to wash gypsum
	Minimum Charge	Job		137		137	drywall or plaster wall surfaces.
Heavy							
	Clean	S.F.		.11		.11	Includes labor and material for heavy
	Minimum Charge	Job		137		137	cleaning (multiple applications) with detergent and solvent.
Prep Walls							
	Paint	S.F.	.03	.14		.17	Includes labor and material to prepare
	Minimum Charge	Job		137		137	for painting including scraping, patching and puttying.
Sand Walls							
	Paint	S.F.		.09		.09	Includes labor to sand gypsum drywall
	Minimum Charge	Job		137		137	wall surfaces.
Seal Walls							
Large Area							
	Paint	S.F.	.04	.14		.18	Includes labor and material to paint
	Minimum Charge	Job		137		137	with primer / sealer by roller.
Spot							
	Paint	S.F.	.04	.24		.28	Includes labor and material to seal
	Minimum Charge	Job		137		137	existing drywall with shellac-based material.

Paint Preparation, Ceilings		Unit	Material	Labor	Equip.	Total	Specification
Cover / Protect Floors							
	Install	S.F.	.01	.11		.12	Includes material and labor to install
	Minimum Charge	Job		137		137	plastic masking sheet on large uninterrupted areas.

Painting

Paint Preparation, Ceilings

Paint Preparation, Ceilings		Unit	Material	Labor	Equip.	Total	Specification
Cover / Protect Walls							
	Install	S.F.	.01	.11		.12	Includes material and labor to install
	Minimum Charge	Job		137		137	plastic masking sheet on walls.
Prep Ceiling							
	Paint	S.F.	.06	.51		.57	Includes labor and material to fill in
	Minimum Charge	Job		137		137	hairline cracks up to 1/8" with filler, sand and prep for paint.
Sand Ceiling							
	Paint	S.F.		.13		.13	Includes labor to sand gypsum drywall
	Minimum Charge	Job		137		137	ceiling surfaces.
Seal Ceiling							
Large Area							
	Paint	S.F.	.06	.21		.27	Includes labor and material to paint
	Minimum Charge	Job		138		138	ceiling, one coat flat latex.
Spot							
	Paint	S.F.	.04	.24		.28	Includes labor and material to seal
	Minimum Charge	Job		138		138	existing drywall with shellac-based material.

Paint / Texture, Walls

Paint / Texture, Walls		Unit	Material	Labor	Equip.	Total	Specification
Texture Walls							
Spray							
	Install	S.F.	.04	.35		.39	Includes labor and material to install
	Minimum Charge	Job		138		138	by spray, texture finish.
Trowel							
	Install	S.F.		.18		.18	Includes labor and material to apply
	Minimum Charge	Job		138		138	hand troweled texture.
Paint Walls 1 Coat							
	Paint	S.F.	.06	.21		.27	Includes labor and material to paint,
	Minimum Charge	Job		138		138	one coat flat latex.
Paint Walls 2 Coats							
	Paint	S.F.	.10	.35		.45	Includes labor and material to paint,
	Minimum Charge	Job		138		138	two coats flat latex.
Paint Walls 3 Coats							
	Paint	S.F.	.15	.42		.57	Includes labor and material to paint,
	Minimum Charge	Job		138		138	three coats flat latex.
Clean and Seal							
	Paint	S.F.	.21	.74		.95	Includes labor and material to do light
	Minimum Charge	Job		138		138	cleaning, sealing with 2 coats of latex paint.

Paint / Texture, Ceilings

Paint / Texture, Ceilings		Unit	Material	Labor	Equip.	Total	Specification
Texture Ceiling							
Spray							
	Install	S.F.	.04	.35		.39	Includes labor and material to install
	Minimum Charge	Job		138		138	by spray, texture finish.
Trowel							
	Install	S.F.		.18		.18	Includes labor and material to apply
	Minimum Charge	Job		138		138	hand troweled texture.

Painting

Paint / Texture, Ceilings		Unit	Material	Labor	Equip.	Total	Specification
Paint Ceiling, 1 Coat							
	Paint	S.F.	.06	.21		.27	Includes labor and material to paint
	Minimum Charge	Job		138		138	ceiling, one coat flat latex.
Paint Ceiling, 2 Coats							
	Paint	S.F.	.10	.35		.45	Includes labor and material to paint
	Minimum Charge	Job		138		138	two coats flat latex paint.
Paint Ceiling, 3 Coats							
	Paint	S.F.	.15	.41		.56	Includes labor and material to paint
	Minimum Charge	Job		138		138	ceiling, three coats flat latex.
Clean and Seal							
	Paint	S.F.	.15	.58		.73	Includes labor and material for light
	Minimum Charge	Job		138		138	cleaning, sealing and paint 3 coats of flat latex paint.

Flooring

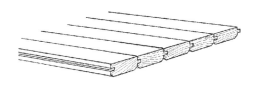

Plank Flooring

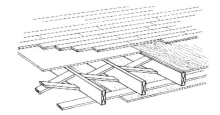

Wood Strip Flooring

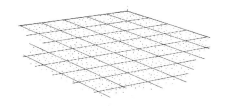

Tile Flooring

Wood Plank Flooring	Unit	Material	Labor	Equip.	Total	Specification
Maple						
2-1/4″						
Demolish	S.F.		1.02		1.02	Includes material and labor to install
Install	S.F.	4.19	1.56		5.75	unfinished T&G maple flooring,
Demolish and Install	S.F.	4.19	2.58		6.77	25/32″ thick, 2-1/4″ wide, random
Reinstall	S.F.		1.25		1.25	3′ to 16′ lengths including felt
Clean	S.F.		.29		.29	underlayment, nailed in place over
Paint	S.F.	1.14	1.77		2.91	prepared subfloor.
Minimum Charge	Job		157		157	
3-1/4″						
Demolish	S.F.		1.02		1.02	Includes material and labor to install
Install	S.F.	3.11	1.56		4.67	unfinished T&G maple flooring,
Demolish and Install	S.F.	3.11	2.58		5.69	33/32″ thick, 3-1/4″ wide, random
Reinstall	S.F.		1.25		1.25	3′ to 16′ lengths including felt
Clean	S.F.		.29		.29	underlayment, nailed in place over
Paint	S.F.	1.14	1.77		2.91	prepared subfloor.
Minimum Charge	Job		157		157	
Sand and Finish						
Install	S.F.	.76	.57		1.33	Includes labor and material to sand (3
Minimum Charge	Job		157		157	passes) and finish and two coats of
						urethane existing floor.
Oak						
2-1/4″						
Demolish	S.F.		1.02		1.02	Includes material and labor to install
Install	S.F.	3.84	1.56		5.40	unfinished T&G oak flooring, 25/32″
Demolish and Install	S.F.	3.84	2.58		6.42	thick, 2-1/4″ wide, random 3′ to 16′
Reinstall	S.F.		1.25		1.25	lengths including felt underlayment,
Clean	S.F.		.29		.29	nailed in place over prepared
Paint	S.F.	1.14	1.77		2.91	subfloor.
Minimum Charge	Job		157		157	
3-1/4″						
Demolish	S.F.		1.02		1.02	Includes material and labor to install
Install	S.F.	5.30	1.11		6.41	unfinished T&G oak flooring, 25/32″
Demolish and Install	S.F.	5.30	2.13		7.43	thick, 3-1/4″ wide, random 3′ to 16′
Reinstall	S.F.		.89		.89	lengths including felt underlayment,
Clean	S.F.		.29		.29	nailed in place over prepared
Paint	S.F.	1.14	1.77		2.91	subfloor.
Minimum Charge	Job		157		157	

Wood Plank Flooring		Unit	Material	Labor	Equip.	Total	Specification
Sand and Finish							
	Install	S.F.	1.14	1.77		2.91	Includes labor and material to sand
	Minimum Charge	Job		157		157	and finish (3 passes) with two coats of urethane on a new floor.
Walnut							
3″ to 7″							
	Demolish	S.F.		1.02		1.02	Includes material and labor to install
	Install	S.F.	9.05	1.11		10.16	unfinished T&G walnut flooring,
	Demolish and Install	S.F.	9.05	2.13		11.18	25/32″ thick, random 3′ to 16′
	Reinstall	S.F.		.89		.89	lengths including felt underlayment,
	Clean	S.F.		.29		.29	nailed in place over prepared
	Paint	S.F.	1.14	1.77		2.91	subfloor.
	Minimum Charge	Job		157		157	
Sand and Finish							
	Install	S.F.	1.14	1.77		2.91	Includes labor and material to sand
	Minimum Charge	Job		157		157	and finish (3 passes) with two coats of urethane on a new floor.
Pine							
2-1/4″							
	Demolish	S.F.		1.02		1.02	Includes material and labor to install
	Install	S.F.	7.35	1.18		8.53	unfinished T&G pine flooring, 25/32″
	Demolish and Install	S.F.	7.35	2.20		9.55	thick, 2-1/4″ wide, random 3′ to 16′
	Reinstall	S.F.		.94		.94	lengths including felt underlayment,
	Clean	S.F.		.29		.29	nailed in place over prepared
	Paint	S.F.	.76	.57		1.33	subfloor.
	Minimum Charge	Job		157		157	
Sand and Finish							
	Install	S.F.	1.14	1.77		2.91	Includes labor and material to sand
	Minimum Charge	Job		157		157	and finish (3 passes) with two coats of urethane on a new floor.
Prefinished Oak							
2-1/4″							
	Demolish	S.F.		1.02		1.02	Includes material and labor to install
	Install	S.F.	5.60	1.56		7.16	prefinished T&G oak flooring, 2-1/4″
	Demolish and Install	S.F.	5.60	2.58		8.18	wide, random 3′ to 16′ lengths
	Reinstall	S.F.		1.25		1.25	including felt underlayment, nailed in
	Clean	S.F.		.29		.29	place over prepared subfloor.
	Minimum Charge	Job		157		157	
3-1/4″							
	Demolish	S.F.		1.02		1.02	Includes material and labor to install
	Install	S.F.	7.15	1.44		8.59	prefinished T&G oak flooring, 25/32″
	Demolish and Install	S.F.	7.15	2.46		9.61	thick, 3″ to 7″ wide, random 3′ to 16′
	Reinstall	S.F.		1.15		1.15	lengths including felt underlayment,
	Clean	S.F.		.29		.29	nailed in place over prepared
	Minimum Charge	Job		157		157	subfloor.
Sand and Finish New							
	Install	S.F.	.76	.78		1.54	Includes labor and material to sand (3
	Minimum Charge	Job		157		157	passes) and finish, two coats of urethane, new floor.
Refinish Existing							
	Install	S.F.	.76	.57		1.33	Includes labor and material to sand (3
	Minimum Charge	Job		157		157	passes) and finish and two coats of urethane existing floor.

Wood Plank Flooring

Wood Plank Flooring	Unit	Material	Labor	Equip.	Total	Specification
Urethane Coat						
Install	S.F.	.13	.15		.28	Includes labor and material to install
Minimum Charge	Job		157		157	two coats of urethane, on existing floor.
Clean and Wax						
Clean	S.F.	.02	.18		.20	Includes labor and material to wash
Minimum Charge	Job		157		157	and wax a hardwood floor.
Underlayment Per S.F.						
Demolish	S.F.		.46		.46	Cost includes material and labor to
Install	S.F.	.53	.42		.95	install 1/4" lauan subfloor, standard
Demolish and Install	S.F.	.53	.88		1.41	interior grade, nailed every 6".
Clean	S.F.		.22		.22	
Minimum Charge	Job		157		157	

Wood Parquet Tile Floor

Wood Parquet Tile Floor	Unit	Material	Labor	Equip.	Total	Specification
Oak						
9" x 9"						
Demolish	S.F.		.77		.77	Includes material and labor to install
Install	S.F.	4.62	2.51		7.13	prefinished oak 9" x 9" parquet block
Demolish and Install	S.F.	4.62	3.28		7.90	flooring, 5/16" thick, installed in
Reinstall	S.F.		2.01		2.01	mastic.
Clean	S.F.		.29		.29	
Paint	S.F.	1.14	1.77		2.91	
Minimum Charge	Job		157		157	
13" x 13"						
Demolish	S.F.		.77		.77	Includes material and labor to install
Install	S.F.	8.75	1.96		10.71	prefinished oak 13" x 13" parquet
Demolish and Install	S.F.	8.75	2.73		11.48	block flooring, 5/16" thick, installed in
Reinstall	S.F.		1.57		1.57	mastic.
Clean	S.F.		.29		.29	
Paint	S.F.	1.14	1.77		2.91	
Minimum Charge	Job		157		157	
Cherry						
9" x 9"						
Demolish	S.F.		.77		.77	Includes material and labor to install
Install	S.F.	8.15	2.51		10.66	prefinished cherry 9" x 9" parquet
Demolish and Install	S.F.	8.15	3.28		11.43	block flooring, 5/16" thick, installed in
Reinstall	S.F.		2.51		2.51	mastic.
Clean	S.F.		.29		.29	
Paint	S.F.	1.14	1.77		2.91	
Minimum Charge	Job		157		157	
13" x 13"						
Demolish	S.F.		.77		.77	Includes material and labor to install
Install	S.F.	9.30	1.96		11.26	prefinished cherry 13" x 13" parquet
Demolish and Install	S.F.	9.30	2.73		12.03	block flooring, 5/16" thick, installed in
Reinstall	S.F.		1.96		1.96	mastic.
Clean	S.F.		.29		.29	
Paint	S.F.	1.14	1.77		2.91	
Minimum Charge	Job		157		157	

Flooring

Wood Parquet Tile Floor		Unit	Material	Labor	Equip.	Total	Specification
Walnut							
9" x 9"							
	Demolish	S.F.		.77		.77	Includes material and labor to install
	Install	S.F.	5	2.51		7.51	prefinished walnut 9" x 9" parquet
	Demolish and Install	S.F.	5	3.28		8.28	flooring in mastic.
	Reinstall	S.F.		2.01		2.01	
	Clean	S.F.		.29		.29	
	Paint	S.F.	1.14	1.77		2.91	
	Minimum Charge	Job		157		157	
13" x 13"							
	Demolish	S.F.		.77		.77	Includes material and labor to install
	Install	S.F.	9.35	1.96		11.31	prefinished walnut 13" x 13" parquet
	Demolish and Install	S.F.	9.35	2.73		12.08	block flooring, 5/16" thick, installed in
	Reinstall	S.F.		1.96		1.96	mastic.
	Clean	S.F.		.29		.29	
	Paint	S.F.	1.14	1.77		2.91	
	Minimum Charge	Job		157		157	
Acrylic Impregnated							
Oak 12" x 12"							
	Demolish	S.F.		.77		.77	Includes material and labor to install
	Install	S.F.	7.55	2.24		9.79	red oak 12" x 12" acrylic
	Demolish and Install	S.F.	7.55	3.01		10.56	impregnated parquet block flooring,
	Reinstall	S.F.		2.24		2.24	5/16" thick, installed in mastic.
	Clean	S.F.		.29		.29	
	Minimum Charge	Job		157		157	
Cherry 12" x 12"							
	Demolish	S.F.		.77		.77	Includes material and labor to install
	Install	S.F.	11.40	2.24		13.64	prefinished cherry 12" x 12" acrylic
	Demolish and Install	S.F.	11.40	3.01		14.41	impregnated parquet block flooring,
	Reinstall	S.F.		2.24		2.24	5/16" thick, installed in mastic.
	Clean	S.F.		.29		.29	
	Minimum Charge	Job		157		157	
Ash 12" x 12"							
	Demolish	S.F.		.77		.77	Includes material and labor to install
	Install	S.F.	10.25	2.24		12.49	ash 12" x 12" acrylic impregnated
	Demolish and Install	S.F.	10.25	3.01		13.26	parquet block flooring, 5/16" thick in
	Reinstall	S.F.		2.24		2.24	mastic.
	Clean	S.F.		.29		.29	
	Minimum Charge	Job		157		157	
Teak							
	Demolish	S.F.		.77		.77	Includes material and labor to install
	Install	S.F.	7.20	1.96		9.16	prefinished teak parquet block
	Demolish and Install	S.F.	7.20	2.73		9.93	flooring, 5/16" thick, and installed in
	Reinstall	S.F.		1.96		1.96	mastic.
	Clean	S.F.		.29		.29	
	Paint	S.F.	1.14	1.77		2.91	
	Minimum Charge	Job		157		157	
Clean and Wax							
	Clean	S.F.	.02	.18		.20	Includes labor and material to wash
	Minimum Charge	Job		157		157	and wax a hardwood floor.

Flooring

Ceramic Tile Flooring

Ceramic Tile Flooring	Unit	Material	Labor	Equip.	Total	Specification
Economy Grade						
Demolish	S.F.		.76		.76	Cost includes material and labor to
Install	S.F.	4.20	5.55		9.75	install 4-1/4" x 4-1/4" ceramic tile in
Demolish and Install	S.F.	4.20	6.31		10.51	mortar bed including grout.
Clean	S.F.		.44		.44	
Minimum Charge	Job		142		142	
Average Grade						
Demolish	S.F.		.76		.76	Cost includes material and labor to
Install	S.F.	5.60	5.70		11.30	install 4-1/4" x 4-1/4" ceramic tile in
Demolish and Install	S.F.	5.60	6.46		12.06	mortar bed including grout.
Clean	S.F.		.44		.44	
Minimum Charge	Job		142		142	
Premium Grade						
Demolish	S.F.		.76		.76	Cost includes material and labor to
Install	S.F.	8.80	5.70		14.50	install 4-1/4" x 4-1/4" ceramic tile in
Demolish and Install	S.F.	8.80	6.46		15.26	mortar bed including grout.
Clean	S.F.		.44		.44	
Minimum Charge	Job		142		142	
Re-grout						
Install	S.F.	.14	2.27		2.41	Includes material and labor to regrout
Minimum Charge	Job		142		142	tile floors.

Hard Tile Flooring

Hard Tile Flooring	Unit	Material	Labor	Equip.	Total	Specification
Thick Set Paver						
Brick						
Demolish	S.F.		1.59		1.59	Cost includes material and labor to
Install	S.F.	.72	13.25		13.97	install 6" x 12" adobe brick paver with
Demolish and Install	S.F.	.72	14.84		15.56	1/2" mortar joints.
Clean	S.F.		.44		.44	
Minimum Charge	Job		142		142	
Mexican Red						
Demolish	S.F.		1.59		1.59	Cost includes material and labor to
Install	S.F.	1.43	5.90		7.33	install 12" x 12" red Mexican paver
Demolish and Install	S.F.	1.43	7.49		8.92	tile in mortar bed with grout.
Clean	S.F.		.44		.44	
Minimum Charge	Job		142		142	
Saltillo						
Demolish	S.F.		.76		.76	Cost includes material and labor to
Install	S.F.	2.24	5.90		8.14	install 12" x 12"saltillo tile in mortar
Demolish and Install	S.F.	2.24	6.66		8.90	bed with grout.
Clean	S.F.		.44		.44	
Minimum Charge	Job		142		142	
Marble						
Premium Grade						
Demolish	S.F.		.76		.76	Cost includes material and labor to
Install	S.F.	18.90	12.35		31.25	install 3/8" x 12" x 12" marble, in
Demolish and Install	S.F.	18.90	13.11		32.01	mortar bed with grout.
Clean	S.F.		.44		.44	
Minimum Charge	Job		142		142	
Terrazzo						
Gray Cement						
Demolish	S.F.		1.01		1.01	Cost includes material and labor to
Install	S.F.	2.68	3.94	1.53	8.15	install 1-3/4" terrazzo , #1 and #2
Demolish and Install	S.F.	2.68	4.95	1.53	9.16	chips in gray Portland cement.
Clean	S.F.		.44		.44	
Minimum Charge	Job		256	99.50	355.50	

Flooring

Hard Tile Flooring

		Unit	Material	Labor	Equip.	Total	Specification
White Cement							
	Demolish	S.F.		1.01		1.01	Cost includes material and labor to
	Install	S.F.	3.06	3.94	1.53	8.53	install 1-3/4" terrazzo, #1 and #2
	Demolish and Install	S.F.	3.06	4.95	1.53	9.54	chips in white Portland cement.
	Clean	S.F.		.44		.44	
	Minimum Charge	Job		256	99.50	355.50	
Non-skid Gray							
	Demolish	S.F.		1.01		1.01	Cost includes material and labor to
	Install	S.F.	4.88	8.55	3.32	16.75	install 1-3/4" terrazzo, #1 and #2
	Demolish and Install	S.F.	4.88	9.56	3.32	17.76	chips in gray Portland cement with
	Clean	S.F.		.44		.44	light non-skid abrasive.
	Minimum Charge	Job		256	99.50	355.50	
Non-skid White							
	Demolish	S.F.		1.01		1.01	Cost includes material and labor to
	Install	S.F.	5.35	8.55	3.32	17.22	install 1-3/4" terrazzo, #1 and #2
	Demolish and Install	S.F.	5.35	9.56	3.32	18.23	chips in white Portland cement with
	Clean	S.F.		.44		.44	light non-skid abrasive.
	Minimum Charge	Job		256	99.50	355.50	
Gray w / Brass Divider							
	Demolish	S.F.		1.02		1.02	Cost includes material and labor to
	Install	S.F.	6.70	9.30	3.62	19.62	install 1-3/4" terrazzo, #1 and #2
	Demolish and Install	S.F.	6.70	10.32	3.62	20.64	chips in gray Portland cement with
	Clean	S.F.		.44		.44	brass strips 2' O.C. each way.
	Minimum Charge	Job		256	99.50	355.50	
Clean							
	Clean	S.F.	.12	.29		.41	Includes labor and material to clean
	Minimum Charge	Job		256	99.50	355.50	terrazzo flooring.
Slate							
	Demolish	S.F.		.91		.91	Includes material and labor to install
	Install	S.F.	4.55	2.80		7.35	slate tile flooring in thin set with grout.
	Demolish and Install	S.F.	4.55	3.71		8.26	
	Clean	S.F.		.44		.44	
	Minimum Charge	Job		142		142	
Granite							
	Demolish	S.F.		.75		.75	Includes material and labor to install
	Install	S.F.	18.55	23.50		42.05	granite tile flooring in thin set with
	Demolish and Install	S.F.	18.55	24.25		42.80	grout.
	Reinstall	S.F.		23.67		23.67	
	Clean	S.F.		.44		.44	
	Minimum Charge	Job		142		142	
Re-grout							
	Install	S.F.	.14	2.27		2.41	Includes material and labor to regrout
	Minimum Charge	Job		142		142	tile floors.

Sheet Vinyl Per S.F.

		Unit	Material	Labor	Equip.	Total	Specification
Economy Grade							
	Demolish	S.F.		.36		.36	Includes material and labor to install
	Install	S.F.	1.44	.89		2.33	resilient sheet vinyl flooring.
	Demolish and Install	S.F.	1.44	1.25		2.69	
	Clean	S.F.		.29		.29	
	Minimum Charge	Job		142		142	

Flooring

Sheet Vinyl Per S.F.

		Unit	Material	Labor	Equip.	Total	Specification
Average Grade							
	Demolish	S.F.		.36		.36	Includes material and labor to install
	Install	S.F.	2.32	.89		3.21	resilient sheet vinyl flooring.
	Demolish and Install	S.F.	2.32	1.25		3.57	
	Clean	S.F.		.29		.29	
	Minimum Charge	Job		142		142	
Premium Grade							
	Demolish	S.F.		.36		.36	Includes material and labor to install
	Install	S.F.	3.66	.89		4.55	resilient sheet vinyl flooring.
	Demolish and Install	S.F.	3.66	1.25		4.91	
	Clean	S.F.		.29		.29	
	Minimum Charge	Job		142		142	
Clean and Wax							
	Clean	S.F.	.02	.14		.16	Includes labor and material to wash
	Minimum Charge	Job		142		142	and wax a vinyl tile floor.

Sheet Vinyl

		Unit	Material	Labor	Equip.	Total	Specification
Economy Grade							
	Demolish	S.Y.		3.29		3.29	Includes material and labor to install
	Install	S.Y.	12.95	8		20.95	no wax sheet vinyl flooring.
	Demolish and Install	S.Y.	12.95	11.29		24.24	
	Clean	S.Y.		2		2	
	Minimum Charge	Job		142		142	
Average Grade							
	Demolish	S.Y.		3.29		3.29	Includes material and labor to install
	Install	S.Y.	21	8		29	no wax sheet vinyl flooring.
	Demolish and Install	S.Y.	21	11.29		32.29	
	Clean	S.Y.		2		2	
	Minimum Charge	Job		142		142	
Premium Grade							
	Demolish	S.Y.		3.29		3.29	Includes material and labor to install
	Install	S.Y.	33	8		41	no wax sheet vinyl flooring.
	Demolish and Install	S.Y.	33	11.29		44.29	
	Clean	S.Y.		2		2	
	Minimum Charge	Job		142		142	
Clean and Wax							
	Clean	S.F.	.02	.14		.16	Includes labor and material to wash
	Minimum Charge	Job		142		142	and wax a vinyl tile floor.

Vinyl Tile

		Unit	Material	Labor	Equip.	Total	Specification
Economy Grade							
	Demolish	S.F.		.51		.51	Includes material and labor to install
	Install	S.F.	1.85	.57		2.42	no wax 12" x 12" vinyl tile.
	Demolish and Install	S.F.	1.85	1.08		2.93	
	Clean	S.F.		.29		.29	
	Minimum Charge	Job		142		142	
Average Grade							
	Demolish	S.F.		.51		.51	Includes material and labor to install
	Install	S.F.	5.05	.57		5.62	no wax 12" x 12" vinyl tile.
	Demolish and Install	S.F.	5.05	1.08		6.13	
	Clean	S.F.		.29		.29	
	Minimum Charge	Job		142		142	

Flooring

Vinyl Tile		Unit	Material	Labor	Equip.	Total	Specification
Premium Grade							
	Demolish	S.F.		.51		.51	Includes material and labor to install
	Install	S.F.	8.80	.57		9.37	no wax 12″ x 12″ vinyl tile.
	Demolish and Install	S.F.	8.80	1.08		9.88	
	Clean	S.F.		.29		.29	
	Minimum Charge	Job		142		142	
Clean and Wax							
	Clean	S.F.	.02	.14		.16	Includes labor and material to wash
	Minimum Charge	Job		142		142	and wax a vinyl tile floor.

Carpeting Per S.F.		Unit	Material	Labor	Equip.	Total	Specification
Economy Grade							
	Demolish	S.F.		.06		.06	Includes material and labor to install
	Install	S.F.	1.06	.55		1.61	carpet including tack strips and hot
	Demolish and Install	S.F.	1.06	.61		1.67	melt tape on seams.
	Clean	S.F.		.40		.40	
	Minimum Charge	Job		142		142	
Average Grade							
	Demolish	S.F.		.06		.06	Includes material and labor to install
	Install	S.F.	1.30	.55		1.85	carpet including tack strips and hot
	Demolish and Install	S.F.	1.30	.61		1.91	melt tape on seams.
	Clean	S.F.		.40		.40	
	Minimum Charge	Job		142		142	
Premium Grade							
	Demolish	S.F.		.06		.06	Includes material and labor to install
	Install	S.F.	2.30	.55		2.85	carpet including tack strips and hot
	Demolish and Install	S.F.	2.30	.61		2.91	melt tape on seams.
	Clean	S.F.		.40		.40	
	Minimum Charge	Job		142		142	
Indoor/Outdoor							
	Demolish	S.F.		.06		.06	Includes material and labor to install
	Install	S.F.	1.01	.67		1.68	indoor-outdoor carpet.
	Demolish and Install	S.F.	1.01	.73		1.74	
	Clean	S.F.		.40		.40	
	Minimum Charge	Job		142		142	
100% Wool							
Average Grade							
	Demolish	S.F.		.06		.06	Includes material and labor to install
	Install	S.F.	5.85	.56		6.41	carpet including tack strips and hot
	Demolish and Install	S.F.	5.85	.62		6.47	melt tape on seams.
	Clean	S.F.		.40		.40	
	Minimum Charge	Job		142		142	
Premium Grade							
	Demolish	S.F.		.06		.06	Includes material and labor to install
	Install	S.F.	6.50	.56		7.06	carpet including tack strips and hot
	Demolish and Install	S.F.	6.50	.62		7.12	melt tape on seams.
	Clean	S.F.		.40		.40	
	Minimum Charge	Job		142		142	
Luxury Grade							
	Demolish	S.F.		.06		.06	Includes material and labor to install
	Install	S.F.	9.20	.56		9.76	carpet including tack strips and hot
	Demolish and Install	S.F.	9.20	.62		9.82	melt tape on seams.
	Clean	S.F.		.40		.40	
	Minimum Charge	Job		142		142	

Flooring

Carpeting Per S.F.

Carpeting Per S.F.	Unit	Material	Labor	Equip.	Total	Specification
Berber						
Average Grade						
Demolish	S.F.		.06		.06	Includes material and labor to install
Install	S.F.	2.72	.67		3.39	carpet including tack strips and hot
Demolish and Install	S.F.	2.72	.73		3.45	melt tape on seams.
Clean	S.F.		.40		.40	
Minimum Charge	Job		142		142	
Premium Grade						
Demolish	S.F.		.06		.06	Includes material and labor to install
Install	S.F.	5.90	.67		6.57	carpet including tack strips and hot
Demolish and Install	S.F.	5.90	.73		6.63	melt tape on seams.
Clean	S.F.		.40		.40	
Minimum Charge	Job		142		142	

Carpeting

Carpeting	Unit	Material	Labor	Equip.	Total	Specification
Indoor/Outdoor						
Demolish	S.Y.		.85		.85	Includes material and labor to install
Install	S.Y.	9.10	6.05		15.15	indoor-outdoor carpet.
Demolish and Install	S.Y.	9.10	6.90		16	
Reinstall	S.Y.		4.83		4.83	
Clean	S.Y.		3.59		3.59	
Minimum Charge	Job		142		142	
Economy Grade						
Demolish	S.Y.		.85		.85	Includes material and labor to install
Install	S.Y.	9.50	4.98		14.48	carpet including tack strips and hot
Demolish and Install	S.Y.	9.50	5.83		15.33	melt tape on seams.
Reinstall	S.Y.		3.99		3.99	
Clean	S.Y.		3.59		3.59	
Minimum Charge	Job		142		142	
Average Grade						
Demolish	S.Y.		.85		.85	Includes material and labor to install
Install	S.Y.	11.70	4.98		16.68	carpet including tack strips and hot
Demolish and Install	S.Y.	11.70	5.83		17.53	melt tape on seams.
Reinstall	S.Y.		3.99		3.99	
Clean	S.Y.		3.59		3.59	
Minimum Charge	Job		142		142	
Premium Grade						
Demolish	S.Y.		.85		.85	Includes material and labor to install
Install	S.Y.	20.50	4.98		25.48	carpet including tack strips and hot
Demolish and Install	S.Y.	20.50	5.83		26.33	melt tape on seams.
Reinstall	S.Y.		3.99		3.99	
Clean	S.Y.		3.59		3.59	
Minimum Charge	Job		142		142	
100% Wool						
Average Grade						
Demolish	S.Y.		.85		.85	Includes material and labor to install
Install	S.Y.	52.50	5.05		57.55	wool carpet including tack strips and
Demolish and Install	S.Y.	52.50	5.90		58.40	hot melt tape on seams.
Reinstall	S.Y.		4.06		4.06	
Clean	S.Y.		3.59		3.59	
Minimum Charge	Job		142		142	

Flooring

Carpeting		Unit	Material	Labor	Equip.	Total	Specification
Premium Grade							
	Demolish	S.Y.		.85		.85	Includes material and labor to install
	Install	S.Y.	59	5.05		64.05	carpet including tack strips and hot
	Demolish and Install	S.Y.	59	5.90		64.90	melt tape on seams.
	Reinstall	S.Y.		4.06		4.06	
	Clean	S.Y.		3.59		3.59	
	Minimum Charge	Job		142		142	
Luxury Grade							
	Demolish	S.Y.		.85		.85	Includes material and labor to install
	Install	S.Y.	82.50	5.05		87.55	carpet including tack strips and hot
	Demolish and Install	S.Y.	82.50	5.90		88.40	melt tape on seams.
	Reinstall	S.Y.		4.06		4.06	
	Clean	S.Y.		3.59		3.59	
	Minimum Charge	Job		142		142	
Berber							
Average Grade							
	Demolish	S.Y.		.85		.85	Includes material and labor to install
	Install	S.Y.	25	6.05		31.05	berber carpet including tack strips and
	Demolish and Install	S.Y.	25	6.90		31.90	hot melt tape on seams.
	Reinstall	S.Y.		4.83		4.83	
	Clean	S.Y.		3.59		3.59	
	Minimum Charge	Job		142		142	
Premium Grade							
	Demolish	S.Y.		.85		.85	Includes material and labor to install
	Install	S.Y.	53	6.05		59.05	carpet including tack strips and hot
	Demolish and Install	S.Y.	53	6.90		59.90	melt tape on seams.
	Reinstall	S.Y.		4.83		4.83	
	Clean	S.Y.		3.59		3.59	
	Minimum Charge	Job		142		142	

Carpet Pad Per S.F.		Unit	Material	Labor	Equip.	Total	Specification
Urethane							
	Demolish	S.F.		.02		.02	Includes material and labor to install
	Install	S.F.	.31	.21		.52	urethane carpet pad.
	Demolish and Install	S.F.	.31	.23		.54	
	Minimum Charge	Job		142		142	
Foam Rubber Slab							
	Demolish	S.F.		.02		.02	Includes material and labor to install
	Install	S.F.	.52	.21		.73	carpet pad, 3/8" thick foam rubber.
	Demolish and Install	S.F.	.52	.23		.75	
	Minimum Charge	Job		142		142	
Waffle							
	Demolish	S.F.		.02		.02	Includes material and labor to install
	Install	S.F.	.33	.02		.35	rubber waffle carpet pad.
	Demolish and Install	S.F.	.33	.04		.37	
	Minimum Charge	Job		142		142	
Jute							
	Demolish	S.F.		.02		.02	Includes material and labor to install
	Install	S.F.	.61	.19		.80	jute hair carpet pad.
	Demolish and Install	S.F.	.61	.21		.82	
	Minimum Charge	Job		142		142	
Rebound							
	Demolish	S.F.		.02		.02	Includes material and labor to install
	Install	S.F.	.29	.16		.45	rebound carpet pad.
	Demolish and Install	S.F.	.29	.18		.47	
	Minimum Charge	Job		142		142	

Flooring

Carpet Pad		Unit	Material	Labor	Equip.	Total	Specification
Urethane							
	Demolish	S.Y.		.15		.15	Includes material and labor to install
	Install	S.Y.	2.74	1.89		4.63	urethane carpet pad.
	Demolish and Install	S.Y.	2.74	2.04		4.78	
	Reinstall	S.Y.		1.51		1.51	
	Minimum Charge	Job		142		142	
Foam Rubber Slab							
	Demolish	S.Y.		.15		.15	Includes material and labor to install
	Install	S.Y.	4.65	1.89		6.54	rubber slab carpet pad.
	Demolish and Install	S.Y.	4.65	2.04		6.69	
	Reinstall	S.Y.		1.51		1.51	
	Minimum Charge	Job		142		142	
Waffle							
	Demolish	S.Y.		.15		.15	Includes material and labor to install
	Install	S.Y.	3.01	.21		3.22	rubber waffle carpet pad.
	Demolish and Install	S.Y.	3.01	.36		3.37	
	Reinstall	S.Y.		.17		.17	
	Minimum Charge	Job		142		142	
Jute							
	Demolish	S.Y.		.15		.15	Includes material and labor to install
	Install	S.Y.	5.45	1.67		7.12	jute hair carpet pad.
	Demolish and Install	S.Y.	5.45	1.82		7.27	
	Reinstall	S.Y.		1.34		1.34	
	Minimum Charge	Job		142		142	
Rebound							
	Demolish	S.Y.		.15		.15	Includes material and labor to install
	Install	S.Y.	2.56	1.42		3.98	rebound carpet pad.
	Demolish and Install	S.Y.	2.56	1.57		4.13	
	Reinstall	S.Y.		1.13		1.13	
	Minimum Charge	Job		142		142	

Stair Components		Unit	Material	Labor	Equip.	Total	Specification
Carpeting							
	Demolish	Ea.		1.55		1.55	Includes material and labor to install
	Install	Riser	29.50	7.45		36.95	carpet and pad on stairs.
	Demolish and Install	Riser	29.50	9		38.50	
	Reinstall	Riser		5.98		5.98	
	Clean	Ea.		3.59		3.59	
	Minimum Charge	Job		142		142	

Finish Mechanical

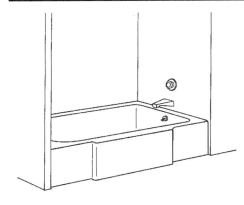

Tub/Shower

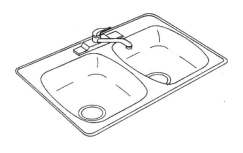

Stainless Steel Double Sink

Water Closet

Commode		Unit	Material	Labor	Equip.	Total	Specification
	Demolish	Ea.		34.50		34.50	Includes material and labor to install a toilet with valve, seat and cover.
	Install	Ea.	176	117		293	
	Demolish and Install	Ea.	176	151.50		327.50	
	Reinstall	Ea.		93.22		93.22	
	Clean	Ea.	.29	11.50		11.79	
	Minimum Charge	Job		172		172	
Wall w / Flush Valve							
	Demolish	Ea.		34.50		34.50	Includes material and labor to install a wall hung toilet with flush valve, seat and cover.
	Install	Ea.	375	106		481	
	Demolish and Install	Ea.	375	140.50		515.50	
	Reinstall	Ea.		85.19		85.19	
	Clean	Ea.	.29	11.50		11.79	
	Minimum Charge	Job		172		172	
Designer Floor Mount							
	Demolish	Ea.		34.50		34.50	Includes material and labor to install a toilet with valve, seat and cover.
	Install	Ea.	535	117		652	
	Demolish and Install	Ea.	535	151.50		686.50	
	Reinstall	Ea.		93.22		93.22	
	Clean	Ea.	.29	11.50		11.79	
	Minimum Charge	Job		172		172	
Urinal							
	Demolish	Ea.		34.50		34.50	Includes material and labor to install a wall hung urinal with flush valve.
	Install	Ea.	320	206		526	
	Demolish and Install	Ea.	320	240.50		560.50	
	Reinstall	Ea.		164.69		164.69	
	Clean	Ea.	.29	11.50		11.79	
	Minimum Charge	Job		172		172	
Bidet							
Vitreous China							
	Demolish	Ea.		34.50		34.50	Includes material and labor to install a vitreous china bidet complete with trim.
	Install	Ea.	615	124		739	
	Demolish and Install	Ea.	615	158.50		773.50	
	Reinstall	Ea.		98.82		98.82	
	Clean	Ea.	.29	11.50		11.79	
	Minimum Charge	Job		172		172	

Finish Mechanical

Commode

	Unit	Material	Labor	Equip.	Total	Specification
Chrome Fittings						
Install	Ea.	150	43		193	Includes labor and materials to install chrome fittings for a bidet.
Brass Fittings						
Install	Ea.	164	43		207	Includes labor and materials to install brass fittings for a bidet.
Seat						
Demolish	Ea.		14.30		14.30	Includes material and labor to install a toilet seat and cover.
Install	Ea.	27.50	14.30		41.80	
Demolish and Install	Ea.	27.50	28.60		56.10	
Reinstall	Ea.		11.44		11.44	
Clean	Ea.	.14	4.78		4.92	
Minimum Charge	Job		172		172	
Rough-in						
Install	Ea.	153	217		370	Includes labor and materials to install water supply pipe with valves and drain, waste and vent pipe with all couplings, hangers and fasteners necessary.

Sink (assembly)

	Unit	Material	Labor	Equip.	Total	Specification
Single						
Porcelain						
Demolish	Ea.		21.50		21.50	Includes material and labor to install a single bowl enamel finished cast iron kitchen sink with faucet and drain.
Install	Ea.	274	110		384	
Demolish and Install	Ea.	274	131.50		405.50	
Reinstall	Ea.		88.23		88.23	
Clean	Ea.	.19	8.20		8.39	
Minimum Charge	Job		172		172	
Stainless Steel						
Demolish	Ea.		21.50		21.50	Includes material and labor to install a single bowl stainless steel kitchen sink with faucet and drain.
Install	Ea.	405	110		515	
Demolish and Install	Ea.	405	131.50		536.50	
Reinstall	Ea.		88.23		88.23	
Clean	Ea.	.19	8.20		8.39	
Minimum Charge	Job		172		172	
Double						
Porcelain						
Demolish	Ea.		24.50		24.50	Includes material and labor to install a double bowl enamel finished cast iron kitchen sink with faucet and drain.
Install	Ea.	525	129		654	
Demolish and Install	Ea.	525	153.50		678.50	
Reinstall	Ea.		102.93		102.93	
Clean	Ea.	.29	11.50		11.79	
Minimum Charge	Job		172		172	
Stainless Steel						
Demolish	Ea.		24.50		24.50	Includes material and labor to install a double bowl stainless steel kitchen sink with faucet and drain.
Install	Ea.	257	129		386	
Demolish and Install	Ea.	257	153.50		410.50	
Reinstall	Ea.		102.93		102.93	
Clean	Ea.	.29	11.50		11.79	
Minimum Charge	Job		172		172	
Bar						
Demolish	Ea.		21.50		21.50	Includes material and labor to install a small single bowl stainless steel bar / vegetable / kitchen sink with faucet and drain.
Install	Ea.	78.50	86		164.50	
Demolish and Install	Ea.	78.50	107.50		186	
Reinstall	Ea.		68.64		68.64	
Clean	Ea.	.19	8.20		8.39	
Minimum Charge	Job		172		172	

Sink (assembly)

Sink (assembly)		Unit	Material	Labor	Equip.	Total	Specification
Floor							
	Demolish	Ea.		68.50		68.50	Includes material and labor to install
	Install	Ea.	540	140		680	an enameled cast iron floor mounted
	Demolish and Install	Ea.	540	208.50		748.50	corner service sink with faucet and
	Reinstall	Ea.		112.29		112.29	drain.
	Clean	Ea.	.29	5.75		6.04	
	Minimum Charge	Job		172		172	
Pedestal							
	Demolish	Ea.		21.50		21.50	Includes material and labor to install a
	Install	Ea.	445	96.50		541.50	vitreous china pedestal lavatory with
	Demolish and Install	Ea.	445	118		563	faucet set and pop-up drain.
	Reinstall	Ea.		77.20		77.20	
	Clean	Ea.	.19	8.20		8.39	
	Minimum Charge	Job		172		172	
Vanity Lavatory							
	Demolish	Ea.		21.50		21.50	Includes material and labor to install a
	Install	Ea.	287	114		401	vitreous china lavatory with faucet set
	Demolish and Install	Ea.	287	135.50		422.50	and pop-up drain.
	Reinstall	Ea.		91.50		91.50	
	Clean	Ea.	.19	8.20		8.39	
	Minimum Charge	Job		172		172	
Rough-in							
	Install	Ea.	78	289		367	Includes labor and materials to install copper supply pipe with valves and drain, waste and vent pipe with all couplings, hangers and fasteners necessary.
Laundry							
	Demolish	Ea.		23		23	Includes material and labor to install a
	Install	Ea.	96.50	95		191.50	laundry sink (wall mounted or with
	Demolish and Install	Ea.	96.50	118		214.50	legs) including faucet and pop up
	Clean	Ea.	.19	8.20		8.39	drain.
	Minimum Charge	Job		172		172	

Sink Only

Sink Only		Unit	Material	Labor	Equip.	Total	Specification
Single							
Porcelain							
	Demolish	Ea.		21.50		21.50	Includes material and labor to install a
	Install	Ea.	204	123		327	single bowl enamel finished cast iron
	Demolish and Install	Ea.	204	144.50		348.50	kitchen sink.
	Reinstall	Ea.		98.06		98.06	
	Clean	Ea.	.19	8.20		8.39	
	Minimum Charge	Job		172		172	
Stainless Steel							
	Demolish	Ea.		21.50		21.50	Includes material and labor to install a
	Install	Ea.	335	86		421	single bowl stainless steel kitchen sink.
	Demolish and Install	Ea.	335	107.50		442.50	
	Reinstall	Ea.		68.64		68.64	
	Clean	Ea.	.19	8.20		8.39	
	Minimum Charge	Job		172		172	

Sink Only		Unit	Material	Labor	Equip.	Total	Specification
Double							
Porcelain							
	Demolish	Ea.		24.50		24.50	Includes material and labor to install a
	Install	Ea.	234	143		377	double bowl enamel finished cast iron
	Demolish and Install	Ea.	234	167.50		401.50	kitchen sink.
	Reinstall	Ea.		114.40		114.40	
	Clean	Ea.	.29	11.50		11.79	
	Minimum Charge	Job		172		172	
Stainless Steel							
	Demolish	Ea.		24.50		24.50	Includes material and labor to install a
	Install	Ea.	500	111		611	double bowl stainless steel kitchen
	Demolish and Install	Ea.	500	135.50		635.50	sink.
	Reinstall	Ea.		88.57		88.57	
	Clean	Ea.	.29	11.50		11.79	
	Minimum Charge	Job		172		172	
Bar							
	Demolish	Ea.		21.50		21.50	Includes material and labor to install a
	Install	Ea.	47.50	86		133.50	small single bowl stainless steel bar /
	Demolish and Install	Ea.	47.50	107.50		155	vegetable / kitchen sink.
	Reinstall	Ea.		68.64		68.64	
	Clean	Ea.	.19	8.20		8.39	
	Minimum Charge	Job		172		172	
Floor							
	Demolish	Ea.		68.50		68.50	Includes material and labor to install
	Install	Ea.	445	156		601	an enameled cast iron floor mounted
	Demolish and Install	Ea.	445	224.50		669.50	corner service sink.
	Reinstall	Ea.		124.80		124.80	
	Clean	Ea.	.29	5.75		6.04	
	Minimum Charge	Job		172		172	
Pedestal							
	Demolish	Ea.		21.50		21.50	Includes material and labor to install a
	Install	Ea.	345	107		452	vitreous china pedestal lavatory.
	Demolish and Install	Ea.	345	128.50		473.50	
	Reinstall	Ea.		85.80		85.80	
	Clean	Ea.	.19	8.20		8.39	
	Minimum Charge	Job		172		172	
Vanity Lavatory							
	Demolish	Ea.		21.50		21.50	Includes material and labor to install a
	Install	Ea.	145	107		252	vitreous china lavatory.
	Demolish and Install	Ea.	145	128.50		273.50	
	Reinstall	Ea.		85.80		85.80	
	Clean	Ea.	.19	8.20		8.39	
	Minimum Charge	Job		172		172	
Rough-in							
	Install	Ea.	69.50	269		338.50	Includes labor and materials to install copper supply pipe with valves and drain, waste and vent pipe with all couplings, hangers and fasteners necessary.
Faucet Sink							
Good Grade							
	Demolish	Ea.		17.15		17.15	Includes material and labor to install a
	Install	Ea.	56.50	23.50		80	sink faucet and fittings.
	Demolish and Install	Ea.	56.50	40.65		97.15	
	Reinstall	Ea.		18.95		18.95	
	Clean	Ea.	.11	7.20		7.31	
	Minimum Charge	Job		172		172	

Finish Mechanical

Sink Only

Sink Only	Unit	Material	Labor	Equip.	Total	Specification
Better Grade						
Demolish	Ea.		17.15		17.15	Includes material and labor to install a
Install	Ea.	95	23.50		118.50	sink faucet and fittings.
Demolish and Install	Ea.	95	40.65		135.65	
Reinstall	Ea.		18.95		18.95	
Clean	Ea.	.11	7.20		7.31	
Minimum Charge	Job		172		172	
Premium Grade						
Demolish	Ea.		17.15		17.15	Includes material and labor to install a
Install	Ea.	186	29.50		215.50	sink faucet and fittings.
Demolish and Install	Ea.	186	46.65		232.65	
Reinstall	Ea.		23.63		23.63	
Clean	Ea.	.11	7.20		7.31	
Minimum Charge	Job		172		172	

Drain & Basket

Drain & Basket	Unit	Material	Labor	Equip.	Total	Specification
Demolish	Ea.		10.75		10.75	Includes material and labor to install a
Install	Ea.	13.35	21.50		34.85	sink drain and basket.
Demolish and Install	Ea.	13.35	32.25		45.60	
Reinstall	Ea.		17.16		17.16	
Clean	Ea.	.11	7.20		7.31	
Minimum Charge	Job		172		172	

Bathtub

Bathtub	Unit	Material	Labor	Equip.	Total	Specification
Enameled						
Steel						
Demolish	Ea.		62		62	Includes material and labor to install a
Install	Ea.	315	112		427	formed steel bath tub with spout,
Demolish and Install	Ea.	315	174		489	mixing valve, shower head and
Reinstall	Ea.		89.83		89.83	pop-up drain.
Clean	Ea.	.57	19.15		19.72	
Minimum Charge	Job		172		172	
Cast Iron						
Demolish	Ea.		62		62	Includes material and labor to install a
Install	Ea.	415	140		555	cast iron bath tub with spout, mixing
Demolish and Install	Ea.	415	202		617	valve, shower head and pop-up drain.
Reinstall	Ea.		112.29		112.29	
Clean	Ea.	.57	19.15		19.72	
Minimum Charge	Job		172		172	
Fiberglass						
Demolish	Ea.		62		62	Includes material and labor to install
Install	Ea.	915	112		1027	an acrylic soaking tub with spout,
Demolish and Install	Ea.	915	174		1089	mixing valve, shower head and
Reinstall	Ea.		89.83		89.83	pop-up drain.
Clean	Ea.	.57	19.15		19.72	
Minimum Charge	Job		172		172	
Institutional						
Demolish	Ea.		62		62	Includes material and labor to install a
Install	Ea.	975	206		1181	hospital / institutional type bathtub
Demolish and Install	Ea.	975	268		1243	with spout, mixing valve, shower head
Reinstall	Ea.		164.69		164.69	and pop-up drain.
Clean	Ea.	.57	19.15		19.72	
Minimum Charge	Job		172		172	

Finish Mechanical

Bathtub		Unit	Material	Labor	Equip.	Total	Specification
Whirlpool (Acrylic)							
	Demolish	Ea.		191		191	Includes material and labor to install a
	Install	Ea.	2100	620		2720	molded fiberglass whirlpool tub.
	Demolish and Install	Ea.	2100	811		2911	
	Reinstall	Ea.		494.08		494.08	
	Clean	Ea.	.57	19.15		19.72	
	Minimum Charge	Job		172		172	
Pump / Motor							
	Demolish	Ea.		47.50		47.50	Includes material and labor to install a
	Install	Ea.	163	143		306	whirlpool tub pump / motor.
	Demolish and Install	Ea.	163	190.50		353.50	
	Reinstall	Ea.		114.40		114.40	
	Clean	Ea.	.14	14.35		14.49	
	Minimum Charge	Job		172		172	
Heater / Motor							
	Demolish	Ea.		47.50		47.50	Includes material and labor to install a
	Install	Ea.	217	153		370	whirlpool tub heater.
	Demolish and Install	Ea.	217	200.50		417.50	
	Reinstall	Ea.		122.03		122.03	
	Clean	Ea.	.14	14.35		14.49	
	Minimum Charge	Job		172		172	
Thermal Cover							
	Demolish	Ea.		7.65		7.65	Includes material and labor to install a
	Install	Ea.	146	21.50		167.50	whirlpool tub thermal cover.
	Demolish and Install	Ea.	146	29.15		175.15	
	Reinstall	Ea.		17.16		17.16	
	Clean	Ea.	1.44	14.35		15.79	
	Minimum Charge	Job		172		172	
Tub / Shower Combination							
	Demolish	Ea.		38		38	Includes material and labor to install a
	Install	Ea.	540	154		694	fiberglass module tub with shower
	Demolish and Install	Ea.	540	192		732	surround with spout, mixing valve,
	Reinstall	Ea.		123.52		123.52	shower head and pop-up drain.
	Clean	Ea.	.73	28.50		29.23	
	Minimum Charge	Job		172		172	
Accessories							
Sliding Door							
	Demolish	Ea.		15.95		15.95	Includes material and labor to install a
	Install	Ea.	365	72		437	48" aluminum framed tempered glass
	Demolish and Install	Ea.	365	87.95		452.95	shower door.
	Reinstall	Ea.		57.73		57.73	
	Clean	Ea.	1.20	9.55		10.75	
	Minimum Charge	Job		172		172	
Shower Head							
	Demolish	Ea.		4.29		4.29	Includes material and labor to install a
	Install	Ea.	71.50	14.30		85.80	water saving shower head.
	Demolish and Install	Ea.	71.50	18.59		90.09	
	Reinstall	Ea.		11.44		11.44	
	Clean	Ea.	.01	2.87		2.88	
	Minimum Charge	Job		172		172	
Faucet Set							
	Demolish	Ea.		21.50		21.50	Includes material and labor to install a
	Install	Ea.	76.50	43		119.50	combination spout / diverter for a
	Demolish and Install	Ea.	76.50	64.50		141	bathtub.
	Reinstall	Ea.		34.32		34.32	
	Clean	Ea.	.11	7.20		7.31	
	Minimum Charge	Job		172		172	

Finish Mechanical

Bathtub

		Unit	Material	Labor	Equip.	Total	Specification
Shower Rod							
	Demolish	Ea.		2.66		2.66	Includes material and labor to install a
	Install	Ea.	13.20	24		37.20	chrome plated shower rod.
	Demolish and Install	Ea.	13.20	26.66		39.86	
	Reinstall	Ea.		19.30		19.30	
	Clean	Ea.		2.87		2.87	
	Minimum Charge	Job		172		172	
Rough-in							
	Install	Ea.	127	298		425	Includes labor and materials to install copper supply pipe with valves and drain, waste and vent pipe with all couplings, hangers and fasteners necessary.

Shower

		Unit	Material	Labor	Equip.	Total	Specification
Fiberglass							
32" x 32"							
	Demolish	Ea.		129		129	Includes material and labor to install a
	Install	Ea.	385	257		642	fiberglass shower stall with door,
	Demolish and Install	Ea.	385	386		771	mixing valve and shower head and
	Reinstall	Ea.		205.87		205.87	drain fitting.
	Clean	Ea.	.57	19.15		19.72	
	Minimum Charge	Job		172		172	
36" x 36"							
	Demolish	Ea.		129		129	Includes material and labor to install a
	Install	Ea.	435	257		692	fiberglass shower stall with door,
	Demolish and Install	Ea.	435	386		821	mixing valve and shower head and
	Reinstall	Ea.		205.87		205.87	drain fitting.
	Clean	Ea.	.57	19.15		19.72	
	Minimum Charge	Job		172		172	
35" x 60"							
	Demolish	Ea.		129		129	Includes material and labor to install a
	Install	Ea.	705	310		1015	handicap fiberglass shower stall with
	Demolish and Install	Ea.	705	439		1144	door & seat, mixing valve and shower
	Reinstall	Ea.		247.04		247.04	head and drain fitting.
	Clean	Ea.	.57	19.15		19.72	
	Minimum Charge	Job		172		172	
Accessories							
Single Door							
	Demolish	Ea.		15.95		15.95	Includes material and labor to install a
	Install	Ea.	108	38		146	shower door.
	Demolish and Install	Ea.	108	53.95		161.95	
	Reinstall	Ea.		30.47		30.47	
	Clean	Ea.	1.20	9.55		10.75	
	Minimum Charge	Job		172		172	
Shower Head							
	Demolish	Ea.		4.29		4.29	Includes material and labor to install a
	Install	Ea.	71.50	14.30		85.80	water saving shower head.
	Demolish and Install	Ea.	71.50	18.59		90.09	
	Reinstall	Ea.		11.44		11.44	
	Clean	Ea.	.01	2.87		2.88	
	Minimum Charge	Job		172		172	

Shower	Unit	Material	Labor	Equip.	Total	Specification
Faucet Set						
Demolish	Ea.		21.50		21.50	Includes material and labor to install a
Install	Ea.	76.50	43		119.50	combination spout / diverter for a
Demolish and Install	Ea.	76.50	64.50		141	bathtub.
Reinstall	Ea.		34.32		34.32	
Clean	Ea.	.11	7.20		7.31	
Minimum Charge	Job		172		172	
Shower Pan						
Demolish	Ea.		15.90		15.90	Includes material and labor to install
Install	Ea.	177	43		220	the base portion of a fiberglass
Demolish and Install	Ea.	177	58.90		235.90	shower unit.
Reinstall	Ea.		34.32		34.32	
Clean	Ea.	.57	19.15		19.72	
Minimum Charge	Job		172		172	
Rough-in						
Install	Ea.	69	300		369	Includes labor and materials to install copper supply pipe with valves and drain, waste and vent pipe with all couplings, hangers and fasteners necessary.

Ductwork	Unit	Material	Labor	Equip.	Total	Specification
Return Grill						
Under 10" Wide						
Demolish	Ea.		7.60		7.60	Includes material and labor to install a
Install	Ea.	34	19.05		53.05	10" x 10" air return register.
Demolish and Install	Ea.	34	26.65		60.65	
Reinstall	Ea.		15.24		15.24	
Clean	Ea.	.14	6.40		6.54	
Minimum Charge	Job		171		171	
12" to 20" Wide						
Demolish	Ea.		7.60		7.60	Includes material and labor to install a
Install	Ea.	62.50	20		82.50	16" x 16" air return register.
Demolish and Install	Ea.	62.50	27.60		90.10	
Reinstall	Ea.		16.13		16.13	
Clean	Ea.	.14	6.40		6.54	
Minimum Charge	Job		171		171	
Over 20" Wide						
Demolish	Ea.		11.55		11.55	Includes material and labor to install a
Install	Ea.	123	31		154	24" x 24" air return register.
Demolish and Install	Ea.	123	42.55		165.55	
Reinstall	Ea.		24.93		24.93	
Clean	Ea.	.14	7.20		7.34	
Minimum Charge	Job		171		171	
Supply Grill						
Baseboard Type						
Demolish	Ea.		5.45		5.45	Includes material and labor to install a
Install	Ea.	17.10	17.15		34.25	14" x 6" baseboard air supply
Demolish and Install	Ea.	17.10	22.60		39.70	register.
Reinstall	Ea.		13.71		13.71	
Clean	Ea.	.14	6.40		6.54	
Minimum Charge	Job		171		171	

Finish Mechanical

Ductwork		Unit	Material	Labor	Equip.	Total	Specification
10" Wide							
	Demolish	Ea.		6.35		6.35	Includes material and labor to install a
	Install	Ea.	25.50	17.15		42.65	10" x 6" air supply register.
	Demolish and Install	Ea.	25.50	23.50		49	
	Reinstall	Ea.		13.71		13.71	
	Clean	Ea.	.14	6.40		6.54	
	Minimum Charge	Job		171		171	
12" to 15" Wide							
	Demolish	Ea.		6.35		6.35	Includes material and labor to install a
	Install	Ea.	35	20		55	14" x 8" air supply register.
	Demolish and Install	Ea.	35	26.35		61.35	
	Reinstall	Ea.		16.13		16.13	
	Clean	Ea.	.14	6.40		6.54	
	Minimum Charge	Job		171		171	
18" to 24" Wide							
	Demolish	Ea.		7.60		7.60	Includes material and labor to install a
	Install	Ea.	49	26.50		75.50	24" x 8" air supply register.
	Demolish and Install	Ea.	49	34.10		83.10	
	Reinstall	Ea.		21.10		21.10	
	Clean	Ea.	.14	7.20		7.34	
	Minimum Charge	Job		171		171	
30" to 36" Wide							
	Demolish	Ea.		9.50		9.50	Includes material and labor to install a
	Install	Ea.	64	24.50		88.50	30" x 8" air supply register.
	Demolish and Install	Ea.	64	34		98	
	Reinstall	Ea.		19.59		19.59	
	Clean	Ea.	.14	14.35		14.49	
	Minimum Charge	Job		171		171	
Diffuser							
12" Louver							
	Demolish	Ea.		7.60		7.60	Includes material and labor to install a
	Install	Ea.	35	24.50		59.50	12" aluminum louvered diffuser.
	Demolish and Install	Ea.	35	32.10		67.10	
	Reinstall	Ea.		19.59		19.59	
	Clean	Ea.	.14	6.40		6.54	
	Minimum Charge	Job		171		171	
14" to 20" Louver							
	Demolish	Ea.		7.60		7.60	Includes material and labor to install a
	Install	Ea.	38.50	34.50		73	14" to 20" aluminum louvered diffuser.
	Demolish and Install	Ea.	38.50	42.10		80.60	
	Reinstall	Ea.		27.42		27.42	
	Clean	Ea.	.14	7.20		7.34	
	Minimum Charge	Job		171		171	
25" to 32" Louver							
	Demolish	Ea.		11.55		11.55	Includes material and labor to install a
	Install	Ea.	47.50	49		96.50	25" to 32" aluminum louvered diffuser.
	Demolish and Install	Ea.	47.50	60.55		108.05	
	Reinstall	Ea.		39.18		39.18	
	Clean	Ea.	.14	9.55		9.69	
	Minimum Charge	Job		171		171	
12" Round							
	Demolish	Ea.		10.60		10.60	Includes material and labor to install a
	Install	Ea.	31.50	28.50		60	12" diameter aluminum diffuser with a
	Demolish and Install	Ea.	31.50	39.10		70.60	butterfly damper.
	Reinstall	Ea.		22.85		22.85	
	Clean	Ea.	.14	6.40		6.54	
	Minimum Charge	Job		171		171	

Finish Mechanical

Ductwork		Unit	Material	Labor	Equip.	Total	Specification
14″ to 20″ Round							
	Demolish	Ea.		14.10		14.10	Includes material and labor to install a
	Install	Ea.	116	38		154	20″ diameter aluminum diffuser with a
	Demolish and Install	Ea.	116	52.10		168.10	butterfly damper.
	Reinstall	Ea.		30.47		30.47	
	Clean	Ea.	.14	7.20		7.34	
	Minimum Charge	Job		171		171	
Flush-out / Sanitize							
	Install	L.F.	.57	2.87		3.44	Includes labor and material costs to
	Minimum Charge	Job		171		171	clean ductwork.

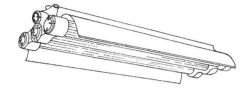

Lighting Fixture

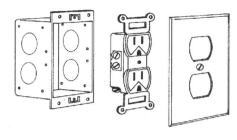

Duplex Receptacle

Electrical Per S.F.	Unit	Material	Labor	Equip.	Total	Specification
Residential						
Light Fixtures						
Install	S.F.	.22	.17		.39	Includes labor and material to replace house light fixtures. Apply S.F. cost to floor area including garages but not basements.

Circuits	Unit	Material	Labor	Equip.	Total	Specification
Outlet with Wiring						
Demolish	Ea.		19.10		19.10	Includes material and labor to install general purpose outlet with cover, up to 20 feet of #12/2 wire.
Install	Ea.	5.80	23.50		29.30	
Demolish and Install	Ea.	5.80	42.60		48.40	
Reinstall	Ea.		18.89		18.89	
Clean	Ea.		1.44		1.44	
Minimum Charge	Job		172		172	
GFCI Outlet w / Wiring						
Demolish	Ea.		19.10		19.10	Cost includes material and labor to install 120V GFCI receptacle and wall plate, with up to 20 feet of #12/2 wire.
Install	Ea.	36.50	28		64.50	
Demolish and Install	Ea.	36.50	47.10		83.60	
Reinstall	Ea.		22.33		22.33	
Clean	Ea.		1.44		1.44	
Minimum Charge	Job		172		172	
Exterior Outlet w / Wiring						
Demolish	Ea.		19.10		19.10	Cost includes material and labor to install 110V waterproof external receptacle and cover plate with up to 20' of #12/2 wire.
Install	Ea.	24	28.50		52.50	
Demolish and Install	Ea.	24	47.60		71.60	
Reinstall	Ea.		22.91		22.91	
Clean	Ea.		1.44		1.44	
Minimum Charge	Job		172		172	
240V Outlet w / Wiring						
Demolish	Ea.		19.10		19.10	Cost includes material and labor to install 240V receptacle and cover plate with up to 20' of #12/2 wire.
Install	Ea.	51.50	53.50		105	
Demolish and Install	Ea.	51.50	72.60		124.10	
Reinstall	Ea.		42.88		42.88	
Clean	Ea.		1.44		1.44	
Minimum Charge	Job		172		172	

Finish Electrical

Circuits

	Unit	Material	Labor	Equip.	Total	Specification
TV / Cable Jack w / Wiring						
Demolish	Ea.		19.10		19.10	Includes material and labor to install
Install	Ea.	12.25	21.50		33.75	TV/cable outlet box and cover plate
Demolish and Install	Ea.	12.25	40.60		52.85	including 20' of wiring.
Reinstall	Ea.		17.18		17.18	
Clean	Ea.		1.44		1.44	
Minimum Charge	Job		172		172	
Phone Jack with Wiring						
Demolish	Ea.		19.10		19.10	Includes material and labor to install
Install	Ea.	7.35	13.20		20.55	phone jack and cover plate including
Demolish and Install	Ea.	7.35	32.30		39.65	up to 20' of telephone cable.
Reinstall	Ea.		10.57		10.57	
Clean	Ea.		.90		.90	
Minimum Charge	Job		172		172	

Light Switch

	Unit	Material	Labor	Equip.	Total	Specification
Single						
Switch						
Demolish	Ea.		3.82		3.82	Includes material and labor to install
Install	Ea.	4.75	16.35		21.10	single switch and wall plate. Wiring is
Demolish and Install	Ea.	4.75	20.17		24.92	not included.
Reinstall	Ea.		13.09		13.09	
Clean	Ea.		1.44		1.44	
Minimum Charge	Job		172		172	
Plate						
Demolish	Ea.		1.91		1.91	Includes material and labor to install a
Install	Ea.	.33	4.30		4.63	one gang brown wall switch plate.
Demolish and Install	Ea.	.33	6.21		6.54	
Reinstall	Ea.		3.44		3.44	
Clean	Ea.	.06	.29		.35	
Minimum Charge	Job		172		172	
Double						
Switch						
Demolish	Ea.		7.65		7.65	Includes material and labor to install
Install	Ea.	6.25	24.50		30.75	double switch and wall plate. Wiring
Demolish and Install	Ea.	6.25	32.15		38.40	is not included.
Reinstall	Ea.		19.63		19.63	
Clean	Ea.	.06	4.59		4.65	
Minimum Charge	Job		172		172	
Plate						
Demolish	Ea.		1.91		1.91	Includes material and labor to install a
Install	Ea.	.85	6.50		7.35	two gang brown wall switch plate.
Demolish and Install	Ea.	.85	8.41		9.26	
Reinstall	Ea.		5.19		5.19	
Clean	Ea.	.06	.29		.35	
Minimum Charge	Job		172		172	
Triple						
Switch						
Demolish	Ea.		11.55		11.55	Includes material and labor to install
Install	Ea.	20.50	28.50		49	triple switch and wall plate. Wiring is
Demolish and Install	Ea.	20.50	40.05		60.55	not included.
Reinstall	Ea.		22.91		22.91	
Clean	Ea.	.06	6.90		6.96	
Minimum Charge	Job		172		172	

Finish Electrical

Light Switch

	Unit	Material	Labor	Equip.	Total	Specification
Plate						
Demolish	Ea.		1.91		1.91	Includes material and labor to install a
Install	Ea.	1.25	10.75		12	three gang brown wall switch plate.
Demolish and Install	Ea.	1.25	12.66		13.91	
Reinstall	Ea.		8.59		8.59	
Clean	Ea.	.06	.29		.35	
Minimum Charge	Job		172		172	
Dimmer						
Switch						
Demolish	Ea.		3.82		3.82	Includes material and labor to install
Install	Ea.	17.60	20		37.60	dimmer switch. Wiring is not included.
Demolish and Install	Ea.	17.60	23.82		41.42	
Reinstall	Ea.		16.17		16.17	
Clean	Ea.		1.44		1.44	
Minimum Charge	Job		172		172	
Plate						
Demolish	Ea.		1.91		1.91	Includes material and labor to install a
Install	Ea.	.33	4.30		4.63	one gang brown wall switch plate.
Demolish and Install	Ea.	.33	6.21		6.54	
Reinstall	Ea.		3.44		3.44	
Clean	Ea.	.06	.29		.35	
Minimum Charge	Job		172		172	
With Wiring						
Single						
Demolish	Ea.		19.10		19.10	Includes material and labor to install
Install	Ea.	7.45	20		27.45	single switch and plate including up to
Demolish and Install	Ea.	7.45	39.10		46.55	20' of romex cable.
Reinstall	Ea.		16.07		16.07	
Clean	Ea.		1.44		1.44	
Minimum Charge	Job		172		172	
Double						
Demolish	Ea.		19.10		19.10	Includes material and labor to install
Install	Ea.	18.20	34.50		52.70	double switch and plate including up
Demolish and Install	Ea.	18.20	53.60		71.80	to 20' of romex cable.
Reinstall	Ea.		27.49		27.49	
Clean	Ea.	.06	4.59		4.65	
Minimum Charge	Job		172		172	
Triple						
Demolish	Ea.		28.50		28.50	Includes material and labor to install
Install	Ea.	25.50	38.50		64	triple switch and plate including up to
Demolish and Install	Ea.	25.50	67		92.50	20' of romex cable.
Reinstall	Ea.		30.92		30.92	
Clean	Ea.	.06	6.90		6.96	
Minimum Charge	Job		172		172	
Dimmer						
Demolish	Ea.		19.10		19.10	Includes material and labor to install
Install	Ea.	20.50	23.50		44	dimmer switch and cover plate
Demolish and Install	Ea.	20.50	42.60		63.10	including up to 20' of romex cable.
Reinstall	Ea.		18.89		18.89	
Clean	Ea.		1.44		1.44	
Minimum Charge	Job		172		172	

Finish Electrical

Cover for Outlet / Switch

	Unit	Material	Labor	Equip.	Total	Specification
High Grade						
Demolish	Ea.		1.91		1.91	Includes material and labor to install
Install	Ea.	1.98	4.30		6.28	wall plate, stainless steel, 1 gang.
Demolish and Install	Ea.	1.98	6.21		8.19	
Reinstall	Ea.		3.44		3.44	
Clean	Ea.	.06	.29		.35	
Minimum Charge	Job		172		172	
Deluxe Grade						
Demolish	Ea.		1.91		1.91	Includes material and labor to install
Install	Ea.	4.24	4.30		8.54	wall plate, brushed brass, 1 gang.
Demolish and Install	Ea.	4.24	6.21		10.45	
Reinstall	Ea.		3.44		3.44	
Clean	Ea.	.06	.29		.35	
Minimum Charge	Job		172		172	

Residential Light Fixture

	Unit	Material	Labor	Equip.	Total	Specification
Ceiling						
Good Quality						
Demolish	Ea.		7.95		7.95	Includes material and labor to install
Install	Ea.	28	8.60		36.60	standard quality incandescent ceiling
Demolish and Install	Ea.	28	16.55		44.55	light fixture. Wiring is not included.
Reinstall	Ea.		6.87		6.87	
Clean	Ea.	.14	9.55		9.69	
Minimum Charge	Job		172		172	
Better Quality						
Demolish	Ea.		7.95		7.95	Includes material and labor to install
Install	Ea.	83.50	18.10		101.60	custom quality incandescent ceiling
Demolish and Install	Ea.	83.50	26.05		109.55	light fixture. Wiring is not included.
Reinstall	Ea.		14.47		14.47	
Clean	Ea.	.14	9.55		9.69	
Minimum Charge	Job		172		172	
Premium Quality						
Demolish	Ea.		7.95		7.95	Includes material and labor to install
Install	Ea.	248	18.10		266.10	deluxe quality incandescent ceiling
Demolish and Install	Ea.	248	26.05		274.05	light fixture. Wiring is not included.
Reinstall	Ea.		14.47		14.47	
Clean	Ea.	.14	9.55		9.69	
Minimum Charge	Job		172		172	
Recessed Ceiling						
Demolish	Ea.		15.90		15.90	Includes material and labor to install
Install	Ea.	38.50	11.45		49.95	standard quality recessed eyeball
Demolish and Install	Ea.	38.50	27.35		65.85	spotlight with housing. Wiring is not
Reinstall	Ea.		9.16		9.16	included.
Clean	Ea.	.03	2.39		2.42	
Minimum Charge	Job		172		172	
Recessed Eyeball						
Demolish	Ea.		15.90		15.90	Includes material and labor to install
Install	Ea.	58	12.25		70.25	custom quality recessed eyeball
Demolish and Install	Ea.	58	28.15		86.15	spotlight with housing. Wiring is not
Reinstall	Ea.		9.82		9.82	included.
Clean	Ea.	.03	2.39		2.42	
Minimum Charge	Job		172		172	

Finish Electrical

Residential Light Fixture	Unit	Material	Labor	Equip.	Total	Specification
Recessed Shower Type						
Demolish	Ea.		15.90		15.90	Includes material and labor to install
Install	Ea.	47.50	11.45		58.95	deluxe quality recessed eyeball
Demolish and Install	Ea.	47.50	27.35		74.85	spotlight with housing. Wiring is not
Reinstall	Ea.		9.16		9.16	included.
Clean	Ea.	.03	2.39		2.42	
Minimum Charge	Job		172		172	
Run Wiring						
Install	Ea.	8.10	13.75		21.85	Includes labor and material to install a
Minimum Charge	Job		172		172	electric fixture utility box with
						non-metallic cable.
Fluorescent						
Good Quality						
Demolish	Ea.		17.35		17.35	Includes material and labor to install
Install	Ea.	46	49		95	standard quality 12" wide by 48" long
Demolish and Install	Ea.	46	66.35		112.35	decorative surface mounted fluorescent
Reinstall	Ea.		39.27		39.27	fixture w/acrylic diffuser.
Clean	Ea.	.03	2.39		2.42	
Minimum Charge	Job		172		172	
Better Quality						
Demolish	Ea.		17.35		17.35	Includes material and labor to install
Install	Ea.	94.50	55.50		150	an interior fluorescent lighting fixture.
Demolish and Install	Ea.	94.50	72.85		167.35	
Reinstall	Ea.		44.34		44.34	
Clean	Ea.	.03	2.39		2.42	
Minimum Charge	Job		172		172	
Premium Quality						
Demolish	Ea.		17.35		17.35	Includes material and labor to install
Install	Ea.	98	65		163	deluxe quality 24" wide by 48" long
Demolish and Install	Ea.	98	82.35		180.35	decorative surface mounted fluorescent
Reinstall	Ea.		51.86		51.86	fixture w/acrylic diffuser.
Clean	Ea.	.03	2.39		2.42	
Minimum Charge	Job		172		172	
Run Wiring						
Install	Ea.	8.10	13.75		21.85	Includes labor and material to install a
Minimum Charge	Job		172		172	electric fixture utility box with
						non-metallic cable.
Exterior						
Good Quality						
Demolish	Ea.		15.25		15.25	Includes material and labor to install
Install	Ea.	34	21.50		55.50	standard quality incandescent outdoor
Demolish and Install	Ea.	34	36.75		70.75	light fixture. Wiring is not included.
Reinstall	Ea.		17.18		17.18	
Clean	Ea.	.14	9.55		9.69	
Minimum Charge	Job		172		172	
Better Quality						
Demolish	Ea.		15.25		15.25	Includes material and labor to install
Install	Ea.	86	21.50		107.50	custom quality incandescent outdoor
Demolish and Install	Ea.	86	36.75		122.75	light fixture. Wiring is not included.
Reinstall	Ea.		17.18		17.18	
Clean	Ea.	.14	9.55		9.69	
Minimum Charge	Job		172		172	
Premium Quality						
Demolish	Ea.		15.25		15.25	Includes material and labor to install
Install	Ea.	164	86		250	deluxe quality incandescent outdoor
Demolish and Install	Ea.	164	101.25		265.25	light fixture. Wiring is not included.
Reinstall	Ea.		68.72		68.72	
Clean	Ea.	.14	9.55		9.69	
Minimum Charge	Job		172		172	

Finish Electrical

Residential Light Fixture		Unit	Material	Labor	Equip.	Total	Specification
Outdoor w / Pole							Includes material and labor to install
	Demolish	Ea.		15.25		15.25	post lantern type incandescent outdoor
	Install	Ea.	121	86		207	light fixture. Wiring is not included.
	Demolish and Install	Ea.	121	101.25		222.25	
	Reinstall	Ea.		68.72		68.72	
	Clean	Ea.	.14	9.55		9.69	
	Minimum Charge	Job		172		172	
Run Wiring							Includes labor and material to install a
	Install	Ea.	8.10	13.75		21.85	electric fixture utility box with
	Minimum Charge	Job		172		172	non-metallic cable.
Track Lighting							Includes material and labor to install
	Demolish	L.F.		5.30		5.30	track lighting.
	Install	L.F.	12.30	14.30		26.60	
	Demolish and Install	L.F.	12.30	19.60		31.90	
	Reinstall	L.F.		11.45		11.45	
	Clean	L.F.	.14	4.59		4.73	
	Minimum Charge	Job		172		172	

Commercial Light Fixture		Unit	Material	Labor	Equip.	Total	Specification
Fluorescent							
2 Bulb, 2'							Includes material and labor to install a
	Demolish	Ea.		15.25		15.25	2' long surface mounted fluorescent
	Install	Ea.	31	43		74	light fixture with two 40 watt bulbs.
	Demolish and Install	Ea.	31	58.25		89.25	
	Reinstall	Ea.		34.36		34.36	
	Clean	Ea.	.73	16.40		17.13	
	Minimum Charge	Job		172		172	
2 Bulb, 4'							Includes material and labor to install a
	Demolish	Ea.		15.25		15.25	4' long surface mounted fluorescent
	Install	Ea.	58	49		107	light fixture with two 40 watt bulbs.
	Demolish and Install	Ea.	58	64.25		122.25	
	Reinstall	Ea.		39.27		39.27	
	Clean	Ea.	.73	16.40		17.13	
	Minimum Charge	Job		172		172	
3 Bulb, 4'							Includes material and labor to install a
	Demolish	Ea.		15.25		15.25	4' long pendant mounted fluorescent
	Install	Ea.	98	49		147	light fixture with three 34 watt bulbs.
	Demolish and Install	Ea.	98	64.25		162.25	
	Reinstall	Ea.		39.27		39.27	
	Clean	Ea.	.73	16.40		17.13	
	Minimum Charge	Job		172		172	
4 Bulb, 4'							Includes material and labor to install a
	Demolish	Ea.		15.25		15.25	4' long pendant mounted fluorescent
	Install	Ea.	89	53		142	light fixture with four 34 watt bulbs.
	Demolish and Install	Ea.	89	68.25		157.25	
	Reinstall	Ea.		42.29		42.29	
	Clean	Ea.	.73	16.40		17.13	
	Minimum Charge	Job		172		172	
2 Bulb, 8'							Includes material and labor to install a
	Demolish	Ea.		19.10		19.10	8' long surface mounted fluorescent
	Install	Ea.	89	65		154	light fixture with two 110 watt bulbs.
	Demolish and Install	Ea.	89	84.10		173.10	
	Reinstall	Ea.		51.86		51.86	
	Clean	Ea.	.73	16.40		17.13	
	Minimum Charge	Job		172		172	

Finish Electrical

Commercial Light Fixture

		Unit	Material	Labor	Equip.	Total	Specification
Run Wiring							
	Install	Ea.	8.10	13.75		21.85	Includes labor and material to install a
	Minimum Charge	Job		172		172	electric fixture utility box with non-metallic cable.
Ceiling, Recessed							
Eyeball Type							
	Demolish	Ea.		15.90		15.90	Includes material and labor to install
	Install	Ea.	58	12.25		70.25	custom quality recessed eyeball
	Demolish and Install	Ea.	58	28.15		86.15	spotlight with housing. Wiring is not
	Reinstall	Ea.		9.82		9.82	included.
	Clean	Ea.	.03	2.39		2.42	
	Minimum Charge	Job		172		172	
Shower Type							
	Demolish	Ea.		15.90		15.90	Includes material and labor to install
	Install	Ea.	47.50	11.45		58.95	deluxe quality recessed eyeball
	Demolish and Install	Ea.	47.50	27.35		74.85	spotlight with housing. Wiring is not
	Reinstall	Ea.		9.16		9.16	included.
	Clean	Ea.	.03	2.39		2.42	
	Minimum Charge	Job		172		172	
Run Wiring							
	Install	Ea.	8.10	13.75		21.85	Includes labor and material to install a
	Minimum Charge	Job		172		172	electric fixture utility box with non-metallic cable.
Ceiling Grid Type							
2 Tube							
	Demolish	Ea.		22.50		22.50	Cost includes material and labor to
	Install	Ea.	53	65		118	install 2 lamp fluorescent light fixture
	Demolish and Install	Ea.	53	87.50		140.50	in 2' x 4' ceiling grid system including
	Reinstall	Ea.		51.86		51.86	lens and tubes.
	Clean	Ea.	.73	16.40		17.13	
	Minimum Charge	Job		172		172	
4 Tube							
	Demolish	Ea.		25.50		25.50	Cost includes material and labor to
	Install	Ea.	60	73		133	install 4 lamp fluorescent light fixture
	Demolish and Install	Ea.	60	98.50		158.50	in 2' x 4' ceiling grid system including
	Reinstall	Ea.		58.49		58.49	lens and tubes.
	Clean	Ea.	.73	16.40		17.13	
	Minimum Charge	Job		172		172	
Acrylic Diffuser							
	Demolish	Ea.		3.98		3.98	Includes material and labor to install
	Install	Ea.	38.50	10.40		48.90	fluorescent light fixture w/acrylic
	Demolish and Install	Ea.	38.50	14.38		52.88	diffuser.
	Reinstall	Ea.		8.33		8.33	
	Clean	Ea.	.46	2.87		3.33	
	Minimum Charge	Job		172		172	
Louver Diffuser							
	Demolish	Ea.		4.77		4.77	Includes material and labor to install
	Install	Ea.	42	10.40		52.40	fluorescent light fixture w/louver
	Demolish and Install	Ea.	42	15.17		57.17	diffuser.
	Reinstall	Ea.		8.33		8.33	
	Clean	Ea.	.46	2.87		3.33	
	Minimum Charge	Job		172		172	
Run Wiring							
	Install	Ea.	8.10	13.75		21.85	Includes labor and material to install a
	Minimum Charge	Job		172		172	electric fixture utility box with non-metallic cable.

Finish Electrical

Commercial Light Fixture		Unit	Material	Labor	Equip.	Total	Specification
Track Lighting							
	Demolish	L.F.		5.30		5.30	Includes material and labor to install
	Install	L.F.	12.30	14.30		26.60	track lighting.
	Demolish and Install	L.F.	12.30	19.60		31.90	
	Reinstall	L.F.		11.45		11.45	
	Clean	L.F.	.14	4.59		4.73	
	Minimum Charge	Job		172		172	
Emergency							
	Demolish	Ea.		27.50		27.50	Includes material and labor to install
	Install	Ea.	121	86		207	battery-powered emergency lighting
	Demolish and Install	Ea.	121	113.50		234.50	unit with two (2) lights, including
	Reinstall	Ea.		68.72		68.72	wiring.
	Clean	Ea.	.36	16.40		16.76	
	Minimum Charge	Job		172		172	
Run Wiring							
	Install	Ea.	8.10	13.75		21.85	Includes labor and material to install a
	Minimum Charge	Job		172		172	electric fixture utility box with
							non-metallic cable.

Fan		Unit	Material	Labor	Equip.	Total	Specification
Ceiling Paddle							
Good Quality							
	Demolish	Ea.		21		21	Includes material and labor to install
	Install	Ea.	82.50	34.50		117	standard ceiling fan.
	Demolish and Install	Ea.	82.50	55.50		138	
	Reinstall	Ea.		27.49		27.49	
	Clean	Ea.	.29	14.35		14.64	
	Minimum Charge	Job		172		172	
Better Quality							
	Demolish	Ea.		21		21	Includes material and labor to install
	Install	Ea.	137	34.50		171.50	better ceiling fan.
	Demolish and Install	Ea.	137	55.50		192.50	
	Reinstall	Ea.		27.49		27.49	
	Clean	Ea.	.29	14.35		14.64	
	Minimum Charge	Job		172		172	
Premium Quality							
	Demolish	Ea.		21		21	Includes material and labor to install
	Install	Ea.	320	43		363	deluxe ceiling fan.
	Demolish and Install	Ea.	320	64		384	
	Reinstall	Ea.		34.36		34.36	
	Clean	Ea.	.29	14.35		14.64	
	Minimum Charge	Job		172		172	
Light Kit							
	Install	Ea.	30.50	43		73.50	Includes labor and material to install a
	Minimum Charge	Job		172		172	ceiling fan light kit.
Bathroom Exhaust							
Ceiling Mounted							
	Demolish	Ea.		17.40		17.40	Cost includes material and labor to
	Install	Ea.	39.50	34.50		74	install 60 CFM ceiling mounted
	Demolish and Install	Ea.	39.50	51.90		91.40	bathroom exhaust fan, up to 10' of
	Reinstall	Ea.		27.41		27.41	wiring with one wall switch.
	Clean	Ea.	.14	5.75		5.89	
	Minimum Charge	Job		172		172	

Finish Electrical

Fan

Fan	Unit	Material	Labor	Equip.	Total	Specification
Wall Mounted						
Demolish	Ea.		17.40		17.40	Cost includes material and labor to
Install	Ea.	53	34.50		87.50	install 110 CFM wall mounted
Demolish and Install	Ea.	53	51.90		104.90	bathroom exhaust fan, up to 10' of
Reinstall	Ea.		27.41		27.41	wiring with one wall switch.
Clean	Ea.	.14	5.75		5.89	
Minimum Charge	Job		172		172	
Lighted						
Demolish	Ea.		17.40		17.40	Cost includes material and labor to
Install	Ea.	38	34.50		72.50	install 50 CFM ceiling mounted lighted
Demolish and Install	Ea.	38	51.90		89.90	exhaust fan, up to 10' of wiring with
Reinstall	Ea.		27.41		27.41	one wall switch..
Clean	Ea.	.14	5.75		5.89	
Minimum Charge	Job		172		172	
Blower Heater w / Light						
Demolish	Ea.		17.40		17.40	Includes material and labor to install
Install	Ea.	86.50	57		143.50	ceiling blower type ventilator with
Demolish and Install	Ea.	86.50	74.40		160.90	switch, light and snap-on grill. Includes
Reinstall	Ea.		45.68		45.68	1300 watt/120 volt heater.
Clean	Ea.	.14	5.75		5.89	
Minimum Charge	Job		172		172	
Fan Heater w / Light						
Demolish	Ea.		17.40		17.40	Includes material and labor to install
Install	Ea.	109	57		166	circular ceiling fan heater unit, 1500
Demolish and Install	Ea.	109	74.40		183.40	watt, surface mounted.
Reinstall	Ea.		45.68		45.68	
Clean	Ea.	.14	5.75		5.89	
Minimum Charge	Job		172		172	
Whole House Exhaust						
36" Ceiling Mounted						
Install	Ea.	505	85.50		590.50	Includes material and labor to install a
Clean	Ea.	.04	4.12		4.16	whole house exhaust fan.
Minimum Charge	Job		172		172	
Kitchen Exhaust						
Standard Model						
Install	Ea.	62	23		85	Includes labor and materials to install
Clean	Ea.	.04	4.12		4.16	a bathroom or kitchen vent fan.
Minimum Charge	Job		172		172	
Low Noise Model						
Install	Ea.	82.50	23		105.50	Includes labor and materials to install
Clean	Ea.	.04	4.12		4.16	a bathroom or kitchen vent fan, low
Minimum Charge	Job		172		172	noise model.

Low Voltage Systems	Unit	Material	Labor	Equip.	Total	Specification
Smoke Detector						
Pre-wired						
Demolish	Ea.		7.95		7.95	Includes material and labor to install
Install	Ea.	19.65	64.50		84.15	AC type smoke detector including
Demolish and Install	Ea.	19.65	72.45		92.10	wiring.
Reinstall	Ea.		51.57		51.57	
Clean	Ea.	.06	7.20		7.26	
Minimum Charge	Job		172		172	

Finish Electrical

Low Voltage Systems		Unit	Material	Labor	Equip.	Total	Specification
Battery-powered							
	Demolish	Ea.		7.95		7.95	Includes material and labor to install
	Install	Ea.	24	21.50		45.50	battery-powered type smoke detector.
	Demolish and Install	Ea.	24	29.45		53.45	
	Reinstall	Ea.		17.18		17.18	
	Clean	Ea.	.06	7.20		7.26	
	Minimum Charge	Job		172		172	
Run Wiring							
	Install	Ea.	8.10	13.75		21.85	Includes labor and material to install a
	Minimum Charge	Job		172		172	electric fixture utility box with
							non-metallic cable.

Door Bell / Chimes
Good Quality

		Unit	Material	Labor	Equip.	Total	Specification
	Demolish	Ea.		24		24	Includes material and labor to install
	Install	Ea.	43	86		129	surface mounted 2 note door bell.
	Demolish and Install	Ea.	43	110		153	Wiring is not included.
	Reinstall	Ea.		68.72		68.72	
	Clean	Ea.	.14	7.20		7.34	
	Minimum Charge	Job		172		172	

Better Quality

		Unit	Material	Labor	Equip.	Total	Specification
	Demolish	Ea.		24		24	Includes material and labor to install
	Install	Ea.	56	86		142	standard quality surface mounted 2
	Demolish and Install	Ea.	56	110		166	note door bell. Wiring is not included.
	Reinstall	Ea.		68.72		68.72	
	Clean	Ea.	.14	7.20		7.34	
	Minimum Charge	Job		172		172	

Premium Quality

		Unit	Material	Labor	Equip.	Total	Specification
	Demolish	Ea.		24		24	Includes material and labor to install
	Install	Ea.	65.50	86		151.50	deluxe quality surface mounted 2 note
	Demolish and Install	Ea.	65.50	110		175.50	door bell. Wiring is not included.
	Reinstall	Ea.		68.72		68.72	
	Clean	Ea.	.14	7.20		7.34	
	Minimum Charge	Job		172		172	

Run Wiring

		Unit	Material	Labor	Equip.	Total	Specification
	Install	Ea.	8.10	13.75		21.85	Includes labor and material to install a
	Minimum Charge	Job		172		172	electric fixture utility box with
							non-metallic cable.

Intercom System
Intercom (master)

		Unit	Material	Labor	Equip.	Total	Specification
	Demolish	Ea.		63.50		63.50	Includes material and labor to install
	Install	Ea.	460	172		632	master intercom station with AM/FM
	Demolish and Install	Ea.	460	235.50		695.50	music and room monitoring. Wiring is
	Reinstall	Ea.		137.44		137.44	not included.
	Clean	Ea.	.14	19.15		19.29	
	Minimum Charge	Job		172		172	

Intercom (remote)

		Unit	Material	Labor	Equip.	Total	Specification
	Demolish	Ea.		38		38	Includes material and labor to install
	Install	Ea.	42.50	43		85.50	remote station for master intercom
	Demolish and Install	Ea.	42.50	81		123.50	system. Wiring is not included.
	Reinstall	Ea.		34.36		34.36	
	Clean	Ea.	.14	7.20		7.34	
	Minimum Charge	Job		172		172	

Low Voltage Systems	Unit	Material	Labor	Equip.	Total	Specification
Security System						
Alarm Panel						
Demolish	Ea.		76.50		76.50	Includes material and labor to install a
Install	Ea.	165	172		337	6 zone burglar alarm panel.
Demolish and Install	Ea.	165	248.50		413.50	
Reinstall	Ea.		137.44		137.44	
Clean	Ea.	.14	7.20		7.34	
Minimum Charge	Job		172		172	
Motion Detector						
Demolish	Ea.		54.50		54.50	Includes material and labor to install a
Install	Ea.	227	149		376	ultrasonic motion detector for a
Demolish and Install	Ea.	227	203.50		430.50	security system.
Reinstall	Ea.		119.51		119.51	
Clean	Ea.	.14	7.20		7.34	
Minimum Charge	Job		172		172	
Door Switch						
Demolish	Ea.		24		24	Includes material and labor to install a
Install	Ea.	34.50	65		99.50	door / window contact for a security
Demolish and Install	Ea.	34.50	89		123.50	system.
Reinstall	Ea.		51.86		51.86	
Clean	Ea.	.14	2.30		2.44	
Minimum Charge	Job		172		172	
Window Switch						
Demolish	Ea.		24		24	Includes material and labor to install a
Install	Ea.	34.50	65		99.50	door / window contact for a security
Demolish and Install	Ea.	34.50	89		123.50	system.
Reinstall	Ea.		51.86		51.86	
Clean	Ea.	.14	2.30		2.44	
Minimum Charge	Job		172		172	
Satellite System						
Dish						
Demolish	Ea.		106		106	Includes material, labor and
Install	Ea.	615	286		901	equipment to install 10' mesh dish.
Demolish and Install	Ea.	615	392		1007	
Reinstall	Ea.		229.07		229.07	
Clean	Ea.	2.88	28.50		31.38	
Minimum Charge	Job		172		172	
Motor						
Demolish	Ea.		53		53	Includes material and labor to install a
Install	Ea.	310	143		453	television satellite dish motor unit.
Demolish and Install	Ea.	310	196		506	
Reinstall	Ea.		114.53		114.53	
Clean	Ea.	.14	7.20		7.34	
Minimum Charge	Job		172		172	
Television Antenna						
40' Pole						
Demolish	Ea.		59.50		59.50	Includes material and labor to install
Install	Ea.	22.50	107		129.50	stand-alone 1"-2" metal pole, 10' long.
Demolish and Install	Ea.	22.50	166.50		189	Cost does not include antenna.
Reinstall	Ea.		85.90		85.90	
Clean	Ea.	.57	28.50		29.07	
Paint	Ea.	6.70	34.50		41.20	
Minimum Charge	Job		172		172	
Rotor Unit						
Demolish	Ea.		47.50		47.50	Includes material and labor to install a
Install	Ea.	107	43		150	television antenna rotor.
Demolish and Install	Ea.	107	90.50		197.50	
Reinstall	Ea.		34.36		34.36	
Clean	Ea.	.57	28.50		29.07	
Minimum Charge	Job		172		172	

Low Voltage Systems	Unit	Material	Labor	Equip.	Total	Specification
Single Booster						Includes material and labor to install
Demolish	Ea.		47.50		47.50	standard grade VHF television signal
Install	Ea.	14.65	43		57.65	booster unit and all hardware required
Demolish and Install	Ea.	14.65	90.50		105.15	for connection to TV cable.
Reinstall	Ea.		34.36		34.36	
Clean	Ea.	.03	3.59		3.62	
Minimum Charge	Job		172		172	

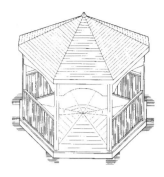

Gazebo   Gutter   Shutter

Window Treatment		Unit	Material	Labor	Equip.	Total	Specification
Drapery							
Good Grade							
	Install	L.F.	33	4.59		37.59	Includes material and labor to install
	Reinstall	L.F.		3.67		3.67	good quality drapery.
	Clean	L.F.	2.31	.11		2.42	
	Minimum Charge	Job		115		115	
Premium Grade							
	Install	L.F.	58	4.59		62.59	Includes material and labor to install
	Reinstall	L.F.		3.67		3.67	premium quality drapery.
	Clean	L.F.	2.31	.13		2.44	
	Minimum Charge	Job		115		115	
Lining							
	Install	L.F.	14.30	4.59		18.89	Includes material and labor to install
	Reinstall	L.F.		3.67		3.67	drapery lining.
	Clean	L.F.	1.16	.07		1.23	
	Minimum Charge	Job		115		115	
Valance Board							
	Demolish	L.F.		1.28		1.28	Includes material and labor to install a
	Install	L.F.	4.22	9.20		13.42	drapery valance board.
	Demolish and Install	L.F.	4.22	10.48		14.70	
	Reinstall	L.F.		7.35		7.35	
	Clean	L.F.	1.16	.23		1.39	
	Minimum Charge	Job		115		115	
Rod							
	Demolish	L.F.		3.19		3.19	Includes material and labor to install
	Install	L.F.	2.09	5.30		7.39	standard traverse drapery hardware.
	Demolish and Install	L.F.	2.09	8.49		10.58	
	Reinstall	L.F.		4.25		4.25	
	Clean	L.F.	.06	.21		.27	
	Paint	L.F.	.06	.53		.59	
	Minimum Charge	Job		115		115	
Blinds							
Mini-type							
	Demolish	S.F.		.68		.68	Includes material and labor to install
	Install	S.F.	2.83	.53		3.36	horizontal 1″ slat blinds.
	Demolish and Install	S.F.	2.83	1.21		4.04	
	Reinstall	S.F.		.43		.43	
	Clean	S.F.	.44	.02		.46	
	Minimum Charge	Job		115		115	

Window Treatment	Unit	Material	Labor	Equip.	Total	Specification
Vertical Type						
Demolish	S.F.		.68		.68	Includes material and labor to install
Install	S.F.	8.65	.68		9.33	vertical 3" - 5" cloth or PVC blinds.
Demolish and Install	S.F.	8.65	1.36		10.01	
Reinstall	S.F.		.55		.55	
Clean	S.F.	.23	.02		.25	
Minimum Charge	Job		115		115	
Vertical Type per L.F.						
Demolish	L.F.		1.06		1.06	Includes material and labor to install
Install	L.F.	51.50	3.06		54.56	vertical 3" - 5" cloth or PVC blinds x
Demolish and Install	L.F.	51.50	4.12		55.62	72" long.
Reinstall	L.F.		2.45		2.45	
Clean	L.F.	1.64	.16		1.80	
Minimum Charge	Job		115		115	
Venetian Type						
Demolish	S.F.		.68		.68	Includes material and labor to install
Install	S.F.	3.27	.53		3.80	horizontal 2" slat blinds.
Demolish and Install	S.F.	3.27	1.21		4.48	
Reinstall	S.F.		.43		.43	
Clean	S.F.	.23	.02		.25	
Minimum Charge	Job		115		115	
Wood Shutter						
27" x 36", 4 Panel						
Demolish	Ea.		22		22	Includes material and labor to install
Install	Pr.	80	18.45		98.45	moveable louver, up to 33" x 36", 4
Demolish and Install	Pr.	80	40.45		120.45	panel shutter.
Reinstall	Pr.		14.76		14.76	
Clean	Ea.	.06	6.85		6.91	
Paint	Ea.	1.07	19.70		20.77	
Minimum Charge	Job		157		157	
33" x 36", 4 Panel						
Demolish	Ea.		22		22	Includes material and labor to install
Install	Pr.	80	18.45		98.45	moveable louver, up to 33" x 36", 4
Demolish and Install	Pr.	80	40.45		120.45	panel shutter.
Reinstall	Pr.		14.76		14.76	
Clean	Ea.	.06	6.85		6.91	
Paint	Ea.	1.11	21		22.11	
Minimum Charge	Job		157		157	
39" x 36", 4 Panel						
Demolish	Ea.		22		22	Includes material and labor to install
Install	Pr.	90	18.45		108.45	moveable louver, up to 47" x 36", 4
Demolish and Install	Pr.	90	40.45		130.45	panel shutter.
Reinstall	Pr.		14.76		14.76	
Clean	Ea.	.06	6.85		6.91	
Paint	Ea.	1.11	21		22.11	
Minimum Charge	Job		157		157	
47" x 36", 4 Panel						
Demolish	Ea.		22		22	Includes material and labor to install
Install	Pr.	90	18.45		108.45	moveable louver, up to 47" x 36", 4
Demolish and Install	Pr.	90	40.45		130.45	panel shutter.
Reinstall	Pr.		14.76		14.76	
Clean	Ea.	.06	6.85		6.91	
Paint	Ea.	1.17	23		24.17	
Minimum Charge	Job		157		157	

Improvements / Appliances / Treatments

Stairs		Unit	Material	Labor	Equip.	Total	Specification
Concrete Steps							
Precast							Includes material and labor to install
	Demolish	Ea.		42.50	29.50	72	precast concrete front entrance stairs,
	Install	Flight	320	107	39.50	466.50	4' wide, with 48" platform and 2
	Demolish and Install	Flight	320	149.50	69	538.50	risers.
	Clean	Ea.		3.19		3.19	
	Paint	Ea.	2.26	34.50		36.76	
	Minimum Charge	Job		279		279	
Metal							
Concrete Filled							Includes material and labor to install
	Demolish	Ea.		12.75		12.75	steel, cement filled metal pan stairs
	Install	Riser	208	52.50	2.76	263.26	with picket rail.
	Demolish and Install	Riser	208	65.25	2.76	276.01	
	Clean	Ea.	.09	1.95		2.04	
	Paint	Ea.	.34	2.04		2.38	
	Minimum Charge	Job		192		192	
Landing							Includes material and labor to install
	Demolish	S.F.		3.19		3.19	pre-erected conventional steel pan
	Install	S.F.	43	6.15	.33	49.48	landing.
	Demolish and Install	S.F.	43	9.34	.33	52.67	
	Clean	S.F.	.09	.41		.50	
	Paint	S.F.	.12	.59		.71	
	Minimum Charge	Job		192		192	
Railing							Includes material and labor to install
	Demolish	L.F.		1.76		1.76	primed steel pipe.
	Install	L.F.	7.15	8.90	.47	16.52	
	Demolish and Install	L.F.	7.15	10.66	.47	18.28	
	Clean	L.F.	.57	1.44		2.01	
	Paint	L.F.	.64	4.60		5.24	
	Minimum Charge	Job		192		192	

Elevator		Unit	Material	Labor	Equip.	Total	Specification
2 Stop Residential							Includes labor, material and
	Install	Ea.	8875	3650		12525	equipment to install 2 stop elevator for
	Clean	Ea.	.66	14.35		15.01	residential use.
	Minimum Charge	Job		365		365	
2 Stop Commercial							Includes labor, material and
	Install	Ea.	32500	7300		39800	equipment to install 2 stop elevator for
	Clean	Ea.	.66	14.35		15.01	apartment or commercial building with
	Minimum Charge	Job		365		365	2,000 lb/13 passenger capacity.
3 Stop Commercial							Includes labor, material and
	Install	Ea.	37300	11200		48500	equipment to install 3 stop elevator for
	Clean	Ea.	.66	14.35		15.01	apartment or commercial building with
	Minimum Charge	Job		365		365	2,000 lb/13 passenger capacity.
4 Stop Commercial							Includes labor, material and
	Install	Ea.	42200	15200		57400	equipment to install 4 stop elevator for
	Clean	Ea.	.66	14.35		15.01	apartment or commercial building with
	Minimum Charge	Job		365		365	2,000 lb/13 passenger capacity.

Elevator		Unit	Material	Labor	Equip.	Total	Specification
5 Stop Commercial							Includes labor, material and
	Install	Ea.	47100	19200		66300	equipment to install 5 stop elevator for
	Clean	Ea.	.66	14.35		15.01	apartment or commercial building with
	Minimum Charge	Job		365		365	2,000 lb/13 passenger capacity.

Rangetop		Unit	Material	Labor	Equip.	Total	Specification
4-burner							Cost includes material and labor to
	Demolish	Ea.		23		23	install 4-burner electric surface unit.
	Install	Ea.	194	57.50		251.50	
	Demolish and Install	Ea.	194	80.50		274.50	
	Reinstall	Ea.		45.81		45.81	
	Clean	Ea.	.04	20.50		20.54	
	Minimum Charge	Job		57.50		57.50	
6-burner							Includes material and labor to install a
	Demolish	Ea.		23		23	6-burner cooktop unit.
	Install	Ea.	450	68.50		518.50	
	Demolish and Install	Ea.	450	91.50		541.50	
	Reinstall	Ea.		54.98		54.98	
	Clean	Ea.	.04	20.50		20.54	
	Minimum Charge	Job		57.50		57.50	
Premium Brand w / Grill							Includes material and labor to install
	Demolish	Ea.		23		23	electric downdraft cooktop with grill.
	Install	Ea.	885	68.50		953.50	
	Demolish and Install	Ea.	885	91.50		976.50	
	Reinstall	Ea.		54.98		54.98	
	Clean	Ea.	.04	20.50		20.54	
	Minimum Charge	Job		57.50		57.50	

Rangehood		Unit	Material	Labor	Equip.	Total	Specification
Vented							
30"							
	Demolish	Ea.		21.50		21.50	Includes material and labor to install a
	Install	Ea.	41.50	68.50		110	vented range hood.
	Demolish and Install	Ea.	41.50	90		131.50	
	Reinstall	Ea.		54.98		54.98	
	Clean	Ea.	.04	5.90		5.94	
	Minimum Charge	Job		57.50		57.50	
36"							
	Demolish	Ea.		21.50		21.50	Includes material and labor to install a
	Install	Ea.	88.50	68.50		157	vented range hood.
	Demolish and Install	Ea.	88.50	90		178.50	
	Reinstall	Ea.		54.98		54.98	
	Clean	Ea.	.04	5.90		5.94	
	Minimum Charge	Job		57.50		57.50	
42"							
	Demolish	Ea.		21.50		21.50	Includes material and labor to install a
	Install	Ea.	248	80		328	vented range hood.
	Demolish and Install	Ea.	248	101.50		349.50	
	Reinstall	Ea.		63.92		63.92	
	Clean	Ea.	.04	5.90		5.94	
	Minimum Charge	Job		57.50		57.50	

Rangehood	Unit	Material	Labor	Equip.	Total	Specification
Ventless						
30"						
Demolish	Ea.		21.50		21.50	Includes material and labor to install a
Install	Ea.	41.50	43		84.50	ventless range hood.
Demolish and Install	Ea.	41.50	64.50		106	
Reinstall	Ea.		34.36		34.36	
Clean	Ea.	.04	5.90		5.94	
Minimum Charge	Job		57.50		57.50	
36"						
Demolish	Ea.		21.50		21.50	Includes material and labor to install a
Install	Ea.	41.50	68.50		110	ventless range hood.
Demolish and Install	Ea.	41.50	90		131.50	
Reinstall	Ea.		54.98		54.98	
Clean	Ea.	.04	5.90		5.94	
Minimum Charge	Job		57.50		57.50	
42"						
Demolish	Ea.		21.50		21.50	Includes material and labor to install a
Install	Ea.	221	57.50		278.50	ventless range hood.
Demolish and Install	Ea.	221	79		300	
Reinstall	Ea.		45.81		45.81	
Clean	Ea.	.04	5.90		5.94	
Minimum Charge	Job		57.50		57.50	

Oven	Unit	Material	Labor	Equip.	Total	Specification
Free Standing						
Demolish	Ea.		23		23	Cost includes material and labor to
Install	Ea.	430	46		476	install 30" free standing electric range
Demolish and Install	Ea.	430	69		499	with self-cleaning oven.
Reinstall	Ea.		36.74		36.74	
Clean	Ea.	.04	20.50		20.54	
Minimum Charge	Job		57.50		57.50	
Single Wall Oven						
Demolish	Ea.		23		23	Includes material and labor to install a
Install	Ea.	425	86		511	built-in oven.
Demolish and Install	Ea.	425	109		534	
Reinstall	Ea.		68.72		68.72	
Clean	Ea.	.04	20.50		20.54	
Minimum Charge	Job		57.50		57.50	
Double Wall Oven						
Demolish	Ea.		23		23	Includes material and labor to install a
Install	Ea.	1025	157		1182	conventional double oven.
Demolish and Install	Ea.	1025	180		1205	
Reinstall	Ea.		125.44		125.44	
Clean	Ea.	.04	20.50		20.54	
Minimum Charge	Job		57.50		57.50	
Drop-in Type w / Range						
Demolish	Ea.		23		23	Includes material and labor to install a
Install	Ea.	490	57.50		547.50	built-in cooking range with oven.
Demolish and Install	Ea.	490	80.50		570.50	
Reinstall	Ea.		45.81		45.81	
Clean	Ea.	.04	20.50		20.54	
Minimum Charge	Job		57.50		57.50	

Improvements / Appliances / Treatments

Oven	Unit	Material	Labor	Equip.	Total	Specification
Hi/Lo w / Range, Microwave						
Demolish	Ea.		23		23	Includes material and labor to install a
Install	Ea.	1600	315		1915	double oven, one conventional, one
Demolish and Install	Ea.	1600	338		1938	microwave.
Reinstall	Ea.		250.88		250.88	
Clean	Ea.	.04	20.50		20.54	
Minimum Charge	Job		57.50		57.50	
Hi/Lo w/ Range / Double Oven						
Demolish	Ea.		23		23	Includes material and labor to install a
Install	Ea.	1025	157		1182	conventional double oven.
Demolish and Install	Ea.	1025	180		1205	
Reinstall	Ea.		125.44		125.44	
Clean	Ea.	.04	20.50		20.54	
Minimum Charge	Job		57.50		57.50	
Countertop Microwave						
Demolish	Ea.		7.95		7.95	Includes material and labor to install a
Install	Ea.	116	28.50		144.50	1.0 cubic foot microwave oven.
Demolish and Install	Ea.	116	36.45		152.45	
Reinstall	Ea.		22.96		22.96	
Clean	Ea.	.04	10.30		10.34	
Minimum Charge	Job		57.50		57.50	
Cabinet / Wall Mounted Microwave						
Demolish	Ea.		15.95		15.95	Includes material and labor to install a
Install	Ea.	550	86		636	space saver microwave oven.
Demolish and Install	Ea.	550	101.95		651.95	
Reinstall	Ea.		68.72		68.72	
Clean	Ea.	.04	10.30		10.34	
Minimum Charge	Job		57.50		57.50	

Dryer	Unit	Material	Labor	Equip.	Total	Specification
Basic Grade						
Demolish	Ea.		10.65		10.65	Includes material and labor to install
Install	Ea.	296	181		477	electric or gas dryer. Vent kit not
Demolish and Install	Ea.	296	191.65		487.65	included.
Reinstall	Ea.		144.64		144.64	
Clean	Ea.	.04	20.50		20.54	
Minimum Charge	Job		57.50		57.50	
Good Grade						
Demolish	Ea.		10.65		10.65	Includes material and labor to install
Install	Ea.	370	181		551	electric or gas dryer. Vent kit not
Demolish and Install	Ea.	370	191.65		561.65	included.
Reinstall	Ea.		144.64		144.64	
Clean	Ea.	.04	20.50		20.54	
Minimum Charge	Job		57.50		57.50	
Premium Grade						
Demolish	Ea.		10.65		10.65	Includes material and labor to install
Install	Ea.	840	271		1111	electric or gas dryer. Vent kit not
Demolish and Install	Ea.	840	281.65		1121.65	included.
Reinstall	Ea.		216.96		216.96	
Clean	Ea.	.04	20.50		20.54	
Minimum Charge	Job		57.50		57.50	

Improvements / Appliances / Treatments

Dryer

	Unit	Material	Labor	Equip.	Total	Specification
Washer / Dryer Combination Unit						Includes material and labor to install
Demolish	Ea.		10.65		10.65	electric washer/dryer combination
Install	Ea.	880	46		926	unit, 27" wide, stackable or unitized.
Demolish and Install	Ea.	880	56.65		936.65	
Reinstall	Ea.		36.74		36.74	
Clean	Ea.	.04	20.50		20.54	
Minimum Charge	Job		57.50		57.50	

Washer

	Unit	Material	Labor	Equip.	Total	Specification
Basic Grade						Includes material and labor to install
Demolish	Ea.		10.65		10.65	washing machine. New hoses not
Install	Ea.	300	114		414	included.
Demolish and Install	Ea.	300	124.65		424.65	
Reinstall	Ea.		91.52		91.52	
Clean	Ea.	.04	20.50		20.54	
Minimum Charge	Job		57.50		57.50	
Good Grade						Includes material and labor to install
Demolish	Ea.		10.65		10.65	washing machine. New hoses not
Install	Ea.	490	172		662	included.
Demolish and Install	Ea.	490	182.65		672.65	
Reinstall	Ea.		137.28		137.28	
Clean	Ea.	.04	20.50		20.54	
Minimum Charge	Job		57.50		57.50	
Premium Grade						Includes material and labor to install
Demolish	Ea.		10.65		10.65	washing machine with digital
Install	Ea.	1125	345		1470	computer readouts and large capacity.
Demolish and Install	Ea.	1125	355.65		1480.65	New hoses not included.
Reinstall	Ea.		274.56		274.56	
Clean	Ea.	.04	20.50		20.54	
Minimum Charge	Job		57.50		57.50	
Washer / Dryer Combination Unit						Includes material and labor to install
Demolish	Ea.		10.65		10.65	electric washer/dryer combination
Install	Ea.	880	46		926	unit, 27" wide, stackable or unitized.
Demolish and Install	Ea.	880	56.65		936.65	
Reinstall	Ea.		36.74		36.74	
Clean	Ea.	.04	20.50		20.54	
Minimum Charge	Job		57.50		57.50	

Refrigerator

	Unit	Material	Labor	Equip.	Total	Specification
12 Cubic Foot						Includes material and labor to install a
Demolish	Ea.		32		32	refrigerator.
Install	Ea.	485	46		531	
Demolish and Install	Ea.	485	78		563	
Reinstall	Ea.		36.74		36.74	
Clean	Ea.	.04	20.50		20.54	
Minimum Charge	Job		57.50		57.50	
16 Cubic Foot						Includes material and labor to install a
Demolish	Ea.		32		32	refrigerator.
Install	Ea.	490	51		541	
Demolish and Install	Ea.	490	83		573	
Reinstall	Ea.		40.82		40.82	
Clean	Ea.	.04	20.50		20.54	
Minimum Charge	Job		57.50		57.50	

Refrigerator		Unit	Material	Labor	Equip.	Total	Specification
18 Cubic Foot							
	Demolish	Ea.		32		32	Includes material and labor to install a
	Install	Ea.	585	57.50		642.50	refrigerator.
	Demolish and Install	Ea.	585	89.50		674.50	
	Reinstall	Ea.		45.92		45.92	
	Clean	Ea.	.04	20.50		20.54	
	Minimum Charge	Job		57.50		57.50	
21 Cubic Foot							
	Demolish	Ea.		32		32	Includes material and labor to install a
	Install	Ea.	710	65.50		775.50	refrigerator.
	Demolish and Install	Ea.	710	97.50		807.50	
	Reinstall	Ea.		52.48		52.48	
	Clean	Ea.	.04	20.50		20.54	
	Minimum Charge	Job		57.50		57.50	
24 Cubic Foot							
	Demolish	Ea.		32		32	Includes material and labor to install a
	Install	Ea.	945	65.50		1010.50	refrigerator.
	Demolish and Install	Ea.	945	97.50		1042.50	
	Reinstall	Ea.		52.48		52.48	
	Clean	Ea.	.04	20.50		20.54	
	Minimum Charge	Job		57.50		57.50	
27 Cubic Foot							
	Demolish	Ea.		32		32	Includes material and labor to install a
	Install	Ea.	1625	46		1671	refrigerator.
	Demolish and Install	Ea.	1625	78		1703	
	Reinstall	Ea.		36.74		36.74	
	Clean	Ea.	.04	20.50		20.54	
	Minimum Charge	Job		57.50		57.50	
Ice Maker							
	Install	Ea.	66	43		109	Includes labor and material to install
	Minimum Charge	Job		57.50		57.50	an automatic ice maker in a refrigerator.

Freezer		Unit	Material	Labor	Equip.	Total	Specification
15 Cubic Foot							
	Demolish	Ea.		32		32	Cost includes material and labor to
	Install	Ea.	430	46		476	install 15 cubic foot freezer.
	Demolish and Install	Ea.	430	78		508	
	Reinstall	Ea.		36.74		36.74	
	Clean	Ea.	.04	20.50		20.54	
	Minimum Charge	Job		57.50		57.50	
18 Cubic Foot							
	Demolish	Ea.		32		32	Cost includes material and labor to
	Install	Ea.	570	46		616	install 18 cubic foot freezer.
	Demolish and Install	Ea.	570	78		648	
	Reinstall	Ea.		36.74		36.74	
	Clean	Ea.	.04	20.50		20.54	
	Minimum Charge	Job		57.50		57.50	
21 Cubic Foot							
	Demolish	Ea.		32		32	Includes material and labor to install a
	Install	Ea.	510	46		556	15 to 23 cubic foot deep freezer.
	Demolish and Install	Ea.	510	78		588	
	Reinstall	Ea.		36.74		36.74	
	Clean	Ea.	.04	20.50		20.54	
	Minimum Charge	Job		57.50		57.50	

Improvements / Appliances / Treatments

Freezer	Unit	Material	Labor	Equip.	Total	Specification
24 Cubic Foot						
Demolish	Ea.		32		32	Includes material and labor to install a
Install	Ea.	565	92		657	24 cubic foot deep freezer.
Demolish and Install	Ea.	565	124		689	
Reinstall	Ea.		73.47		73.47	
Clean	Ea.	.04	20.50		20.54	
Minimum Charge	Job		57.50		57.50	
Chest-type						
Demolish	Ea.		32		32	Cost includes material and labor to
Install	Ea.	430	46		476	install 15 cubic foot freezer.
Demolish and Install	Ea.	430	78		508	
Clean	Ea.	.04	20.50		20.54	
Minimum Charge	Job		57.50		57.50	

Dishwasher	Unit	Material	Labor	Equip.	Total	Specification
4 Cycle						
Demolish	Ea.		25.50		25.50	Includes material and labor to install
Install	Ea.	288	107		395	good grade automatic dishwasher.
Demolish and Install	Ea.	288	132.50		420.50	
Reinstall	Ea.		85.82		85.82	
Clean	Ea.	.04	20.50		20.54	
Minimum Charge	Job		57.50		57.50	

Garbage Disposal	Unit	Material	Labor	Equip.	Total	Specification
In Sink Unit						
Demolish	Ea.		24		24	Cost includes material and labor to
Install	Ea.	88.50	43		131.50	install 1/2 HP custom disposal unit.
Demolish and Install	Ea.	88.50	67		155.50	
Reinstall	Ea.		34.33		34.33	
Clean	Ea.	.02	5.90		5.92	
Minimum Charge	Job		57.50		57.50	

Trash Compactor	Unit	Material	Labor	Equip.	Total	Specification
4 to 1 Compaction						
Demolish	Ea.		24		24	Includes material and labor to install a
Install	Ea.	605	62.50		667.50	trash compactor.
Demolish and Install	Ea.	605	86.50		691.50	
Reinstall	Ea.		50.18		50.18	
Clean	Ea.	.04	14.35		14.39	
Minimum Charge	Job		57.50		57.50	

Kitchenette Unit	Unit	Material	Labor	Equip.	Total	Specification
Range, Refrigerator and Sink						
Demolish	Ea.		95.50		95.50	Includes material and labor to install
Install	Ea.	3000	360		3360	range/refrigerator/sink unit, 72"
Demolish and Install	Ea.	3000	455.50		3455.50	wide, base unit only.
Reinstall	Ea.		286.07		286.07	
Clean	Ea.	.73	10.30		11.03	
Minimum Charge	Job		57.50		57.50	

Improvements / Appliances / Treatments

Bath Accessories		Unit	Material	Labor	Equip.	Total	Specification
Towel Bar							
	Demolish	Ea.		5.30		5.30	Includes material and labor to install
	Install	Ea.	28.50	13.05		41.55	towel bar up to 24" long.
	Demolish and Install	Ea.	28.50	18.35		46.85	
	Reinstall	Ea.		10.45		10.45	
	Clean	Ea.		2.87		2.87	
	Minimum Charge	Job		157		157	
Toothbrush Holder							
	Demolish	Ea.		5.30		5.30	Includes material and labor to install a
	Install	Ea.	23.50	15.70		39.20	surface mounted tumbler and
	Demolish and Install	Ea.	23.50	21		44.50	toothbrush holder.
	Reinstall	Ea.		12.54		12.54	
	Clean	Ea.		2.87		2.87	
	Minimum Charge	Job		157		157	
Grab Rail							
	Demolish	Ea.		7.95		7.95	Includes material and labor to install a
	Install	Ea.	50.50	13.65		64.15	24" long grab bar. Blocking is not
	Demolish and Install	Ea.	50.50	21.60		72.10	included.
	Reinstall	Ea.		10.91		10.91	
	Clean	Ea.	.29	4.78		5.07	
	Minimum Charge	Job		157		157	
Soap Dish							
	Demolish	Ea.		5.30		5.30	Includes material and labor to install a
	Install	Ea.	21	15.70		36.70	surface mounted soap dish.
	Demolish and Install	Ea.	21	21		42	
	Reinstall	Ea.		12.54		12.54	
	Clean	Ea.		2.87		2.87	
	Minimum Charge	Job		157		157	
Toilet Paper Roller							
	Demolish	Ea.		5.30		5.30	Includes material and labor to install a
	Install	Ea.	12.25	10.45		22.70	surface mounted toilet tissue dispenser.
	Demolish and Install	Ea.	12.25	15.75		28	
	Reinstall	Ea.		8.36		8.36	
	Clean	Ea.	.29	7.20		7.49	
	Minimum Charge	Job		157		157	
Dispenser							
Soap							
	Demolish	Ea.		7.10		7.10	Includes material and labor to install a
	Install	Ea.	45.50	15.70		61.20	surface mounted soap dispenser.
	Demolish and Install	Ea.	45.50	22.80		68.30	
	Reinstall	Ea.		12.54		12.54	
	Clean	Ea.	.29	7.20		7.49	
	Minimum Charge	Job		157		157	
Towel							
	Demolish	Ea.		7.95		7.95	Includes material and labor to install a
	Install	Ea.	40	19.60		59.60	surface mounted towel dispenser.
	Demolish and Install	Ea.	40	27.55		67.55	
	Reinstall	Ea.		15.68		15.68	
	Clean	Ea.	.29	9.55		9.84	
	Minimum Charge	Job		157		157	
Seat Cover							
	Demolish	Ea.		5.30		5.30	Includes material and labor to install a
	Install	Ea.	26	21		47	surface mounted toilet seat cover
	Demolish and Install	Ea.	26	26.30		52.30	dispenser.
	Reinstall	Ea.		16.73		16.73	
	Clean	Ea.	.29	7.20		7.49	
	Minimum Charge	Job		157		157	

Bath Accessories

Bath Accessories		Unit	Material	Labor	Equip.	Total	Specification
Sanitary Napkin							
	Demolish	Ea.		13.45		13.45	Includes material and labor to install a
	Install	Ea.	22	33		55	surface mounted sanitary napkin
	Demolish and Install	Ea.	22	46.45		68.45	dispenser.
	Reinstall	Ea.		26.41		26.41	
	Clean	Ea.	.14	9.55		9.69	
	Minimum Charge	Job		157		157	
Partition							
Urinal							
	Demolish	Ea.		15.95		15.95	Includes material and labor to install a
	Install	Ea.	178	78.50		256.50	urinal screen.
	Demolish and Install	Ea.	178	94.45		272.45	
	Reinstall	Ea.		62.72		62.72	
	Clean	Ea.	.43	6.40		6.83	
	Minimum Charge	Job		157		157	
Toilet							
	Demolish	Ea.		47		47	Includes material and labor to install
	Install	Ea.	430	125		555	baked enamel standard size partition,
	Demolish and Install	Ea.	430	172		602	floor or ceiling mounted with one wall
	Reinstall	Ea.		100.35		100.35	and one door.
	Clean	Ea.	2.88	14.35		17.23	
	Minimum Charge	Job		157		157	
Handicap							
	Demolish	Ea.		47		47	Includes material and labor to install
	Install	Ea.	730	125		855	baked enamel handicap partition,
	Demolish and Install	Ea.	730	172		902	floor or ceiling mounted with one wall
	Reinstall	Ea.		100.35		100.35	and one door.
	Clean	Ea.	2.88	14.35		17.23	
	Minimum Charge	Job		157		157	

Fire Prevention / Protection

Fire Prevention / Protection		Unit	Material	Labor	Equip.	Total	Specification
Fire Extinguisher							
5# Carbon Dioxide							
	Install	Ea.	111			111	Includes labor and material to install one wall mounted 5 lb. carbon dioxide factory charged unit, complete with hose, horn and wall mounting bracket.
10# Carbon Dioxide							
	Install	Ea.	165			165	Includes labor and material to install one wall mounted 10 lb. carbon dioxide factory charged unit, complete with hose, horn and wall mounting bracket.
15# Carbon Dioxide							
	Install	Ea.	189			189	Includes labor and material to install one wall mounted 15 lb. carbon dioxide factory charged unit, complete with hose, horn and wall mounting bracket.
5# Dry Chemical							
	Install	Ea.	44			44	Includes labor and material to install one wall mounted 5 lb. dry chemical factory charged unit, complete with hose, horn and wall mounting bracket.

Improvements / Appliances / Treatments

Fire Prevention / Protection		Unit	Material	Labor	Equip.	Total	Specification
20# Dry Chemical							Includes labor and material to install one wall mounted 20 lb. dry chemical factory charged unit, complete with hose, horn and wall mounting bracket.
	Install	Ea.	99			99	

Gutters		Unit	Material	Labor	Equip.	Total	Specification
Galvanized							Cost includes material and labor to install 5" galvanized steel gutter with enamel finish, fittings, hangers, and corners.
	Demolish	L.F.		1.06		1.06	
	Install	L.F.	1.19	2.86		4.05	
	Demolish and Install	L.F.	1.19	3.92		5.11	
	Reinstall	L.F.		2.29		2.29	
	Clean	L.F.		.31		.31	
	Paint	L.F.	.18	.85		1.03	
	Minimum Charge	Job		171		171	
Aluminum							Cost includes material and labor to install 5" aluminum gutter with enamel finish, fittings, hangers, and corners.
	Demolish	L.F.		1.06		1.06	
	Install	L.F.	1.34	2.86		4.20	
	Demolish and Install	L.F.	1.34	3.92		5.26	
	Reinstall	L.F.		2.29		2.29	
	Clean	L.F.		.31		.31	
	Paint	L.F.	.18	.85		1.03	
	Minimum Charge	Job		171		171	
Copper							Cost includes material and labor to install 5" half-round copper gutter with fittings, hangers, and corners.
	Demolish	L.F.		1.06		1.06	
	Install	L.F.	5.65	2.86		8.51	
	Demolish and Install	L.F.	5.65	3.92		9.57	
	Reinstall	L.F.		2.29		2.29	
	Clean	L.F.		.31		.31	
	Minimum Charge	Job		171		171	
Plastic							Cost includes material and labor to install 4" white vinyl gutter fittings, hangers, and corners.
	Demolish	L.F.		1.06		1.06	
	Install	L.F.	.94	2.85		3.79	
	Demolish and Install	L.F.	.94	3.91		4.85	
	Reinstall	L.F.		2.28		2.28	
	Clean	L.F.		.31		.31	
	Paint	L.F.	.18	.85		1.03	
	Minimum Charge	Job		171		171	

Downspouts		Unit	Material	Labor	Equip.	Total	Specification
Galvanized							Cost includes material and labor to install 4" diameter downspout with enamel finish.
	Demolish	L.F.		.73		.73	
	Install	L.F.	1.11	2.36		3.47	
	Demolish and Install	L.F.	1.11	3.09		4.20	
	Reinstall	L.F.		1.89		1.89	
	Clean	L.F.		.31		.31	
	Paint	L.F.	.18	.85		1.03	
	Minimum Charge	Job		171		171	
Aluminum							Includes material and labor to install aluminum enameled downspout.
	Demolish	L.F.		.73		.73	
	Install	L.F.	1.75	2.45		4.20	
	Demolish and Install	L.F.	1.75	3.18		4.93	
	Reinstall	L.F.		1.96		1.96	
	Clean	L.F.		.31		.31	
	Paint	L.F.	.18	.85		1.03	
	Minimum Charge	Job		171		171	

Improvements / Appliances / Treatments

Downspouts

	Unit	Material	Labor	Equip.	Total	Specification
Copper						
Demolish	L.F.		.73		.73	Includes material and labor to install
Install	L.F.	4.96	1.80		6.76	round copper downspout.
Demolish and Install	L.F.	4.96	2.53		7.49	
Reinstall	L.F.		1.44		1.44	
Clean	L.F.		.31		.31	
Paint	L.F.	.18	.85		1.03	
Minimum Charge	Job		171		171	
Plastic						
Demolish	L.F.		.73		.73	Cost includes material and labor to
Install	L.F.	.79	1.63		2.42	install 2" x 3" vinyl rectangular
Demolish and Install	L.F.	.79	2.36		3.15	downspouts.
Reinstall	L.F.		1.31		1.31	
Clean	L.F.		.31		.31	
Paint	L.F.	.18	.85		1.03	
Minimum Charge	Job		171		171	

Shutters

	Unit	Material	Labor	Equip.	Total	Specification
Exterior						
25" High						
Demolish	Ea.		7.95		7.95	Includes material and labor to install
Install	Pr.	89	31.50		120.50	pair of exterior shutters 25" x 16",
Demolish and Install	Pr.	89	39.45		128.45	unpainted, installed as fixed shutters
Reinstall	Pr.		31.36		31.36	on concrete or wood frame.
Clean	Ea.	.06	6.85		6.91	
Paint	Ea.	1.42	14.85		16.27	
Minimum Charge	Job		157		157	
39" High						
Demolish	Ea.		7.95		7.95	Includes material and labor to install
Install	Pr.	96.50	31.50		128	pair of shutters 39" x 16", unpainted,
Demolish and Install	Pr.	96.50	39.45		135.95	installed as fixed shutters on concrete
Reinstall	Pr.		31.36		31.36	or wood frame.
Clean	Ea.	.06	6.85		6.91	
Paint	Ea.	2.20	20		22.20	
Minimum Charge	Job		157		157	
51" High						
Demolish	Ea.		7.95		7.95	Includes material and labor to install
Install	Pr.	107	31.50		138.50	pair of shutters 51" x 16", unpainted,
Demolish and Install	Pr.	107	39.45		146.45	installed as fixed shutters on concrete
Reinstall	Pr.		31.36		31.36	or wood frame.
Clean	Ea.	.06	6.85		6.91	
Paint	Ea.	2.86	21		23.86	
Minimum Charge	Job		157		157	
59" High						
Demolish	Ea.		7.95		7.95	Includes material and labor to install
Install	Pr.	151	31.50		182.50	pair of shutters 59" x 16", unpainted,
Demolish and Install	Pr.	151	39.45		190.45	installed as fixed shutters on concrete
Reinstall	Pr.		31.36		31.36	or wood frame.
Clean	Ea.	.06	6.85		6.91	
Paint	Ea.	3.30	23		26.30	
Minimum Charge	Job		157		157	

Shutters		Unit	Material	Labor	Equip.	Total	Specification
67" High							
	Demolish	Ea.		7.95		7.95	Includes material and labor to install
	Install	Pr.	152	35		187	pair of shutters 67" x 16", unpainted,
	Demolish and Install	Pr.	152	42.95		194.95	installed as fixed shutters on concrete
	Reinstall	Pr.		34.84		34.84	or wood frame.
	Clean	Ea.	.06	6.85		6.91	
	Paint	Ea.	3.77	25.50		29.27	
	Minimum Charge	Job		157		157	
Interior Wood							
23" x 24", 2 Panel							
	Demolish	Ea.		7.95		7.95	Cost includes material and labor to
	Install	Ea.	81	18.45		99.45	install 23" x 24", 4 panel interior
	Demolish and Install	Ea.	81	26.40		107.40	moveable pine shutter, 1-1/4" louver,
	Reinstall	Ea.		14.76		14.76	knobs and hooks and hinges.
	Clean	Ea.	.06	6.85		6.91	
	Paint	Ea.	1.07	19.70		20.77	
	Minimum Charge	Job		157		157	
27" x 24", 4 Panel							
	Demolish	Ea.		7.95		7.95	Cost includes material and labor to
	Install	Ea.	94	18.45		112.45	install 27" x 24", 4 panel interior
	Demolish and Install	Ea.	94	26.40		120.40	moveable pine shutter, 1-1/4" louver,
	Reinstall	Ea.		14.76		14.76	knobs and hooks and hinges.
	Clean	Ea.	.06	6.85		6.91	
	Paint	Ea.	1.07	19.70		20.77	
	Minimum Charge	Job		157		157	
31" x 24", 4 Panel							
	Demolish	Ea.		7.95		7.95	Cost includes material and labor to
	Install	Ea.	108	18.45		126.45	install 31" x 24", 4 panel interior
	Demolish and Install	Ea.	108	26.40		134.40	moveable pine shutter, 1-1/4" louver,
	Reinstall	Ea.		14.76		14.76	knobs and hooks and hinges.
	Clean	Ea.	.06	6.85		6.91	
	Paint	Ea.	1.07	19.70		20.77	
	Minimum Charge	Job		157		157	
35" x 24", 4 Panel							
	Demolish	Ea.		7.95		7.95	Cost includes material and labor to
	Install	Ea.	108	18.45		126.45	install 35" x 24", 4 panel interior
	Demolish and Install	Ea.	108	26.40		134.40	moveable pine shutter, 1-1/4" louver,
	Reinstall	Ea.		14.76		14.76	knobs and hooks and hinges.
	Clean	Ea.	.06	6.85		6.91	
	Paint	Ea.	1.07	19.70		20.77	
	Minimum Charge	Job		157		157	
39" x 24", 4 Panel							
	Demolish	Ea.		7.95		7.95	Cost includes material and labor to
	Install	Ea.	124	18.45		142.45	install 39" x 24", 4 panel interior
	Demolish and Install	Ea.	124	26.40		150.40	moveable pine shutter, 1-1/4" louver,
	Reinstall	Ea.		14.76		14.76	knobs and hooks and hinges.
	Clean	Ea.	.06	6.85		6.91	
	Paint	Ea.	1.07	19.70		20.77	
	Minimum Charge	Job		157		157	
47" x 24", 4 Panel							
	Demolish	Ea.		7.95		7.95	Cost includes material and labor to
	Install	Ea.	138	18.45		156.45	install 47" x 24", 4 panel interior
	Demolish and Install	Ea.	138	26.40		164.40	moveable pine shutter, 1-1/4" louver,
	Reinstall	Ea.		14.76		14.76	knobs and hooks and hinges.
	Clean	Ea.	.06	6.85		6.91	
	Paint	Ea.	1.11	21		22.11	
	Minimum Charge	Job		157		157	

Awning	Unit	Material	Labor	Equip.	Total	Specification
Standard Aluminum						
36″ Wide x 30″ Deep						
Demolish	Ea.		7.95		7.95	Includes material and labor to install
Install	Ea.	169	26		195	one aluminum awning with
Demolish and Install	Ea.	169	33.95		202.95	weather-resistant finish.
Reinstall	Ea.		20.91		20.91	
Clean	Ea.		9.55		9.55	
Minimum Charge	Job		157		157	
48″ Wide x 30″ Deep						
Demolish	Ea.		7.95		7.95	Includes material and labor to install
Install	Ea.	195	31.50		226.50	one aluminum awning with
Demolish and Install	Ea.	195	39.45		234.45	weather-resistant finish.
Reinstall	Ea.		25.09		25.09	
Clean	Ea.		9.55		9.55	
Minimum Charge	Job		157		157	
60″ Wide x 30″ Deep						
Demolish	Ea.		15.95		15.95	Includes material and labor to install
Install	Ea.	219	35		254	one aluminum awning with
Demolish and Install	Ea.	219	50.95		269.95	weather-resistant finish.
Reinstall	Ea.		27.88		27.88	
Clean	Ea.		9.55		9.55	
Minimum Charge	Job		157		157	
72″ Wide x 30″ Deep						
Demolish	Ea.		15.95		15.95	Includes material and labor to install
Install	Ea.	243	39		282	one aluminum awning with
Demolish and Install	Ea.	243	54.95		297.95	weather-resistant finish.
Reinstall	Ea.		31.36		31.36	
Clean	Ea.		9.55		9.55	
Minimum Charge	Job		157		157	
Roll-up Type						
Demolish	S.F.		.57		.57	Includes material and labor to install
Install	S.F.	6.40	3.14		9.54	an aluminum roll-up awning.
Demolish and Install	S.F.	6.40	3.71		10.11	
Reinstall	S.F.		2.51		2.51	
Clean	S.F.	.03	.48		.51	
Minimum Charge	Job		157		157	
Canvas						
30″ Wide						
Demolish	Ea.		4.78		4.78	Includes material and labor to install a
Install	Ea.	182	26		208	canvas window awning.
Demolish and Install	Ea.	182	30.78		212.78	
Reinstall	Ea.		20.91		20.91	
Clean	Ea.		2.87		2.87	
Minimum Charge	Job		157		157	
36″ Wide						
Demolish	Ea.		4.78		4.78	Includes material and labor to install a
Install	Ea.	193	31.50		224.50	canvas window awning.
Demolish and Install	Ea.	193	36.28		229.28	
Reinstall	Ea.		25.09		25.09	
Clean	Ea.		3.15		3.15	
Minimum Charge	Job		157		157	
42″ Wide						
Demolish	Ea.		4.78		4.78	Includes material and labor to install a
Install	Ea.	204	39		243	canvas window awning.
Demolish and Install	Ea.	204	43.78		247.78	
Reinstall	Ea.		31.36		31.36	
Clean	Ea.		3.73		3.73	
Minimum Charge	Job		157		157	

Improvements / Appliances / Treatments

Awning

	Unit	Material	Labor	Equip.	Total	Specification
48″ Wide						Includes material and labor to install a canvas window awning.
Demolish	Ea.		4.78		4.78	
Install	Ea.	215	45		260	
Demolish and Install	Ea.	215	49.78		264.78	
Reinstall	Ea.		35.84		35.84	
Clean	Ea.		4.33		4.33	
Minimum Charge	Job		157		157	

Cloth Canopy Cover

	Unit	Material	Labor	Equip.	Total	Specification
8′ x 10′						Includes material and labor to install one 8′ x 10′ cloth canopy patio cover with front bar and tension support rafters and 9″ valance.
Demolish	Ea.		260		260	
Install	Ea.	305	157		462	
Demolish and Install	Ea.	305	417		722	
Reinstall	Ea.		125.44		125.44	
Clean	Ea.	2.31	38.50		40.81	
Minimum Charge	Job		315		315	
8′ x 15′						Includes material and labor to install one 8′ x 15′ cloth canopy patio cover with front bar and tension support rafters and 9″ valance.
Demolish	Ea.		260		260	
Install	Ea.	415	157		572	
Demolish and Install	Ea.	415	417		832	
Reinstall	Ea.		125.44		125.44	
Clean	Ea.	2.31	57.50		59.81	
Minimum Charge	Job		315		315	

Vent

	Unit	Material	Labor	Equip.	Total	Specification
Small Roof Turbine						Includes material and labor to install spinner ventilator, wind driven, galvanized, 4″ neck diam, 180 CFM.
Demolish	Ea.		7.95		7.95	
Install	Ea.	57	31		88	
Demolish and Install	Ea.	57	38.95		95.95	
Reinstall	Ea.		24.69		24.69	
Clean	Ea.		5.10		5.10	
Minimum Charge	Job		172		172	
Large Roof Turbine						Includes material and labor to install spinner ventilator, wind driven, galvanized, 8″ neck diam, 360 CFM.
Demolish	Ea.		6.40		6.40	
Install	Ea.	66.50	44		110.50	
Demolish and Install	Ea.	66.50	50.40		116.90	
Reinstall	Ea.		35.27		35.27	
Clean	Ea.		5.75		5.75	
Minimum Charge	Job		172		172	
Roof Ventilator						Includes material and labor to install a roof ventilator.
Demolish	Ea.		7.95		7.95	
Install	Ea.	163	38.50		201.50	
Demolish and Install	Ea.	163	46.45		209.45	
Reinstall	Ea.		30.86		30.86	
Clean	Ea.	.57	5.75		6.32	
Paint	Ea.	1.34	6.90		8.24	
Minimum Charge	Job		172		172	
Gable Louvered						Includes material and labor to install a vinyl gable end vent.
Demolish	Ea.		15.95		15.95	
Install	Ea.	31	10.45		41.45	
Demolish and Install	Ea.	31	26.40		57.40	
Reinstall	Ea.		8.36		8.36	
Clean	Ea.		3.06		3.06	
Paint	Ea.	.89	4.25		5.14	
Minimum Charge	Job		157		157	

Improvements / Appliances / Treatments

Vent		Unit	Material	Labor	Equip.	Total	Specification
Ridge Vent							
	Demolish	L.F.		.82		.82	Includes material and labor to install a
	Install	L.F.	2.53	2.21		4.74	mill finish aluminum ridge vent strip.
	Demolish and Install	L.F.	2.53	3.03		5.56	
	Minimum Charge	Job		157		157	
Foundation							
	Demolish	Ea.		5.50		5.50	Includes material and labor to install
	Install	Ea.	19.80	10.65		30.45	galvanized foundation block vent.
	Demolish and Install	Ea.	19.80	16.15		35.95	
	Reinstall	Ea.		8.51		8.51	
	Clean	Ea.	.14	5.75		5.89	
	Minimum Charge	Job		157		157	
Frieze							
	Demolish	L.F.		.57		.57	Includes material and labor to install
	Install	L.F.	.61	.71		1.32	pine frieze board.
	Demolish and Install	L.F.	.61	1.28		1.89	
	Reinstall	L.F.		.57		.57	
	Clean	L.F.	.03	.19		.22	
	Minimum Charge	Job		157		157	
Water Heater Cap							
	Demolish	Ea.		11.90		11.90	Includes material and labor to install a
	Install	Ea.	23	43		66	galvanized steel water heater cap.
	Demolish and Install	Ea.	23	54.90		77.90	
	Reinstall	Ea.		34.28		34.28	
	Clean	Ea.	1.01	14.35		15.36	

Cupolas		Unit	Material	Labor	Equip.	Total	Specification
22″ x 22″							
	Demolish	Ea.		151		151	Includes material and labor to install a
	Install	Ea.	274	78.50		352.50	cedar cupola with aluminum roof
	Demolish and Install	Ea.	274	229.50		503.50	covering.
	Reinstall	Ea.		62.72		62.72	
	Minimum Charge	Job		157		157	
29″ x 29″							
	Demolish	Ea.		151		151	Includes material and labor to install a
	Install	Ea.	360	78.50		438.50	cedar cupola with aluminum roof
	Demolish and Install	Ea.	360	229.50		589.50	covering.
	Reinstall	Ea.		62.72		62.72	
	Minimum Charge	Job		157		157	
35″ x 35″							
	Demolish	Ea.		151		151	Includes material and labor to install a
	Install	Ea.	650	105		755	cedar cupola with aluminum roof
	Demolish and Install	Ea.	650	256		906	covering.
	Reinstall	Ea.		83.63		83.63	
	Minimum Charge	Job		157		157	
47″ x 47″							
	Demolish	Ea.		151		151	Includes material and labor to install a
	Install	Ea.	1050	157		1207	cedar cupola with aluminum roof
	Demolish and Install	Ea.	1050	308		1358	covering.
	Reinstall	Ea.		125.44		125.44	
	Minimum Charge	Job		157		157	
Weather Vane							
	Install	Ea.	44	39		83	Includes minimum labor and
	Minimum Charge	Job		157		157	equipment to install residential type weathervane.

Improvements / Appliances / Treatments

Aluminum Carport		Unit	Material	Labor	Equip.	Total	Specification
Natural Finish							
	Demolish	S.F.		1.59		1.59	Includes material, labor and
	Install	S.F.	19.35	5.20	.88	25.43	equipment to install an aluminum
	Demolish and Install	S.F.	19.35	6.79	.88	27.02	carport. Foundations are not included.
	Reinstall	S.F.		4.17	.71	4.88	
	Clean	S.F.	.03	.17		.20	
	Minimum Charge	Job		315		315	
Enamel Finish							
	Demolish	S.F.		1.59		1.59	Includes material, labor and
	Install	S.F.	19.35	5.20	.88	25.43	equipment to install an aluminum
	Demolish and Install	S.F.	19.35	6.79	.88	27.02	carport. Foundations are not included.
	Reinstall	S.F.		4.17	.71	4.88	
	Clean	S.F.	.03	.17		.20	
	Minimum Charge	Job		315		315	
20' x 20' Complete							
	Demolish	Ea.		159		159	Includes material, labor and
	Install	Ea.	6075	520	88	6683	equipment to install an aluminum
	Demolish and Install	Ea.	6075	679	88	6842	carport. Foundations are not included.
	Reinstall	Ea.		417.31	70.56	487.87	
	Clean	Ea.	.15	57.50		57.65	
	Minimum Charge	Job		315		315	
24' x 24' Complete							
	Demolish	Ea.		159		159	Includes material, labor and
	Install	Ea.	8250	695	118	9063	equipment to install an aluminum
	Demolish and Install	Ea.	8250	854	118	9222	carport. Foundations are not included.
	Reinstall	Ea.		556.42	94.08	650.50	
	Clean	Ea.	.15	115		115.15	
	Minimum Charge	Job		315		315	
Metal Support Posts							
	Demolish	Ea.		2.13		2.13	Includes material and labor to install
	Install	Ea.	13.50	39		52.50	metal support posts for a carport.
	Demolish and Install	Ea.	13.50	41.13		54.63	
	Reinstall	Ea.		31.36		31.36	
	Clean	Ea.	.15	2.39		2.54	
	Paint	Ea.	.12	5.50		5.62	
	Minimum Charge	Job		315		315	

Storage Shed		Unit	Material	Labor	Equip.	Total	Specification
Aluminum							
Pre-fab (8' x 10')							
	Demolish	Ea.		46.50		46.50	Includes material and labor to install
	Install	Ea.	273	450		723	an aluminum storage shed.
	Demolish and Install	Ea.	273	496.50		769.50	
	Clean	Ea.	.96	28.50		29.46	
	Minimum Charge	Ea.		385		385	
Pre-fab (10' x 12')							
	Demolish	Ea.		87		87	Includes material and labor to install
	Install	Ea.	460	450		910	an aluminum storage shed.
	Demolish and Install	Ea.	460	537		997	
	Clean	Ea.	.96	32		32.96	
	Minimum Charge	Ea.		385		385	

Improvements / Appliances / Treatments

Storage Shed		Unit	Material	Labor	Equip.	Total	Specification
Wood							
Pre-fab (8' x 10')							
	Demolish	Ea.		46.50		46.50	Includes material and labor to install a
	Install	Ea.	760	450		1210	pre-fabricated wood storage shed.
	Demolish and Install	Ea.	760	496.50		1256.50	
	Clean	Ea.	.96	28.50		29.46	
	Minimum Charge	Ea.		385		385	
Pre-fab (10' x 12')							
	Demolish	Ea.		87		87	Includes material and labor to install a
	Install	Ea.	990	450		1440	pre-fabricated wood storage shed.
	Demolish and Install	Ea.	990	537		1527	
	Clean	Ea.	.96	32		32.96	
	Minimum Charge	Ea.		385		385	

Gazebo		Unit	Material	Labor	Equip.	Total	Specification
Good Grade							
	Demolish	S.F.		2.55		2.55	Includes material and labor to install a
	Install	S.F.	18.90	23.50		42.40	gazebo.
	Demolish and Install	S.F.	18.90	26.05		44.95	
	Reinstall	S.F.		18.82		18.82	
	Clean	S.F.	.11	1.53		1.64	
	Minimum Charge	Job		157		157	
Custom Grade							
	Demolish	S.F.		2.55		2.55	Includes material and labor to install a
	Install	S.F.	30.50	23.50		54	gazebo.
	Demolish and Install	S.F.	30.50	26.05		56.55	
	Reinstall	S.F.		18.82		18.82	
	Clean	S.F.	.11	1.53		1.64	
	Minimum Charge	Job		157		157	
Bench Seating							
	Demolish	L.F.		4.25		4.25	Includes material and labor to install
	Install	L.F.	2.15	15.70		17.85	pressure treated bench seating.
	Demolish and Install	L.F.	2.15	19.95		22.10	
	Reinstall	L.F.		12.54		12.54	
	Clean	L.F.	.11	1.15		1.26	
	Paint	L.F.	.50	2.21		2.71	
	Minimum Charge	Job		157		157	
Re-screen							
	Install	S.F.	.53	1.25		1.78	Includes labor and material to
	Minimum Charge	Job		157		157	re-screen wood frame.

Pool		Unit	Material	Labor	Equip.	Total	Specification
Screen Enclosure							
Re-screen							
	Install	SF Wall		3.85		3.85	Includes material and labor to install a
	Minimum Charge	Job		157		157	screen enclosure for a pool.
Screen Door							
	Demolish	Ea.		6.20		6.20	Includes material and labor to install
	Install	Ea.	132	52.50		184.50	wood screen door with aluminum cloth
	Demolish and Install	Ea.	132	58.70		190.70	screen. Frame and hardware not
	Reinstall	Ea.		41.81		41.81	included.
	Clean	Ea.	.11	8.25		8.36	
	Paint	Ea.	2.96	11.05		14.01	
	Minimum Charge	Job		157		157	

Improvements / Appliances / Treatments

Pool		Unit	Material	Labor	Equip.	Total	Specification
Heater / Motor							
	Demolish	Ea.		30		30	Includes material and labor to install
	Install	Ea.	1850	385		2235	swimming pool heater, not incl. base
	Demolish and Install	Ea.	1850	415		2265	or pad, elec., 12 kW, 4,800 gal
	Reinstall	Ea.		308.51		308.51	pool, incl. pump.
	Clean	Ea.	2.88	28.50		31.38	
	Minimum Charge	Job		172		172	
Filter System							
	Demolish	Ea.		51		51	Includes material, labor and
	Install	Total	560	275		835	equipment to install sand filter system
	Demolish and Install	Total	560	326		886	tank including hook-up to existing
	Reinstall	Total		219.65		219.65	equipment. Pump and motor not
	Minimum Charge	Job		172		172	included.
Pump / Motor							
	Demolish	Ea.		30		30	Includes material and labor to install a
	Install	Ea.	515	68.50		583.50	22 GPM pump.
	Demolish and Install	Ea.	515	98.50		613.50	
	Reinstall	Ea.		54.91		54.91	
	Clean	Ea.	2.88	28.50		31.38	
	Minimum Charge	Job		172		172	
Pump Water							
	Install	Day		305	73.50	378.50	Includes labor and equipment to pump
	Minimum Charge	Job		143		143	water from a swimming pool.
Replaster							
	Install	SF Surf	.91	7.40		8.31	Includes labor and material to
	Minimum Charge	Job		143		143	replaster gunite pool with a surface area of 500 to 600 S.F.
Solar Panels							
	Demolish	S.F.		1.27		1.27	Includes material and labor to install
	Install	S.F.	18.75	2.26		21.01	solar panel.
	Demolish and Install	S.F.	18.75	3.53		22.28	
	Reinstall	S.F.		1.81		1.81	
	Minimum Charge	Ea.		385		385	

Construction Clean-up	Unit	Material	Labor	Equip.	Total	Specification
Final						
Clean	S.F.		.20		.20	Includes labor for final site/job
Minimum Charge	Job		115		115	clean-up including detail cleaning.

Location Factors

Costs shown in *Means cost data publications* are based on National Averages for materials and installation. To adjust these costs to a specific location, simply multiply the base cost by the factor for that city. The data is arranged alphabetically by state and postal zip code numbers. For a city not listed, use the factor for a nearby city with similar economic characteristics.

STATE	CITY	Residential
ALABAMA		
350-352	Birmingham	.86
354	Tuscaloosa	.73
355	Jasper	.70
356	Decatur	.76
357-358	Huntsville	.84
359	Gadsden	.73
360-361	Montgomery	.75
362	Anniston	.68
363	Dothan	.74
364	Evergreen	.71
365-366	Mobile	.79
367	Selma	.72
368	Phenix City	.73
369	Butler	.71
ALASKA		
995-996	Anchorage	1.28
997	Fairbanks	1.30
998	Juneau	1.28
999	Ketchikan	1.30
ARIZONA		
850,853	Phoenix	.87
852	Mesa/Tempe	.85
855	Globe	.81
856-857	Tucson	.85
859	Show Low	.84
860	Flagstaff	.85
863	Prescott	.82
864	Kingman	.82
865	Chambers	.82
ARKANSAS		
716	Pine Bluff	.77
717	Camden	.66
718	Texarkana	.71
719	Hot Springs	.64
720-722	Little Rock	.82
723	West Memphis	.75
724	Jonesboro	.73
725	Batesville	.72
726	Harrison	.73
727	Fayetteville	.62
728	Russellville	.71
729	Fort Smith	.77
CALIFORNIA		
900-902	Los Angeles	1.07
903-905	Inglewood	1.04
906-908	Long Beach	1.03
910-912	Pasadena	1.04
913-916	Van Nuys	1.07
917-918	Alhambra	1.08
919-921	San Diego	1.03
922	Palm Springs	1.00
923-924	San Bernardino	1.01
925	Riverside	1.07
926-927	Santa Ana	1.05
928	Anaheim	1.08
930	Oxnard	1.08
931	Santa Barbara	1.07
932-933	Bakersfield	1.05
934	San Luis Obispo	1.07
935	Mojave	1.04
936-938	Fresno	1.11
939	Salinas	1.12
940-941	San Francisco	1.22
942,956-958	Sacramento	1.12
943	Palo Alto	1.17
944	San Mateo	1.21
945	Vallejo	1.14
946	Oakland	1.19
947	Berkeley	1.22
948	Richmond	1.23
949	San Rafael	1.21
950	Santa Cruz	1.15
951	San Jose	1.20
952	Stockton	1.10

STATE	CITY	Residential
CALIFORNIA (CONT'D)		
953	Modesto	1.09
954	Santa Rosa	1.15
955	Eureka	1.11
959	Marysville	1.11
960	Redding	1.12
961	Susanville	1.11
COLORADO		
800-802	Denver	.95
803	Boulder	.94
804	Golden	.91
805	Fort Collins	.90
806	Greeley	.79
807	Fort Morgan	.94
808-809	Colorado Springs	.92
810	Pueblo	.93
811	Alamosa	.90
812	Salida	.92
813	Durango	.94
814	Montrose	.89
815	Grand Junction	.93
816	Glenwood Springs	.92
CONNECTICUT		
060	New Britain	1.08
061	Hartford	1.07
062	Willimantic	1.07
063	New London	1.07
064	Meriden	1.07
065	New Haven	1.08
066	Bridgeport	1.08
067	Waterbury	1.08
068	Norwalk	1.08
069	Stamford	1.09
D.C.		
200-205	Washington	.92
DELAWARE		
197	Newark	1.01
198	Wilmington	1.01
199	Dover	1.01
FLORIDA		
320,322	Jacksonville	.78
321	Daytona Beach	.85
323	Tallahassee	.72
324	Panama City	.66
325	Pensacola	.76
326,344	Gainesville	.76
327-328,347	Orlando	.79
329	Melbourne	.86
330-332,340	Miami	.83
333	Fort Lauderdale	.84
334,349	West Palm Beach	.83
335-336,346	Tampa	.87
337	St. Petersburg	.76
338	Lakeland	.84
339,341	Fort Myers	.81
342	Sarasota	.85
GEORGIA		
300-303,399	Atlanta	.89
304	Statesboro	.67
305	Gainesville	.73
306	Athens	.74
307	Dalton	.70
308-309	Augusta	.76
310-312	Macon	.77
313-314	Savannah	.79
315	Waycross	.71
316	Valdosta	.71
317	Albany	.73
318-319	Columbus	.76
HAWAII		
967	Hilo	1.23
968	Honolulu	1.24

Location Factors

STATE	CITY	Residential
STATES & POSS.		
969	Guam	1.60
IDAHO		
832	Pocatello	.89
833	Twin Falls	.73
834	Idaho Falls	.72
835	Lewiston	.99
836-837	Boise	.90
838	Coeur d'Alene	.85
ILLINOIS		
600-603	North Suburban	1.11
604	Joliet	1.13
605	South Suburban	1.10
606	Chicago	1.15
609	Kankakee	1.03
610-611	Rockford	1.04
612	Rock Island	.97
613	La Salle	1.02
614	Galesburg	.99
615-616	Peoria	1.02
617	Bloomington	.98
618-619	Champaign	.99
620-622	East St. Louis	.99
623	Quincy	.98
624	Effingham	.99
625	Decatur	.99
626-627	Springfield	.99
628	Centralia	.97
629	Carbondale	.96
INDIANA		
460	Anderson	.92
461-462	Indianapolis	.96
463-464	Gary	1.02
465-466	South Bend	.92
467-468	Fort Wayne	.92
469	Kokomo	.94
470	Lawrenceburg	.88
471	New Albany	.86
472	Columbus	.93
473	Muncie	.92
474	Bloomington	.96
475	Washington	.91
476-477	Evansville	.91
478	Terre Haute	.91
479	Lafayette	.92
IOWA		
500-503,509	Des Moines	.93
504	Mason City	.78
505	Fort Dodge	.77
506-507	Waterloo	.81
508	Creston	.83
510-511	Sioux City	.87
512	Sibley	.74
513	Spencer	.76
514	Carroll	.76
515	Council Bluffs	.82
516	Shenandoah	.74
520	Dubuque	.86
521	Decorah	.77
522-524	Cedar Rapids	.94
525	Ottumwa	.84
526	Burlington	.88
527-528	Davenport	.98
KANSAS		
660-662	Kansas City	.95
664-666	Topeka	.78
667	Fort Scott	.85
668	Emporia	.73
669	Belleville	.74
670-672	Wichita	.81
673	Independence	.74
674	Salina	.73
675	Hutchinson	.68
676	Hays	.74
677	Colby	.76
678	Dodge City	.74
679	Liberal	.68
KENTUCKY		
400-402	Louisville	.92
403-405	Lexington	.84

STATE	CITY	Residential
KENTUCKY (CONT'D)		
406	Frankfort	.82
407-409	Corbin	.67
410	Covington	.95
411-412	Ashland	.93
413-414	Campton	.68
415-416	Pikeville	.78
417-418	Hazard	.67
420	Paducah	.89
421-422	Bowling Green	.89
423	Owensboro	.83
424	Henderson	.92
425-426	Somerset	.67
427	Elizabethtown	.88
LOUISIANA		
700-701	New Orleans	.86
703	Thibodaux	.82
704	Hammond	.80
705	Lafayette	.78
706	Lake Charles	.80
707-708	Baton Rouge	.80
710-711	Shreveport	.79
712	Monroe	.75
713-714	Alexandria	.74
MAINE		
039	Kittery	.81
040-041	Portland	.89
042	Lewiston	.89
043	Augusta	.84
044	Bangor	.87
045	Bath	.82
046	Machias	.82
047	Houlton	.87
048	Rockland	.80
049	Waterville	.76
MARYLAND		
206	Waldorf	.84
207-208	College Park	.85
209	Silver Spring	.85
210-212	Baltimore	.90
214	Annapolis	.86
215	Cumberland	.85
216	Easton	.69
217	Hagerstown	.86
218	Salisbury	.75
219	Elkton	.83
MASSACHUSETTS		
010-011	Springfield	1.05
012	Pittsfield	1.00
013	Greenfield	1.00
014	Fitchburg	1.09
015-016	Worcester	1.10
017	Framingham	1.11
018	Lowell	1.12
019	Lawrence	1.12
020-022, 024	Boston	1.17
023	Brockton	1.11
025	Buzzards Bay	1.09
026	Hyannis	1.08
027	New Bedford	1.11
MICHIGAN		
480,483	Royal Oak	1.05
481	Ann Arbor	1.05
482	Detroit	1.10
484-485	Flint	.99
486	Saginaw	.96
487	Bay City	.96
488-489	Lansing	.98
490	Battle Creek	.94
491	Kalamazoo	.93
492	Jackson	.94
493,495	Grand Rapids	.84
494	Muskegon	.90
496	Traverse City	.83
497	Gaylord	.85
498-499	Iron Mountain	.92
MINNESOTA		
550-551	Saint Paul	1.14
553-555	Minneapolis	1.18

STATE	CITY	Residential
MINNESOTA (CONT'D)		
556-558	Duluth	1.10
559	Rochester	1.05
560	Mankato	1.02
561	Windom	.85
562	Willmar	.86
563	St. Cloud	1.10
564	Brainerd	.98
565	Detroit Lakes	.99
566	Bemidji	.96
567	Thief River Falls	.94
MISSISSIPPI		
386	Clarksdale	.60
387	Greenville	.69
388	Tupelo	.63
389	Greenwood	.63
390-392	Jackson	.72
393	Meridian	.66
394	Laurel	.62
395	Biloxi	.75
396	McComb	.73
397	Columbus	.64
MISSOURI		
630-631	St. Louis	1.01
633	Bowling Green	.91
634	Hannibal	.88
635	Kirksville	.81
636	Flat River	.94
637	Cape Girardeau	.87
638	Sikeston	.84
639	Poplar Bluff	.85
640-641	Kansas City	1.01
644-645	St. Joseph	.95
646	Chillicothe	.86
647	Harrisonville	.96
648	Joplin	.83
650-651	Jefferson City	.89
652	Columbia	.89
653	Sedalia	.87
654-655	Rolla	.89
656-658	Springfield	.85
MONTANA		
590-591	Billings	.88
592	Wolf Point	.85
593	Miles City	.87
594	Great Falls	.89
595	Havre	.82
596	Helena	.89
597	Butte	.84
598	Missoula	.84
599	Kalispell	.83
NEBRASKA		
680-681	Omaha	.90
683-685	Lincoln	.79
686	Columbus	.69
687	Norfolk	.78
688	Grand Island	.78
689	Hastings	.76
690	Mccook	.69
691	North Platte	.75
692	Valentine	.66
693	Alliance	.65
NEVADA		
889-891	Las Vegas	1.01
893	Ely	.92
894-895	Reno	.97
897	Carson City	.97
898	Elko	.93
NEW HAMPSHIRE		
030	Nashua	.91
031	Manchester	.91
032-033	Concord	.88
034	Keene	.73
035	Littleton	.81
036	Charleston	.71
037	Claremont	.72
038	Portsmouth	.85

STATE	CITY	Residential
NEW JERSEY		
070-071	Newark	1.13
072	Elizabeth	1.15
073	Jersey City	1.12
074-075	Paterson	1.13
076	Hackensack	1.12
077	Long Branch	1.12
078	Dover	1.12
079	Summit	1.12
080,083	Vineland	1.10
081	Camden	1.11
082,084	Atlantic City	1.14
085-086	Trenton	1.12
087	Point Pleasant	1.11
088-089	New Brunswick	1.12
NEW MEXICO		
870-872	Albuquerque	.86
873	Gallup	.86
874	Farmington	.86
875	Santa Fe	.86
877	Las Vegas	.86
878	Socorro	.86
879	Truth/Consequences	.84
880	Las Cruces	.83
881	Clovis	.85
882	Roswell	.86
883	Carrizozo	.86
884	Tucumcari	.86
NEW YORK		
100-102	New York	1.37
103	Staten Island	1.30
104	Bronx	1.32
105	Mount Vernon	1.18
106	White Plains	1.21
107	Yonkers	1.22
108	New Rochelle	1.23
109	Suffern	1.15
110	Queens	1.30
111	Long Island City	1.33
112	Brooklyn	1.34
113	Flushing	1.32
114	Jamaica	1.32
115,117,118	Hicksville	1.22
116	Far Rockaway	1.31
119	Riverhead	1.23
120-122	Albany	.96
123	Schenectady	.96
124	Kingston	1.04
125-126	Poughkeepsie	1.08
127	Monticello	1.05
128	Glens Falls	.88
129	Plattsburgh	.93
130-132	Syracuse	.96
133-135	Utica	.93
136	Watertown	.92
137-139	Binghamton	.92
140-142	Buffalo	1.06
143	Niagara Falls	1.04
144-146	Rochester	.99
147	Jamestown	.91
148-149	Elmira	.89
NORTH CAROLINA		
270,272-274	Greensboro	.74
271	Winston-Salem	.74
275-276	Raleigh	.75
277	Durham	.74
278	Rocky Mount	.64
279	Elizabeth City	.62
280	Gastonia	.74
281-282	Charlotte	.75
283	Fayetteville	.72
284	Wilmington	.72
285	Kinston	.62
286	Hickory	.62
287-288	Asheville	.72
289	Murphy	.66
NORTH DAKOTA		
580-581	Fargo	.81
582	Grand Forks	.76
583	Devils Lake	.81
584	Jamestown	.75

STATE	CITY	Residential
NORTH DAKOTA (CONT'D)		
585	Bismarck	.81
586	Dickinson	.78
587	Minot	.81
588	Williston	.78
OHIO		
430-432	Columbus	.96
433	Marion	.94
434-436	Toledo	1.02
437-438	Zanesville	.91
439	Steubenville	.96
440	Lorain	1.03
441	Cleveland	1.03
442-443	Akron	1.00
444-445	Youngstown	.97
446-447	Canton	.95
448-449	Mansfield	.97
450	Hamilton	.96
451-452	Cincinnati	.96
453-454	Dayton	.93
455	Springfield	.94
456	Chillicothe	.97
457	Athens	.88
458	Lima	.91
OKLAHOMA		
730-731	Oklahoma City	.81
734	Ardmore	.79
735	Lawton	.83
736	Clinton	.78
737	Enid	.78
738	Woodward	.77
739	Guymon	.67
740-741	Tulsa	.80
743	Miami	.83
744	Muskogee	.72
745	Mcalester	.75
746	Ponca City	.78
747	Durant	.77
748	Shawnee	.77
749	Poteau	.78
OREGON		
970-972	Portland	1.02
973	Salem	1.02
974	Eugene	1.01
975	Medford	1.00
976	Klamath Falls	1.00
977	Bend	1.02
978	Pendleton	.99
979	Vale	.98
PENNSYLVANIA		
150-152	Pittsburgh	.98
153	Washington	.94
154	Uniontown	.92
155	Bedford	.88
156	Greensburg	.95
157	Indiana	.91
158	Dubois	.90
159	Johnstown	.91
160	Butler	.94
161	New Castle	.94
162	Kittanning	.95
163	Oil City	.91
164-165	Erie	.97
166	Altoona	.90
167	Bradford	.90
168	State College	.93
169	Wellsboro	.89
170-171	Harrisburg	.93
172	Chambersburg	.89
173-174	York	.89
175-176	Lancaster	.91
177	Williamsport	.86
178	Sunbury	.90
179	Pottsville	.90
180	Lehigh Valley	.99
181	Allentown	1.02
182	Hazleton	.90
183	Stroudsburg	.93
184-185	Scranton	.96
186-187	Wilkes-Barre	.92
188	Montrose	.90

STATE	CITY	Residential
PENNSYLVANIA (CONT'D)		
189	Doylestown	1.04
190-191	Philadelphia	1.13
193	Westchester	1.07
194	Norristown	1.06
195-196	Reading	.95
PUERTO RICO		
009	San Juan	.84
RHODE ISLAND		
028	Newport	1.07
029	Providence	1.07
SOUTH CAROLINA		
290-292	Columbia	.73
293	Spartanburg	.72
294	Charleston	.72
295	Florence	.67
296	Greenville	.71
297	Rock Hill	.65
298	Aiken	.84
299	Beaufort	.67
SOUTH DAKOTA		
570-571	Sioux Falls	.77
572	Watertown	.73
573	Mitchell	.75
574	Aberdeen	.76
575	Pierre	.76
576	Mobridge	.74
577	Rapid City	.76
TENNESSEE		
370-372	Nashville	.84
373-374	Chattanooga	.77
375,380-381	Memphis	.83
376	Johnson City	.72
377-379	Knoxville	.74
382	Mckenzie	.70
383	Jackson	.71
384	Columbia	.73
385	Cookeville	.68
TEXAS		
750	Mckinney	.77
751	Waxahackie	.78
752-753	Dallas	.84
754	Greenville	.70
755	Texarkana	.75
756	Longview	.69
757	Tyler	.76
758	Palestine	.69
759	Lufkin	.74
760-761	Fort Worth	.84
762	Denton	.77
763	Wichita Falls	.80
764	Eastland	.73
765	Temple	.76
766-767	Waco	.78
768	Brownwood	.69
769	San Angelo	.73
770-772	Houston	.86
773	Huntsville	.69
774	Wharton	.71
775	Galveston	.84
776-777	Beaumont	.83
778	Bryan	.74
779	Victoria	.75
780	Laredo	.73
781-782	San Antonio	.79
783-784	Corpus Christi	.78
785	Mc Allen	.76
786-787	Austin	.80
788	Del Rio	.66
789	Giddings	.69
790-791	Amarillo	.79
792	Childress	.76
793-794	Lubbock	.77
795-796	Abilene	.76
797	Midland	.77
798-799,885	El Paso	.76
UTAH		
840-841	Salt Lake City	.84

STATE	CITY	Residential
UTAH (CONT'D)		
842,844	Ogden	.82
843	Logan	.83
845	Price	.73
846-847	Provo	.84
VERMONT		
050	White River Jct.	.74
051	Bellows Falls	.75
052	Bennington	.74
053	Brattleboro	.75
054	Burlington	.80
056	Montpelier	.81
057	Rutland	.82
058	St. Johnsbury	.75
059	Guildhall	.74
VIRGINIA		
220-221	Fairfax	.86
222	Arlington	.87
223	Alexandria	.90
224-225	Fredericksburg	.76
226	Winchester	.72
227	Culpeper	.78
228	Harrisonburg	.68
229	Charlottesville	.73
230-232	Richmond	.82
233-235	Norfolk	.79
236	Newport News	.77
237	Portsmouth	.75
238	Petersburg	.80
239	Farmville	.69
240-241	Roanoke	.73
242	Bristol	.68
243	Pulaski	.66
244	Staunton	.69
245	Lynchburg	.70
246	Grundy	.68
WASHINGTON		
980-981,987	Seattle	1.01
982	Everett	1.03
983-984	Tacoma	.99
985	Olympia	.99
986	Vancouver	.98
988	Wenatchee	.92
989	Yakima	.95
990-992	Spokane	1.00
993	Richland	.99
994	Clarkston	.98
WEST VIRGINIA		
247-248	Bluefield	.89
249	Lewisburg	.89
250-253	Charleston	.97
254	Martinsburg	.85
255-257	Huntington	.96
258-259	Beckley	.90
260	Wheeling	.92
261	Parkersburg	.92
262	Buckhannon	.91
263-264	Clarksburg	.91
265	Morgantown	.92
266	Gassaway	.92
267	Romney	.87
268	Petersburg	.89
WISCONSIN		
530,532	Milwaukee	1.06
531	Kenosha	1.06
534	Racine	1.04
535	Beloit	1.02
537	Madison	1.01
538	Lancaster	1.00
539	Portage	.98
540	New Richmond	1.00
541-543	Green Bay	1.03
544	Wausau	.96
545	Rhinelander	.97
546	La Crosse	.96
547	Eau Claire	1.00
548	Superior	.99
549	Oshkosh	.96

STATE	CITY	Residential
WYOMING		
820	Cheyenne	.76
821	Yellowstone Nat. Pk.	.71
822	Wheatland	.72
823	Rawlins	.70
824	Worland	.69
825	Riverton	.70
826	Casper	.77
827	Newcastle	.69
828	Sheridan	.74
829-831	Rock Springs	.74
ALBERTA		
	Calgary	1.05
	Edmonton	1.05
	Fort McMurray	1.03
	Lethbridge	1.04
	Lloydminster	1.03
	Medicine Hat	1.03
	Red Deer	1.03
BRITISH COLUMBIA		
	Kamloops	1.01
	Prince George	1.02
	Vancouver	1.07
	Victoria	1.02
MANITOBA		
	Brandon	.99
	Portage la Prairie	.99
	Winnipeg	.99
NEW BRUNSWICK		
	Bathurst	.91
	Dalhousie	.91
	Fredericton	.97
	Moncton	.91
	Newcastle	.91
	Saint John	.97
NEWFOUNDLAND		
	Corner Brook	.92
	St. John's	.93
NORTHWEST TERRITORIES		
	Yellowknife	.99
NOVA SCOTIA		
	Dartmouth	.93
	Halifax	.94
	New Glasgow	.93
	Sydney	.92
	Yarmouth	.93
ONTARIO		
	Barrie	1.10
	Brantford	1.11
	Cornwall	1.10
	Hamilton	1.11
	Kingston	1.11
	Kitchener	1.05
	London	1.09
	North Bay	1.08
	Oshawa	1.10
	Ottawa	1.11
	Owen Sound	1.08
	Peterborough	1.09
	Sarnia	1.12
	Sudbury	1.03
	Thunder Bay	1.07
	Toronto	1.14
	Windsor	1.08
PRINCE EDWARD ISLAND		
	Charlottetown	.88
	Summerside	.88
QUEBEC		
	Cap-de-la-Madeleine	1.10
	Charlesbourg	1.10
	Chicoutimi	1.10
	Gatineau	1.09
	Laval	1.09
	Montreal	1.09
	Quebec	1.11

Location Factors

STATE	CITY	Residential
QUEBEC **(CONT'D)**		
	Sherbrooke	1.09
	Trois Rivieres	1.10
SASKATCHEWAN		
	Moose Jaw	.91
	Prince Albert	.90
	Regina	.91
	Saskatoon	.90
YUKON		
	Whitehorse	.89

Abbreviations

A	Area
ASTM	American Society for Testing and Materials
B.F.	Board feet
Carp.	Carpenter
C.F.	Cubic feet
CWJ	Composite wood joist
C.Y.	Cubic yard
Ea.	Each
Equip.	Equipment
Exp.	Exposure
Ext.	Exterior
F	Fahrenheit
Ft.	Foot, feet
Gal.	Gallon
Hr.	Hour
in.	Inch, inches
Inst.	Installation
Int.	Interior
Lb.	Pound
L.F.	Linear feet
LVL	Laminated veneer lumber
Mat.	Material
Max.	Maximum
MBF	Thousand board feet
MBM	Thousand feet board measure
MSF	Thousand square feet
Min.	Minimum
O.C.	On center
O&P	Overhead and profit
OWJ	Open web wood joist
Oz.	Ounce
Pr.	Pair
Quan.	Quantity
S.F.	Square foot
Sq.	Square, 100 square feet
S.Y.	Square yard
V.L.F.	Vertical linear feet
'	Foot, feet
"	Inch, inches
°	Degrees

Index

Index

Index

Notes

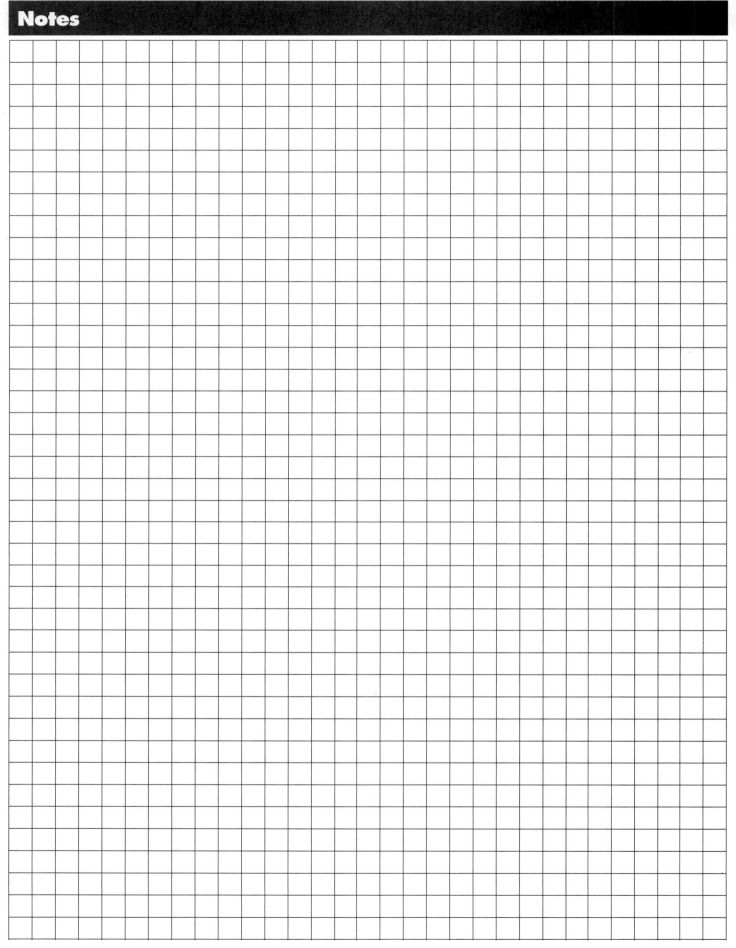

Reed Construction Data, a leading worldwide provider of total construction information solutions, is comprised of three main product groups designed specifically to help construction professionals advance their businesses with timely, accurate and actionable project, product, and cost data. Reed Construction Data is a division of Reed Business Information, a member of the Reed Elsevier plc group of companies.

The *Project, Product, and Cost & Estimating* divisions offer a variety of innovative products and services designed for the full spectrum of design, construction, and manufacturing professionals. Through it's *International* companies, Reed Construction Data's reputation for quality construction market data is growing worldwide.

Cost Information
RSMeans, the undisputed market leader and authority on construction costs, publishes current cost and estimating information in annual cost books and on the CostWorks CD-ROM. RSMeans furnishes the construction industry with a rich library of complementary reference books and a series of professional seminars that are designed to sharpen professional skills and maximize the effective use of cost estimating and management tools. RSMeans also provides construction cost consulting for Owners, Manufacturers, Designers, and Contractors.

Project Data
Reed Construction Data provides complete, accurate and relevant project information through all stages of construction. Customers are supplied industry data through leads, project reports, contact lists, plans and specifications surveys, market penetration analyses and sales evaluation reports. Any of these products can pinpoint a county, look at a state, or cover the country. Data is delivered via paper, e-mail, CD-ROM or the Internet.

Building Product Information
The First Source suite of products is the only integrated building product information system offered to the commercial construction industry for comparing and specifying building products. These print and online resources include *First Source,* CSI's SPEC-DATA™, CSI's MANU-SPEC™, First Source CAD, and Manufacturer Catalogs. Written by industry professionals and organized using CSI's MasterFormat™, construction professionals use this information to make better design decisions.

FirstSourceONL.com combines Reed Construction Data's project, product and cost data with news and information from Reed Business Information's *Building Design & Construction* and *Consulting-Specifying Engineer,* this industry-focused site offers easy and unlimited access to vital information for all construction professionals.

International
BIMSA/Mexico provides construction project news, product information, cost-data, seminars and consulting services to construction professionals in Mexico. Its subsidiary, PRISMA, provides job costing software.

Byggfakta Scandinavia AB, founded in 1936, is the parent company for the leaders of customized construction market data for Denmark, Estonia, Finland, Norway and Sweden. Each company fully covers the local construction market and provides information across several platforms including subscription, ad-hoc basis, electronically and on paper.

Reed Construction Data Canada serves the Canadian construction market with reliable and comprehensive project and product information services that cover all facets of construction. Core services include: *BuildSource, BuildSpec, BuildSelect,* product selection and specification tools available in print and on the Internet; Building Reports, a national construction project lead service; CanaData, statistical and forecasting information; *Daily Commercial News,* a construction newspaper reporting on news and projects in Ontario; and *Journal of Commerce,* reporting news in British Columbia and Alberta.

Cordell Building Information Services, with its complete range of project and cost and estimating services, is Australia's specialist in the construction information industry. Cordell provides in-depth and historical information on all aspects of construction projects and estimation, including several customized reports, construction and sales leads, and detailed cost information among others.

For more information, please visit our Web site at www.reedconstructiondata.com.

Reed Construction Data Corporate Office
30 Technology Parkway South
Norcross, GA 30092-2912
(800) 322-6996
(800) 895-8661 (fax)
info@reedbusiness.com
www.reedconstructiondata.com

Contractor's Pricing Guides

For more information
visit Means Web Site
at www.rsmeans.com

Contractor's Pricing Guide:
Residential Detailed Costs 2004

Every aspect of residential construction, from overhead costs to residential lighting and wiring, is in here. All the detail you need to accurately estimate the costs of your work with or without markups—labor-hours, typical crews and equipment are included as well. When you need a detailed estimate, this publication has all the costs to help you come up with a complete, on the money, price you can rely on to win profitable work.

$39.95 per copy
Over 300 pages, with charts and tables, 8-1/2 x 11
Catalog No. 60334   ISBN 0-87629-718-1

Contractor's Pricing Guide:
Residential Repair &
Remodeling Costs 2004

This book provides total unit price costs for every aspect of the most common repair & remodeling projects. Organized in the order of construction by component and activity, it includes demolition and installation, cleaning, painting, and more.

With simplified estimating methods; clear, concise descriptions; and technical specifications for each component, the book is a valuable tool for contractors who want to speed up their estimating time, while making sure their costs are on target.

$39.95 per copy
Over 250 pages, illustrated, 8-1/2 x 11
Catalog No. 60344   ISBN 0-87629-717-3

Contractor's Pricing Guide:
Residential Square Foot
Costs 2004

Now available in one concise volume, all you need to know to plan and budget the cost of new homes. If you are looking for a quick reference, the model home section contains costs for over 250 different sizes and types of residences, with hundreds of easily applied modifications. If you need even more detail, the Assemblies Section lets you build your own costs or modify the model costs further. Hundreds of graphics are provided, along with forms and procedures to help you get it right.

$39.95 per copy
Over 250 pages, illustrated, 8-1/2 x 11
Catalog No. 60324   ISBN 0-87629-719-X

Means Electrical Estimating
Methods
3rd Edition

Expanded new edition includes sample estimates and cost information in keeping with the latest version of the CSI MasterFormat and UNIFORMAT II. Complete coverage of Fiber Optic and Uninterruptible Power Supply electrical systems, broken down by components and explained in detail. Includes a new chapter on computerized estimating methods. A practical companion to *Means Electrical Cost Data.*

$64.95 per copy
Over 325 pages, hardcover
Catalog No. 67230A   ISBN 0-87629-701-7

Means Repair & Remodeling
Estimating Methods

New 4th Edition
By Edward B. Wetherill and RSMeans Engineering Staff

This updated edition focuses on the unique problems of estimating renovations in existing structures—using the latest cost resources and construction methods. The book helps you determine the true costs of remodeling, and includes:

    Part I–The Estimating Process
    Part II–Estimating by CSI Division
    Part III–Two Complete Sample
        Estimates–Unit Price & Assemblies
    Part IV–Disaster Reconstruction

$69.95 per copy
Over 450 pages, illustrated, hardcover
Catalog No. 67265B   ISBN 0-87629-661-4

Means Landscape Estimating
Methods New 4th Edition
By Sylvia H. Fee

Professional Methods for Estimating and Bidding Landscaping Projects and Grounds Maintenance Contracts

- Easy-to-understand text. Clearly explains the estimating process and how to use *Means Site Work & Landscape Cost Data.*
- Sample forms and worksheets to save you time and avoid errors.
- Tips on best techniques for saving money and winning jobs.
- **Updated cost and estimate examples** help you control your equipment costs and bid landscape maintenance projects.

$62.95 per copy
Over 300 pages, illustrated, hardcover
Catalog No. 67295B   ISBN 0-87629-633-0